Emergence in Context

Praise for *Emergence in Context*

At present there is little agreement on what exactly emergence is, whether it in fact occurs, and, if it does, what its scientific and metaphysical implications might be. Robert Bishop and Michael Silberstein offer an impressive, historically informed and scientifically detailed account of a phenomenon deserving to be called emergence, an account that advances the discussion by moving beyond the familiar categories of 'strong' and 'weak' emergence: emergence occurs when multiscale 'contextual' factors play a role in the behaviour of a population of particles, cells, organisms, societies, and the universe as a whole. Thus construed, emergence is both ubiquitous and definitive of natural processes at every level of organisation.

John Heil, *Washington University, Durham University, Monash University*

Bishop, Silberstein and Pexton offer, in a Contextual Emergence Tour-de-Force, a highly persuasive view of the material world, and the models of it that the sciences build. Refusing and refuting brute dualisms, either of strong ontological vs. weak emergence, or indeed of mind vs. matter, their radical third way is demonstrated to be faithful to a rich variety of examples from the physical and life sciences. It is the combination of detailed and scientifically-mastered examples, together with a powerful formal and philosophical approach, that makes their case so compelling, and so readable. Lifting one's eyes from their pages to the world around, it also seems clearly true.

Tom McLeish, *Professor of Natural Philosophy in the Department of Physics, University of York, FRS*

Discussions of reduction and emergence have long used scientific examples for illustration but more rarely analyzed scientific practice productively. Bishop, Silberstein, and Pexton give us an exemplar of this productivity in a systematic account of what contexts are—not merely background or boundary conditions—and how they yield a distinctive conception of emergence via forms of stabilization. This makes explicit common patterns of successful reasoning across the sciences and yields a deeply relational ontology that challenges tenets in contemporary metaphysics of science. Its strength lies in the cumulative case built from a close examination of many natural sciences and scrutiny of the consequences for topics such as determinism, mental causation, modality, and realization. This book has the potential to reshape future inquiry, moving us away from false dichotomies of "reduction versus emergence" and toward a practice-grounded scientific metaphysics.

Alan Love, *Professor of Philosophy, Winton Chair in the Liberal Arts, University of Minnesota, Director of the Minnesota Center for Philosophy of Science*

Bishop, Silberstein and Pexton have written a book of compelling value to both philosophers and scientists. The authors help readers understand and conceptualise the difficult phenomenon of emergence, and they put it into context with a rich array of examples from different scientific fields, ranging from quantum physics to ecology. The book is a must-read for anybody interested in the topic.

Roman Frigg (*Professor, London School of Economics, Department of Philosophy Logic and Scientific Method and Deputy Director CPNSS*)

This impressive and fascinating book offers a new approach to understanding reduction and emergence. Avoiding the usual options of weak or strong emergence, it proposes contextual emergence to show that higher-level phenomena are neither adequately explained by micro-reduction nor arise by miracle in nature. Detailed case studies from physics, biology and neuroscience illustrate that multi-scale constraints are crucial to address the reduction problems of the sciences. The book challenges foundationalist scientific explanations and proposes a non-reductive, pluralistic unity of nature. By emphasizing that multi-scale modelling is more important than inter-theoretical relations, it gives a new turn to philosophy of science. It also discusses the far-reaching consequences of contextual emergence for a metaphysics of nature that is up to contemporary science.

Prof. Dr. Dr. Brigitte Falkenburg, *Institut für Philosophie und Politikwissenschaft, Technische Universität Dortmund*

Emergence in Context

A Treatise in Twenty-First Century Natural Philosophy

Robert C. Bishop
Wheaton College, USA

Michael Silberstein
Elizabethtown College and University of Maryland, USA

Mark Pexton
University of Durham, UK

OXFORD
UNIVERSITY PRESS

OXFORD
UNIVERSITY PRESS

Great Clarendon Street, Oxford, OX2 6DP,
United Kingdom

Oxford University Press is a department of the University of Oxford.
It furthers the University's objective of excellence in research, scholarship,
and education by publishing worldwide. Oxford is a registered trade mark of
Oxford University Press in the UK and in certain other countries

Published in the United States of America by Oxford University Press
198 Madison Avenue, New York, NY 10016, United States of America

British Library Cataloguing in Publication Data
Data available

Library of Congress Control Number: 2021952581

ISBN 978–0–19–284978–6

DOI: 10.1093/oso/9780192849786.001.0001

Printed and bound by
CPI Group (UK) Ltd, Croydon, CR0 4YY

Robert Bishop: For my parents, Ed and Gloria Bishop, and my brothers, Jerry (deceased) and Terry, and Hans Primas, who showed a bright light down the path

Michael Silberstein: For my wife Elizabeth Newell, my son Robin Silberstein, my parents Fred and Jean Silberstein, and to fireflies everywhere, with much love and gratitude

Mark Pexton: For Douglas Pexton, John Pexton, and Ann Brook

Contents

Preface xiv
Acknowledgments xxvi

1 Introduction 1
- 1.1 Setting the scene 1
- 1.2 Emergence and metaphysics 8
- 1.3 Introducing contextual emergence 14
- 1.4 Where we go from here 19

2 Contextual Emergence 20
- 2.1 Introduction 20
- 2.2 Preliminaries 20
- 2.3 Contextual emergence 24
  - 2.3.1 The basic framework 24
  - 2.3.2 Contexts and stability conditions 27
  - 2.3.3 Stability conditions and physical possibility space 30
  - 2.3.4 Summarizing and situating contextual emergence 33
- 2.4 Case study: Rayleigh–Bénard convection 37
  - 2.4.1 Basic setup 37
  - 2.4.2 Transition to convection 38
  - 2.4.3 Rayleigh–Bénard convection and contextual emergence 39
- 2.5 Summary and moving forward 43

3 Intertheoretic Relations and Multiscale Modeling 44
- 3.1 Introduction 44
- 3.2 Scientific theories, observables, and contextual emergence 48
  - 3.2.1 States, observables, and topologies 48
  - 3.2.2 Contextual topologies 50

3.2.3 Contextual topologies and intertheoretic relations 56
3.2.4 Abstractions, idealizations, and contextual topologies 59
3.3 Intertheoretic relations and contextual emergence 64
3.3.1 Intertheoretic reduction and context enrichment: comparison with Nagel and Nickels 65
3.3.2 Comparison with Butterfield's neo-Nagelian intertheoretical reductionism 71
3.3.3 Schaffner, interlevel theories in biology, and intertheoretic reduction 76
3.3.4 Wimsatt and functional reduction 78
3.4 Multiscale modeling 83
3.4.1 Contexts in multiscale modeling 85
3.4.2 Multiscale data analysis 90
3.4.3 Stability conditions and multiscale models 93
3.4.4 Multiscale modeling and contextual emergence 94
3.5 Summary and moving forward 97
Appendix 99

4 Classical and Quantum Physics 100
4.1 Introduction 100
4.2 Contexts and interactions 101
4.3 Property emergence and global stability conditions: temperature, chemical potential, and pressure 102
4.3.1 The ontological origin of temperature 103
4.3.2 Chemical potential 106
4.3.3 Pressure 107
4.4 Emergent causal relations: white dwarf stars 108
4.4.1 White dwarfs as contextually emergent 109
4.5 The contextual emergence of the macroscopic world 113
4.6 Contextual property emergence: molecular shape 115
4.6.1 The problem of molecular structure 117
4.6.2 Ontological emergence of molecular shape 120
4.6.3 The molecular environment 122
4.6.4 Molecular structure and the contextual emergence pattern 123
4.7 Quantum superpositions, entanglement, and contextual emergence 124
4.7.1 Quantum entanglement and nonlocality 126
4.7.2 The no-go theorems of Bell and Kochen-Specker 131

4.7.3 Quantum mechanics and contextual emergence 135
4.8 Renormalization, universality, and contextual emergence 140
4.8.1 Renormalization group methods 141
4.8.2 Universality 143
4.8.3 Renormalization, universality, and contextual emergence 149
4.9 Further brief examples 152
4.9.1 Lasers 152
4.9.2 Coherent electrons in scanning tunneling microscopy 153
4.9.3 Topological invariants in ideal fluids 153
4.9.4 Coalescence in metastable liquids 154
4.10 Conclusion 154
Appendix 1: Classifying algebras 155
Appendix 2: The Pauli exclusion principle 155
Appendix 3: Degeneracy pressure in a white dwarf 156

5 Contextual Emergence in Condensed-Matter Physics 158
5.1 Introduction 158
5.2 Condensed-matter physics and contexts 159
5.3 Basic concepts in condensed-matter physics 162
5.3.1 Compartmentalization of phase space and contextual boundaries 163
5.3.2 Order 164
5.3.3 Symmetry 165
5.3.4 Generalized rigidity 167
5.4 Contextual aggregate environments 170
5.4.1 Superconductivity and gauge symmetry breaking 170
5.4.1.1 Morrison's argument for emergent superconductivity 172
5.4.1.2 The physical significance of gauge 174
5.4.2 The fractional quantum Hall effect and topological entanglement 179
5.4.2.1 Topology and FQH states 181
5.4.2.2 The problem with a microscale description of correlated electron physics 183

5.4.2.3 Multiple realizability, stability, and topological computation 183
5.4.2.4 Long-range entanglement 185
5.4.2.5 Summary of the key physics 186
5.4.2.6 Different ways of articulating emergence in the fractional quantum Hall effect 187
5.4.2.6.1 Quasiparticle realism 187
5.4.2.6.2 Good coordinates 189
5.4.2.6.3 Failure of mereological supervenience 191
5.4.3 Contextual emergence and the fractional quantum Hall effect 193
5.5 Contextual object/environment distinctions and effective field theories 198
5.5.1 Objections to giving up on fundamental theories 201
5.5.2 Effective field theories and approximations 204
5.5.3 Effective field theories and fundamentality 208
5.6 Conclusion 210

6 Contextual Emergence in Biology and Neuroscience 211
6.1 Introduction 211
6.2 Systems biology 214
6.3 Contextual emergence in neuroscience and cognitive science 221
6.3.1 Systems neuroscience and social neuroscience 223
6.3.2 Organizational features of the brain and related complex biological systems 226
6.3.3 Network neuroscience 228
6.3.4 The return of localization and decomposition? 235
6.3.5 Broader implications of contextual emergence 239
6.4 Biological and ecological examples of contextual emergence 242
6.4.1 Microbial community assembly 242
6.4.2 Appendage regeneration 247
6.4.3 Bee colonies 249
6.4.4 Ecology 249
6.5 Conclusion 251

7 Ontological and Scientific Implications of Contextual Emergence 253
7.1 Introduction 253
7.2 Context, possibility spaces, and modal constraint 256
7.3 Modalities and dispositions 257
7.4 Properties 261
7.5 Property incompatibility relations 265
7.6 Nested possibility spaces 267
7.7 Realization and multiple realizability 268
7.8 Laws 278
7.9 Causation, laws, and context 283
7.10 Determinism as contextually emergent 286
7.11 Fundamental ontology? 288
7.12 Neutral monism 303
7.13 Objections and frequently asked questions 311
7.13.1 Ontology or epistemology? 311
7.13.2 Contextual emergence compared to transformational emergence and ontological structural realism 315
7.14 Conclusion 326

References 327
Index 354

Preface

Whisky, fireflies, and fireworks. It sounds like the beginning of a bad country song. For several summers now, around late June and early July, Bishop and Silberstein have been hunkered down in Silberstein's home in Lancaster, Pennsylvania, working on this book. Festooned with fireflies every summer, Silberstein is lucky enough to live in a small wooded enclave with large dense trees that surround several fields and yards. An ideal arboreal grotto for our friendly flying bioluminescent beetles, as it happens. The closed canopy allows fireflies to congregate together in safety—they like to hug the tree line, and the open fields allow them to clearly see one another's flashing mating call (i.e., "mate-search-communication"). We later learned that such a setting is well known as a key environmental constraint for increasing the probability that fireflies will congregate, be able to see each other's flashing, and grow dense in number as a result. All of which makes sense, because every summer, ranging from the ground to treetops, we have been treated to an amazing private display of fireflies so dense in every spatial dimension that we can only describe the experience *à la* Leibnitz, as being like a tiny cell-sized observer trapped inside a brain, watching, perhaps both synchronously and chaotically, countless flying neurons firing.

We say "perhaps," because given that most of the flashing is asynchronous, it is not always easy to see the synchronous activity (being "on" and "off" at the same time) or to distinguish it from periodic behavior (patterns of flashing and darkness that occur at specific repeated intervals). Furthermore, the property of being synchronous is essentially on a dimmer switch as it ranges from intermittent to nearly perfect.

Like so many before us, we became mesmerized and enchanted by the nightly display and wanted to understand it as we would any other natural phenomena that we had studied as philosophers of science. And of course, synchronous fireflies are commonly cited as an exemplary case of "Sync" in nature, which is supposedly a paradigmatic example of "emergence" (Strogatz 2004). The key question driving scientists who study synchronization is: how and why does the synchronous pattern in question emerge?

Needless to say, neither Bishop or Silberstein are entomologists or experts in Lampyridae, so we began to engage in "data collection," (i.e., recovering from the day's writing session by spending long evenings drinking scotch or bourbon outdoors in the company of our glowing aerobatic visitors). Being highly skilled researchers, through arduous exploration, the first thing we learned is that fireflies are the state bug of Pennsylvania. But could we really be witnessing synchronous fireflies, which are allegedly rare? According to the experts there are only a few places on Earth that reliably produce synchronous fireflies: Great Smoky Mountains, Tennessee; Elkmont, Tennessee; Congaree National Park, South Carolina; Allegheny National Forest, Pennsylvania; Kampung Kuantan fireflies park, Malaysia; Kuala Selangor District, Malaysia; Bhandardara, Maharashtra, India; Setagaya, Japan; and Nanacamilpa, Mexico, to name a few. So, in addition to the mystery of how and why firefly synchronization occurs, there is the mystery of why it is so uncommon. Therefore, if such synchronous behavior were occurring in Silberstein's backyard every summer, that would be news indeed.

Again, synchronous behavior such as fireflies flashing, birds and fish flocking, and the like, has long surrounded the discussion of emergence, but especially since the 1990s with the appearance of Steven Strogatz's formal account, wherein synchronous fireflies are modeled as weakly coupled self-sustaining oscillators, explicable by nearest-neighbor effects. Sync is everywhere, of course, including, perhaps most famously, neurons and cardiac pacemaker cells. It is well known that under particular conditions people's heart rates and breathing, bodily movements, emotions, and even neural oscillations become Synced (Valencia and Froese 2020). Phase synchronization is thus defined as the continuous adjustment of rhythms of self-sustained oscillators as the result of weak coupling. Synchronous patterns are not only multiply instantiated in nature and out, not only amenable to mathematical modeling, but are striking to watch in their novelty, sometimes hard to predict or anticipate, and have no top-down explanation such as an external coordinator (i.e., they are "self-organizing"). These are properties often associated with the term "emergence." But what kind of emergence are such cases, and how interesting and representative are such cases? We shall return to these questions shortly.

We note for now that the standard way of thinking about such "complex systems" (Bishop and Silberstein 2019) is to view some complex macropattern of behavior emerging from the interactions of the atomistic-like components, such as particles, proteins, cells, organisms, via some sort of algorithm or formal rule—indeed, often physically and biologically distinct systems allegedly following the same rule. We think this conception of emergence confuses reality with the model.

Bishop and Silberstein took a deep dive into the mysteries of how and why such synchronous behavior occurs in fireflies and beyond. Unsurprisingly, the goal for researchers here is to create a science of synchronization that would apply to any number of different cases. In addition to being mathematically modeled as weakly coupled oscillators, synchronous fireflies have been modeled with algorithms or rules in a computer simulation not unlike Conway's Game of Life (Stewart and Strogatz 1993), with tools from dynamical systems theory such as attractors in phase space (Strogatz 2004), and with tools from graph theory (Strogatz 2001). Researchers into Sync in the late 1990s and early 2000s made the following advances:

1. The tendency to synchronize is ubiquitous for multiple entities at all scales, ranging from electrons to neurons to woman's menstrual cycles (Strogatz 2004, p. 14). All such Sync was dubbed "self-organization that emerges spontaneously from chaos." The basic thinking at the time was that in most cases Sync is best explained as a local nearest-neighbor effect that cascades over time like a wave.
2. This tendency suggests a kind of "universality" as the entities, mechanisms, and structural dynamics differ in each case, but there appear to be common principles, or rules, at work. For instance, such mutual synchronization can be modeled as a phase transition, like the freezing of water into ice. This allowed researchers to bring in mathematical methods from statistical mechanics (such as renormalization group methods) as well as dynamical systems theory. For example, synchronous fireflies not only flash together in unison, but also flash periodically at fixed intervals (Strogatz 2004, p. 30).
3. All (biological) oscillators send and receive signals. These properties change through an oscillator's cycle. The ability to send and receive signals, and to send information (the "influence function") and absorb it (the "receiver function"), is a function of phase. For instance, the fireflies' cycle consists of a sudden flash followed by an interval of darkness, then another flash. Female fireflies notice only the flash and ignore the dark period. Depending on where in space relative to other fireflies in their line of sight and when in their cycle they receive information or fail to receive it, their own behavior, such as advances or delays in their own oscillations, is affected. Such general time- and position-dependent behavior can be modeled with differential equations, as random or non-random networks in graph theory, or as information-theoretic rules or computational algorithms (Strogatz 2004, p. 47).

Certainly, all of this represents an advance in the description, understanding, and prediction of various classes of synchronous systems, but have our questions of *how* and *why* some fireflies of various genera and species behave synchronously been answered? While Strogatz acknowledges that mathematical modeling efforts have achieved success at describing and predicting self-organization in such biological systems, he notes: "Still, a more tough-minded question is whether the models describe reality faithfully. Do they predict phenomena that agree with data from real fireflies, heart cells, or neurons?" His answer: "We don't know. There have been no tests so far" (Strogatz 2004 p. 65). It is notoriously hard to do relevant experiments (e.g., performing measurements at the level of individual fireflies or heart cells with their differing natural frequencies and stimulus response).

In the intervening years there has been growing evidence that formal models of synchronization do capture the behavior of fireflies in the wild (Peleg 2021). In our view, however, the deeper concerns are as follows: 1) one can have predictive and descriptive formal models that do not necessarily provide the deepest *explanation* for a phenomenon (e.g., why did the "phase transition" occur in the physical world for this particular system?), and thus 2) whether or not key formal features of those models, such as universality of one sort or another, get at the essence of emergence. We will discuss the question of how universality, phase transitions, and asymptotic singularities relate to emergence in subsequent chapters, but suffice it to say that while we think such things are often symptomatic of emergence, they are neither necessary nor sufficient. We will also make clear why the neomechanistic model of emergence is not the alternative for which we are looking (Silberstein, 2021).

Philosophers of science sometimes focus on the logical and mathematical relationships between theories as ends in themselves (e.g., whether some sort of intertheoretic reduction obtains between one theory and another, or whether a particular phenomenon can be mathematically modeled as a well-known phase transition). We will argue throughout that although these questions have merit, they are rarely the most important or most interesting. Rather, the main goal is multiscale modeling properly understood and all it can tell us about a system or phenomenon in the real world.

Of course, all of this depends on what one means by "emergence," which by now has many, many meanings and covers many different cases. However, most cases of emergence tend to get categorized as some variant of "strong" or "weak" emergence (see Chapter 1 for details). Most researchers would agree that while striking, synchronous behavior, however we model it mathematically in terms of attractors, order parameters, or some other way, is some species of weak or epistemic emergence at best. In other words, synchrony is not like some completely new ontological phenomenon or order-of-being emerging, such as chemistry from physics, life from chemistry, or mind from life. Synchrony requires no new fundamental laws, entities, or causes, whereas some people believe that the other cases just mentioned represent radical categorical "jumps" in nature's great chain of being.

If so, these would be cases of strong emergence with important ontological implications.

One major question looming over the discussion of emergence—the question of how order and stability arise, and how novel phenomena come into being from more "fundamental" physical processes—is whether there really is any strong emergence in the world. That is, starting with the Big Bang (the initial conditions of the universe) and given whatever dynamical laws "govern" "basic" microphysical entities, does everything else, such as life and mind, come for free? Is all the complexity, variety, and stability of order in the world a consequence of fundamental physics, the way seemingly self-replicating complex life-like patterns such as Gliders emerge in Conway's finite cellular automaton known as The Game of Life? At its base, The Game of Life has only cells, upon which updating the state of the system based on three simple rules, either live or die. Yet as the cellular automaton is time-evolved, very complex and impossible-to-predict stable patterns emerge, such as Glider-Guns that keep "reproducing" more Gliders. Such patterns, being nothing but resultants from the basic dynamical rules plus the initial conditions, are said to be only weakly emergent. Is the real world just a much more complex Game of Life? That is, are there really some fundamental physical entities (i.e., fully autonomous at some smallest spatial and temporal length scale) and some fully autonomous dynamical law that evolve such entities in time, which is sufficient to explain in principle the emergence of everything else in the universe? We argue that this model of the universe is false and does not best reflect what the sciences are showing us.

For instance, the relatively early artificial life program, Boids,[1] gets virtual birds to approximate flocking behavior by each bird (really the system in this case simulating individual birds) following three simple rules:

1. Separation: steer to avoid crowding local flock mates.
2. Alignment: steer towards the average heading of local flock mates.
3. Cohesion: steer to move toward the average position (center of mass) of local flock mates.

Since the original software, more complex rules have been added, such as obstacle avoidance and goal-seeking. Keep in mind that no one bird is in charge (decentralized control), and many local interactions between birds are required. You can also read about things such as "Applying Ant Colony Optimization Algorithms to Solve the Traveling Salesman Problem."[2] Much

[1] http://www.red3d.com/cwr/boids/

[2] http://www.codeproject.com/Articles/644067/Applying-Ant-Colony-Optimization-Algorithms-to-Sol

of this knowledge about "swarm intelligence" has been implemented into relatively smart swarm-bots of various kinds (coming soon to a drone near you!). So, from all this analysis we have improved machine intelligence of both the embodied artificial intelligence (AI) kind, such as swarm-bots, and the good old-fashioned AI kind, such as genetic algorithms.

Nonetheless, do such mathematical models and simulations exemplifying weak emergence tell us most deeply how nature works, how these tricks are accomplished in the real world? Do birds, fish, and fireflies really "follow" simple rules in some computational sense from which larger-scale patterns of behavior emerge? Probably not, and unlike the rules of finite automata, the behavioral propensities of complex organisms are generally probabilistic and context-dependent. We want to challenge the idea that novelty, order, and stability come primarily from rules or laws viewed as (internal or external) "governing" principles. There would be nothing truly self-organizing or spontaneous about such systems, except the absence of a top-down agent. We will argue that the rules, laws, principles, and modalities of all varieties emerge from inherently multiscale interactions involving stability conditions defining concrete contexts. All of this goes for the entities themselves and not just the laws; indeed, we will argue that this is a misleading duality to begin with. Taking the sciences as our guide, the universe is best viewed as a multiscale network of extrinsic dispositions. Only this conception of reality allows us to explain emergence in a scientific fashion.

As we discuss in Chapter 1, the consensus these days is that all phenomena are at best weakly emergent, with experience as the possible exception. It is true that people discuss quantum entanglement as an emergent phenomenon that is a real-life example of something's being "more than the sum of the parts," but not, obviously, in the sense of transcending fundamental microphysical laws. If the only cases of emergence that might transcend being of the weak variety are quantum weirdness, perhaps the emergence of life, and experience, this would be a very strange disunified world, and ontological emergence would be a topic with a very limited and relatively rare subject matter. However, if the only choice is between the very common cases of weak or epistemic emergence on the one hand and spooky strong emergence on the other, with its inexplicable brute bridge laws and causal powers popping into being (e.g., psychophysical bridge laws pertaining to the emergence of experience from essentially insensate matter), then we understand why the preferred choice by default is weak emergence. Is there no other alternative to weak or strong emergence?

Let us return to the how and why of synchronous fireflies. Set aside the helpful idealization and abstraction of firefly Sync being like a phase transition or finite automata, and look at some of the real-world complexity involved. There are roughly 2,000 known species of fireflies, with countless genera. They are often taxonomized according to geography, habitat, morphology, flashing related behavior, time of year, time of night, flying- and movement-related behavior such as altitude and flight path, and other lifestyle-related factors. To focus on the flash-related behavior, different species tend to flash in different colors, such as yellow, green, or amber. The general consensus is that the flashing of males especially is all about finding a mate. Most experts also believe that the synchronous flashing, in particular, is also about attracting a mate, as generally only males are synchronous.

However, this does raise the question of how an individual firefly can stand out if they are all synced. Strogatz notes some further possibilities for the synchronous strategy: "A second possible advantage of synchrony is that you might get lucky—a female with eyes for your look-alike neighbor might become confused and mate with you instead." In other words, luck! Synchrony could also be a means of avoiding predators: "It's always safest to blend in with a crowd." Or, as he continues: "The latest theory is that synchrony reflects competition, not cooperation. Every firefly is trying to be the first to flash (because females seem to prefer that), but if everyone follows that strategy, sync automatically ensues" (Strogatz 2004, p. 35).

Another possibility is that synchronous flashing creates more of a light stimulus for the females to notice. Still another possibility is that since there are often multiple species in one place, synchronous flashing is a strategy for fireflies of the same species to find each other. Females generally stay on the ground, flashing dimly while males move towards them. All males are definitely advertising their availability to all females, but females select a specific individual male and then privately signal back. It must be noted that there are probably other social functions of different flash patterns, such as competition or showing off, rejection as in "I'm not interested!," cooperation in the face of specific predators (it is hard to pick out an individual firefly to eat when they are synced), and to send an alarm signal when threatened (Frierson Faust 2017, chapter 3). In any case, such behaviors are definitely social.

The flash patterns themselves can vary in countless ways (Frierson Faust 2017, chapter 3). Here are just a few:

- Flash, flicker, fade, or glow.
- Slow or fast flash.
- Longer versus shorter flash.
- Varying intensity or brightness of flash.
- Intermittently versus continuous flash.
- Single flash versus flash-trains.

- Rhythmic or periodic at constant intervals (e.g., flash every five versus every sixty seconds).
- Flash while stationary versus flash while flying at some variable altitude.
- Make a characteristic shape or tracer with the flash while flying, like a child with a sparkler.
- Synchronous flashes can move as a localized wave throughout the group or appear more global.
- Synchrony can be more or less perfect, or partial.

In principle, any of these choice points or combinations thereof could be used to send information. While mate attraction might be the most common reason why synchronous flashing occurs, there is no reason to think that such behavior has only one biological or social function. Indeed, it need not have any direct function at all; it could be fireflies at play.

Back to the main question. What makes some fireflies synchronous and others behave in a more asynchronous fashion? The answer turns out not be some unique genetic or neural mechanism. So, which factors turn out to be the most common among species of synchronous fireflies, and which common factors are the most predictive of Sync? The leading answer is that density of fireflies as engendered by the appropriate environment is the greatest predictive factor (Sarfati, Hayes, and Peleg 2021): "At low firefly density, flashes appear uncorrelated. *At high density, the swarm produces synchronous flashes within periodic bursts*... Our results suggest that fireflies interact locally through a dynamic network of visual connections defined by visual occlusion from terrain and vegetation. *This model illuminates the importance of the environment in shaping self-organization and collective behavior*" (2021, p. 1, emphasis added).

Here are the important lessons learned in this recent study involving both hours of footage collected in the wild and computer modeling of *P. carolinus* swarms (2021, pp. 2-4, emphasis added):

1. While density-dependent transitions from disorder to order have been observed in various animal systems exhibiting collective behavior, even only a few fireflies in isolation in a sufficiently small space attempt to flash in synchrony.
2. A high density is necessary to receive and relay the flash information across the swarm, because fireflies interact at relatively short range.
3. Flash propagation among fireflies does not always follow the orientation of the terrain (contrary to some hypotheses).
4. Flash propagation occurs through a "relay-like process" to carry information.

5. Information propagation in swarms is linear on average. While the time-distance relationship in flocking birds and human crowds is very narrow, for fireflies, "simultaneous flashes, both later and earlier ones, can be relatively spatially distant." So, "local interactions extend beyond each individual firefly's immediate geometric vicinity."
6. Fireflies determine "local" through line of sight.
7. Nevertheless, "motion asymmetry suggests that at least some fireflies, such as "pacemaker males," are able to perceive the global state of the swarm (i.e., relative delay in the burst), and not simply their local environment." This is the source of collective behavior disparities during bursts.
8. Hence, firefly behavior is multiscale with different distances involved.
9. In dense populations of animals, visual occlusion occurs due to nearest neighbors, whereas in dilute populations there is little occlusion. *P. carolinus* swarms present a situation wherein the swarms are dilute, but the environment creates significant visual occlusion. For example, fireflies at higher elevations have a fuller view of the swarm, but density drops with altitude. In contrast: "Most fireflies are located in the vegetation, and thus their line of sight is obstructed... *Terrain and vegetation also create large variations in the number and range of accessible interactions depending on the firefly's orientation*" in space.
10. Even though *P. carolinus* males synchronize their rhythmic flashing with their "local" peers, there must be a sufficient number of active fireflies to transport collective pace information to form global swarm synchronization. So-called local firefly interactions have more to do with a dynamic network of visual connections defined by relative orientations and visual occlusion from terrain and vegetation, thus resulting in a mixture of short-range and long-range interactions, rather than being primarily metric or topological in nature.

In addition to terrain and vegetation, there are other environmental factors across the globe that are universal to synchronous fireflies, such as humidity, heat, moisture, and darkness (e.g., low light pollution). What we see is that there are many contextual factors at work in producing firefly synchrony, including the existence of three spatial dimensions. For instance, at 84° F the same male firefly producing the same flash-train flashes twice as fast as he does at 60° F. At 50° F or below there is no flashing at all (Frierson Faust 2017, chapter 3). These external, environmental conditions— part of the context shaping firefly behavior—might easily be relegated to nothing but background conditions or boundary conditions, but the view of ontological emergence we advocate in this book—contextual emergence—indicates that this move underestimates the importance of contexts and relationality.

Back to Silberstein's backyard. Could we have seen, could there be synchronous firefly activity present? Yes indeed, so long as all the relevant conditions are met. We need a sufficient number of fireflies per unit area, enough of whom have line of sight with other fireflies in three-dimensional space, so as to produce relatively short-, medium-, and longer-range statistically stable connections. We need a local terrain and vegetation setting that enables the density necessary as well as structuring firefly interaction and orientations. That same setting must be a good breeding ground and have all the necessary features to attract and keep fireflies at the appropriate density, and so forth. In short, you do not have to go to the Great Smoky Mountains National Park in Tennessee, or Malaysia, and you do not have to find a special firefly with special biomechanisms to experience synchronous firefly behavior.

We could codify or model all these factors with differential equations, causal graphs, algorithms, and so forth, but the actual explanation for synchronous behavior lies in a specific set of multiscale contexts and relations that constrain behavior, thus enabling fireflies to synchronize. Even in this relatively simple case, explaining synchronous behavior requires environmental conditions involving temperature, humidity, moisture, darkness, terrain, vegetation, anthropogenic behavior (e.g., low light levels and the absence of other forms of pollution), and so on. Also required are the "biopsychosocial" mating and communicating factors that the fireflies only bring to the table given appropriate contexts that prohibit some firefly behaviors while promoting others to produce synchrony. This is contextual emergence *par excellence*, and we will argue that it is the alternative to strong and weak emergence that some people have been seeking. We begin our journey with fireflies to illustrate that the deepest understanding of emergence does not require going to the realms of quantum mechanics and experience, though we will get there.

Explaining synchronous behavior for any system at any scale seems mysterious only when one assumes that the individual unit in question (e.g., the firefly, the neuron, the particle, etc.) is *the fundamental entity* in the sense of having a relatively intrinsic nature and behavioral autonomy. For example, in the case of synchronous fireflies, the most up-to-date analysis, dubbed "the first ever spatiotemporal description of how firefly synchronization begins," shows the following:

> While even a small number of interacting males flash, periodic bursts occur only in groups larger than 15 males. At high density, the swarm produces synchronous flashes within periodic bursts, like a dancefloor strobe light that's turned on and off at certain intervals. While the swarm of fireflies flashed in linked sequences of light bursts and dark inactivity, solitary males did not maintain a steady rhythm of illumination. This contrasts with a major assumption of the [older] Kuramoto

> model, which is that individual agents perform a periodic behaviour even when they are by themselves, insulated from the swarm. This creates a coupling or causal loop that links the individuals, the group, and the environment in a perpetual cycle of affecting and being affected.
>
> (Peleg 2021)

We see here that there is a real sense in firefly Sync that the individual firefly is not the fundamental unit or explainer in this process. Such Sync is an essentially dynamical multiscale process that ranges across the voluntary action of the organism regulating the release of the neurotransmitter octopamine that then triggers a complex chemical reaction in the fireflies' "lantern" to produce cold light, to the relative positions of the fireflies in space, to every key geographical, topological, and ecological feature of the surrounding environment, to the social norms of various fireflies.

Examining several cases from across the sciences that range far beyond Sync, we will argue that multiscale contexts and relationality are as fundamental as anything else in nature. We will often refer to these as stability conditions, of which there are many different types, as well as case-specific varieties. That is, reality is nothing like the Game of Life; there are no fully autonomous or fully determining entities and laws at any scale.

Let us not forget the whisky, as if we could, which, by the way, pairs well with the watching of fireflies. Whisky, whether bourbon or scotch, is often present at academic discussions of emergence, but rarely the topic itself. Yet it turns out that the flavor profiles of specific whiskies is a nice case of contextual emergence. The flavor profile of any particular bottle of whisky is a complex function of multiscale factors such as size, shape, material, and make of still; how long the whisky was aged; mash bill; water source; type of yeast; particular cuts or distillates put into the barrel; geographic location such as proximity to (specific) oceans or mountains; elevation relative to sea level; temperature and weather variations over time; type of warehouse, such as building materials and style of shelving; internal elevation and location in warehouse, such as which side the barrel is on or directions faced by the barrel; the way the barrels are stored and rotated; the variable airflow, humidity, sunlight, and temperature in the warehouse; the type and size of barrel being utilized and how it is prepared, such as what kind of wood it is, how old the wood is, and from what region the wood is; the particular barrel charr or seasoning; the properties of staves that were used, wherein all the previous barrel choice points apply; the presence and specifics of peat and smoke. No doubt we could name more factors, but you get the point.

Whisky is an historical entity, and each bottle is a multifaceted reflection of a thousand things from across its multiscale spatiotemporal environment; a specific time and place reflected in each and every bottle. One is reminded of Indra's Net. Just imagine an infinite number of whisky bottles all reflecting one another into infinity—or maybe that is the whisky talking. The point, of course, is that whisky is a much better metaphor for emergence and reality than finite cellular automata. And as it turns out, contextual emergence provides a much better account of the emergence of quantum entanglement, life, and even mind, even if we cannot give you all the details for the emergence of life and mind in this book.

We know that some readers will immediately be concerned that contextual emergence is too common to be interesting. And some readers will be thinking that, for example, the case of whisky is easily discharged because each factor can be reduced in principle to some more basic physical or chemical process. After all, every concrete contextual feature at every scale, whatever it may be, just reduces to physics, right? Wrong. This is a book-length argument spanning the sciences that multiscale contexts and stability conditions are at the very nature of reality and are ineliminable in principle in almost every scientific explanation, including what is considered to be fundamental physics.

As we shall see, the pattern of contextual emergence arises no matter what length and time scale of the universe one chooses to attend to. So, while contextual emergence is common, it turns out to be terribly interesting. Unlike the allegedly rare and definitely spooky strong emergence, contextual emergence is always scientifically explicable. And unlike weak emergence, contextual emergence is not merely about epistemic, computational, or other limits. It is because of the contextual nature of reality that emergence is possible and fully explainable by science and yields a unified, coherent account of nature. We think that is a virtue and not a bug.

Cue the fireflies! The show is about to begin.

Acknowledgments

While all three authors have published a good deal on the topic of emergence over the last few years, this project began in earnest roughly eight years ago at the University of Munich (LMU) during the summer of 2013, where professors Bishop and Silberstein were Scholars in Residence at the Munich Center for Mathematical Philosophy (MCMP). Bishop and Silberstein have been working on the project ever since. Starting in July 2015, the manuscript has been helped by the Emergence Project at Durham University, England, at the Workshop on Emergence in Physics, where the manuscript was part of a reading group where professors Bishop and Silberstein gave multiple related talks. In April 2016, Bishop and Silberstein returned to Durham for the Concluding Conference on Emergence at Durham University, where multiple talks were given again, and many comments received. It was there that we found our third coauthor, post-doctoral researcher Mark Pexton—an expert on emergence in condensed-matter physics, among other things. We should also formally note the special assistance from Professor Tom Lancaster (Department of Physics at Durham University) on Chapter 5, which relates to condensed-matter physics. In the interim, parts of the manuscript were vetted at several other venues, including:

April 2016: The Problems of Scale Workshop, University of Wollongong, City of Wollongong, New South Wales, Australia.

May 2017: Department of Physics, University of Cape Town.

February 2018: Philosophy Department Colloquia, Dartmouth College.

May 2018: Reductionism, Emergence and Critical Thinking Workshop, ETH Zürich.

July 2018: 19th UK and European Conference on Foundations of Physics, Utrecht University, Utrecht, The Netherlands.

October 2019: Re-Thinking Emergence Workshop, University of Lisbon, jointly hosted by the Centre of Philosophy and the Centre for Philosophy of Science of the University of Lisbon. The event inaugurated their new project *Emergence in the Natural Sciences*, funded by the Fundação para a Ciência e a Tecnologia. The reading group on emergence at the University of Lisbon was also kind enough to provide us with detailed feedback on an earlier draft of the manuscript.

We would like to thank several anonymous reviewers for their very thoughtful and helpful comments. Finally, we would also like to thank our editor at Oxford University Press, Sonke Adlung (Senior Commissioning Editor, Science), for his support and guidance.

1

Introduction

> In Nature we never see anything isolated, but everything in connection with something else which is before it, beside it, under it, and over it.
>
> Johann Wolfgang von Goethe

1.1 Setting the scene

In the West at least, the topic of emergence has been going in and out of fashion at least since the time of Galen (Caston 1997, 2001) and apparently even earlier in the East (Ganeri 2011). Why has this topic persisted for so long? The answer probably depends a great deal on the place and time, and the community involved. For us, the answer to the question will be a mix of metaphysical and scientific concerns. These days there is a large intersection between those two sets of concerns. It is worth noting that naturalistic metaphysics and science, which were of course once historically co-extensional in natural philosophy, are in general both concerned with the question of how order, stability, and novelty are possible and how they happen. How, on multiple scales, do new and novel robust systems and patterns arise and maintain their stability? How can order come out of disorder? Where does organization come from? These questions are perhaps especially pressing if one thinks physics alone is fundamental or brute, and that basic physical processes determine all the other facts at larger spatiotemporal scales.

Scientists, when confronted with questions about order and novelty with respect to some specific phenomenon in their discipline, will be inclined to invoke various laws, constraints, mechanisms, or other determining factors to answer these questions, whereas many in analytic metaphysics are more concerned about the nature of laws, mechanisms, and properties themselves. Metaphysics is interested in questions such as: what is a law of nature? For instance, a biologist might say that a particular mechanism X explains why a biological pattern or product Y is reliably produced, and a metaphysician

Emergence in Context. Robert Bishop, Michael Silberstein, and Mark Pexton, Oxford University Press.
 DOI: 10.1093/oso/9780192849786.003.0001

might ask: but what makes that mechanism or any mechanism continue to be stable and robust? Metaphysicians are asking: how are such stable and robust mechanisms possible at all? Thus, they tend to seek the answer in the realm of the *meta*-physical—something second-order and yet not supernatural; something such as a form of causal or nomological metaphysical glue, or modal necessity. Scientists might wonder at least in principle if, for example, chemistry reduces to quantum mechanics, in the actual world, whereas metaphysicians might wonder if there are any possible worlds where chemistry could fail to be modally determined by physics. We can see that what is explanans for the scientist is often explanandum for the metaphysician.

Scientists and philosophers of science often have little patience for such metaphysical endeavors, as they seem scholastic, beyond our ability to know in principle, and insufficiently grounded in actual sciences—hence the charge that analytic metaphysics never makes any real progress, and that scientists do not need philosophy. This is a very public charge we have heard over the years from physicists such as Steven Weinberg, Stephen Hawking, and Lawrence Krauss (Holt 2012). Recently, a multidisciplinary cohort of scientists, including Carlo Rovelli and Margaret McFall-Ngai, as well as key historians and philosophers of science, offered a defense of philosophy of science as valuable for

> The clarification of scientific concepts, the critical assessment of scientific assumptions or methods, the formulation of new concepts and theories, and the fostering of dialogue between different sciences, as well as between science and society.
>
> (*PNAS*, March 5, 2019, vol. 116, no. 10, p. 3948)

We agree, and we hope this book is a testament to their assessment.

We often talk about emergent properties—but obviously, talk about emergence or reduction is not limited to properties. Other examples include entities such as molecular structure, behaviors such as flocking, swarming or Sync, mechanisms such as circadian clocks, laws such as those of thermodynamics, theories more generally such as classical mechanics or evolution, models such as agent-based models, processes such as protein folding, levels of organization such as neural networks, spatiotemporal scales such as the molecular, "causal powers" such as mental causation, substances such as the chemical elements, agents such as ourselves, and robust and novel patterns of all varieties. The list could go on.

Moreover, emergence is recently a trendy topic, in part because of its resurrection across the sciences of and foundations for "complexity" (another loaded word, see Bishop and Silberstein 2019) which ranges from condensed matter theory (CMT) to network theory, and beyond. Just as with thermodynamics, statistical mechanics, and quantum field theory, the sciences involving complexity and many-bodied systems now share the same bag of techniques with CMT. These include formal modeling methods from dynamical systems theory, renormalization group methods, universality and phase transitions, and symmetry-breaking.

Nothing perhaps offers more confusion or creates more disagreement when thinking about order, stability, and novelty than talk of emergence. Some of the confusion in discussions of emergence can be seen in how prominent physicists writing on the topic have been interpreted. Take the following example from P. W. Anderson:

> That I saw the "theory of everything" as the theory of almost nothing. The actual universe is the consequence of layer upon layer of emergence, and the concepts and laws necessary to understand it are complicated, subtle and, in some cases, as universal as anything the particle folks are likely to come up with.
>
> (P. W. Anderson, 2011, p. 7)

Whenever a claim is made about X being emergent with respect to Y, we need to ascertain what senses of emergence the speaker has in mind, and then we can begin to critically evaluate their claim. The passage just quoted was not chosen at random, as Nobel Prize winner Anderson is a leading figure in the recent debates on emergence. Many would argue that his well-known 1972 paper "More is Different" (MID), was the key impetus for the renewed interest in emergence in the last few decades.

Anderson has been viewed by some as advocating ontological reductionism, while others view him as advocating an ontological form of emergence. What really is his view? He begins MID by acknowledging that "the great majority of active scientists" accept reductionism (1972, p. 393), and nowhere in the article does he challenge that acceptance. What he does argue is that "The ability to reduce everything to simple fundamental laws does not imply the ability to start from those laws and reconstruct the universe" (1972, p. 393). For example, even if quantum-mechanical processes completely determine all the essential facts about chemistry, such as molecular structure, that does not mean we can derive, deduce, or predict molecular structures *ab initio* starting with nothing but non-relativistic quantum mechanics. In other words, Anderson is arguing primarily for methodological or epistemic emergence.

Anderson's claims rest on two observations. First: "At each stage, entirely new laws, concepts, and generalizations are necessary, requiring inspiration and creativity to just as great a degree as in the previous one" (1972, p. 393). Second, there is a hierarchical ordering of scientific disciplines, where "The elementary entities of science X obey the laws of science Y," with Y belonging to the lower-level description and X belonging to the higher (1972, p. 393). Anderson is not explicit in his 1972 paper as to whether "obey" means *determined by* or merely *dependent on*, and this is a crucial difference, as we shall see. The former expresses ontological reductionism while the latter gloss does not. Nonetheless, Anderson does admit that he fully accepts reductionism in the ontological sense (1972, p. 394).

Anderson reaffirmed this view when reflecting on his 1972 paper: "I was perhaps, at that time, no less a reductionist, nor less willing to mystify" (2011, p. 135). His point in MID was to argue that the descriptions of high-energy physics (HEP) plus a computer of even infinite computational power would never yield an understanding of phenomena such as superconductivity or self-replicating molecules (2011, pp. 136–137). New concepts are always needed at larger special science scales because, as he puts it, such concepts "allow enormous compression of the brute-force calculational algorithm, down to a set of ideas which the human mind can grasp as a whole" (2011, p. 136). The emphasis in MID was twofold. First, "the idea of intellectual autonomy of the two levels of understanding," HEP and CMT, respectively (2011, p. 137). Second, that symmetry-breaking—particularly spontaneous symmetry-breaking—is an "emergent" property in the sense that broken symmetry is distinct from the most fundamental dynamical laws governing the system (i.e., not found in the relevant dynamical equations).

Anderson's motivations for embracing ontological reductionism are revealing: "It is primarily by means of such a reduction that we can rescue ourselves from the possibility, or at least from the accusation, that science may be purely a socially constructed consensus with no objective external reality to which it corresponds" (2011, p. 138). He goes on to say that "when one succeeds in finding a reductionist explanation for a given phenomenon, one embeds it into the entire web of internally consistent scientific knowledge" (2011, p. 138). This means that it becomes harder to modify one piece of scientific knowledge without affecting the rest. His motivation then speaks to objectivity, realism, and unification in science. Anderson clearly is no mere instrumentalist.

Anderson and other condensed-matter theorists are frequently interpreted as arguing for stronger or more metaphysical notions of emergence, where X is not merely *dependent* on and *distinct* from Y, but not even fully *determined* by Y. Take Robert Laughlin, yet another vocal Nobel Prize winner in CMT who writes about emergence. In their well-known 2000 paper Laughlin and David Pines say the following: "Low-energy excited quantum states in such systems as superfluids in Bose liquids, atomic condensates, superconductors, band insulators, ferromagnets, and quantum Hall states are particles in exactly the same sense that the electron in the vacuum of quantum electrodynamics is a particle." Laughlin and Pines note that these phenomena "carry momentum, energy, spin, and charge, scatter off one another according to simple rules, obey Fermi or Bose statistics depending on their nature, and in some cases are even "relativistic," in the sense of being described quantitatively by Dirac or Klein-Gordon equations at low energy scales." But these look more like systems than elementary particles. These phenomena do not "exist outside the context of the stable state of matter in which they live." Laughlin and Pines dub these "quantum protectorates" and go on to say that such "protectorates, with their associated emergent behavior, provide us with explicit demonstrations that the underlying microscopic theory can easily have no measurable consequences whatsoever at low energies." One does not even know what that underlying theory is "until one raises the energy scale sufficiently to escape protection" (p. 29).

Laughlin and Pines go on to claim that "self-organization and protection are not inherently quantum phenomena" and that "quantum and classical protectorates ... are governed by emergent rules" (p. 29). These passages might suggest to some that Laughlin and Pines are advocating a more ontological kind of emergence. For instance, Carl Gillett, in a volume edited by Corradini and O'Connor, interprets Laughlin as claiming that "we need to posit further fundamental laws, or "organizing principles," to explain the novel behaviors and powers of the components" and "the composed entity ... synchronously, influences the features of its components" (2010, p. 32), whereas in the same volume Mark Bedau claims that Laughlin and Pines are only defending epistemic or "weak" emergence (2010, p. 59). These sorts of confusions illustrate why conceptual analysis is useful.

There actually is very little daylight between Laughlin and Anderson, as Laughlin says:

> There are many notorious failures of alleged *ab initio* computation methods, including the phase diagram of liquid ^{3}He and the entire phenomenology of high-temperature super-conductors. Predicting protein functionality or the behavior of the human brain from these equations [Schrödinger equation or Dirac equation] is patently absurd. So the triumph of the reductionism of the Greeks is a pyrrhic victory: we have succeeded in reducing all of ordinary physical behavior to a simple correct TOE [i.e., quantum mechanics] only to discover that it has revealed nothing about many things of great importance.
>
> (2000, p. 28)

In case there is any interpretive doubt remaining, in his 2005 book on emergence Laughlin says, "All physicists are reductionists at heart, myself included. I do not wish to impugn reductionism so much as establish its proper place in the grand scheme of things" (2005, p. xv).

Now that we have a sense of what Anderson and Laughlin are claiming about the emergence of CMT and other larger-scale phenomena from quantum mechanics, we can begin to critically assess it. Thus, in Chapter 5 on CMT, emergence, and universality, we will evaluate Anderson's and

Laughlin's claims and those of other notable figures in CMT. To do so, in addition to more rigorous conceptual analysis, we will have to look at detailed case studies.

As useful as conceptual analysis is for bringing clarity, we acknowledge that philosophers can be overzealous in the quest for clarity and classification. Indeed, in both science and philosophy, one oft-remarked problem with writing about emergence is that the word is used in a mind-numbingly heterogeneous number of ways. The use of the term "emergence" has proliferated wildly in the last few decades, as have different meanings for the word. Philosophers have tried to bring order to all this by concocting various taxonomic schemas, unfortunately often using the exact same expression such as "strong emergence" or "weak emergence" to mean different things. No doubt many scientists find this practice to be another exercise in scholasticism.

Consider Fig. 1.1, meant to be a map of the space of different types of emergence. Each ellipse of a specific color represents a continuum ranging between its specific pair of "opposites" named at each pole (e.g., Ontic versus Epistemic or Strong versus Weak). Each such pair represents a dimension in the possibility space of types of emergence. We doubt the diagram is exhaustive, but one could in principle plot the various types of or definitions of emergence that have come into being over time if found worthwhile.

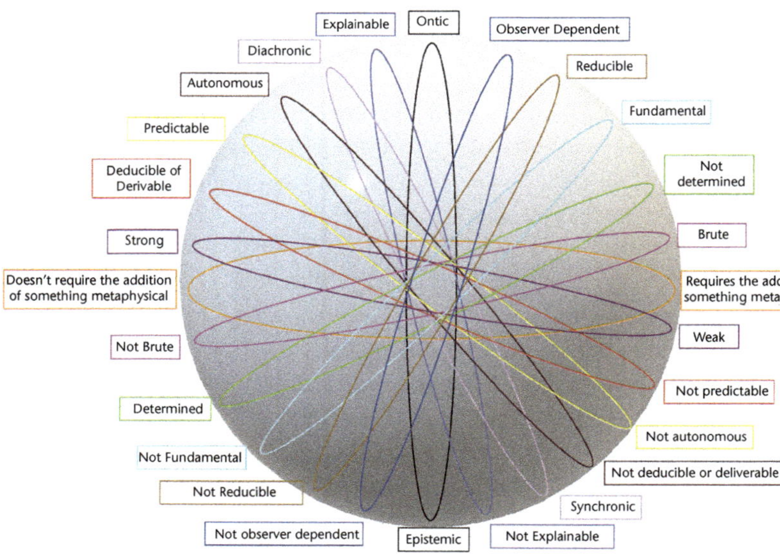

Figure 1.1 Types of emergence.

The task would also be complicated by the fact that, again, different people sometimes use the same name to mean different things. Plotting the positions would also be complicated by the fact that there are combinations in the literature that might seem counterintuitive. For instance, there is Jessica Wilson's "weak metaphysical [ontic] emergence" (2013, p. 225). In contrast, Bedau's "weak emergence" is *mutually exclusive* with ontological emergence (2008).

Or take the "opposing" pair diachronic/synchronic. Diachronic emergence is supposed to inherently involve dynamical or causal processes and take time for the emergent property to appear, whereas with synchronic emergence the larger scale property P is said to emerge at a time *t*, as soon as its smaller scale emergent base B is present at a time *t*. But are diachronic and synchronic emergence really mutually exclusive? Consider quantum entanglement, which many of us have argued is a case of emergence wherein an entangled state is, as the expression goes, definitely "more than the sum of its parts" (more on this in Chapters 4 and 5). There is certainly a synchronic aspect to this case of emergence if we consider an entangled state at time *t* with respect to the particles participating in that state at a time *t*, but also a diachronic aspect, as the entanglement presumably exists as the result of the temporal evolution of the Schrödinger equation (Hawthorne and Silberstein 1995; Silberstein 1999, 2002; Kronz and Tiehen 2002; Hütteman 2005; Silberstein 2009a, 2017).

The "opposing" pair ontic/epistemic is particularly troubling. There is a sense in which the reduction/emergence debates are driven by concerns along this axis. In the case of Anderson and Laughlin, they are not bothered by the idea that there is ontological reductionism in the physical world, but there probably will never be *wholly reductive explanations*. We are stuck with epistemic emergence as our lot in life. An important reason why so many in the debates are comfortable with this kind of position is that they think the opposite of ontological reductionism is an ontological emergence of the most radical type (what some call "strong" emergence), i.e., there are some kinds of brute bridge laws or causal processes linking various domains such as the physical, the chemical, the biological, and the mental. This entails that there are brute novel properties and processes independent of fundamental physics. For scientists, and many philosophers, such a radical ontological emergence is highly problematic for making sense of the order and stability of our world, as well as seeming irrelevant to the sciences.

Granted, if the only options for ontology were ontological reductionism or a radical ontological emergence, then the norms of order and coherence push strongly in the direction of choosing a reductionist world. We want to argue, however, that there is another choice that is well supported by science. In this book, we will be discussing and defending a type of emergence called *contextual emergence* that is both epistemic and ontological while not being radical, and illuminates the coherence and unified structure of the world without being reductionist. Comparisons to other conceptions of emergence will be made throughout the book, but we can say that contextual emergence will transcend and subsume many of the alleged dichotomies discussed previously.

For now, note that contextual emergence is: 1) explanatory and not brute, meaning it is grounded in and meant to subsume many cases from across the sciences; 2) explanatory without being reductionist, but is often consistent with *some notions of reductive explanation*; 3) unifying both with respect to relating different scientific disciplines and with respect to other accounts of emergence; and 4) not rare or exotic. Emergence would be a much less interesting and much more puzzling topic if the only potentially real cases were quantum weirdness and conscious experience.

1.2 Emergence and metaphysics

Metaphysicians often complain that science and philosophy of science, while great at generating detailed case studies such as the emergence and stability of superconductivity, never get around to drawing deeper or more universal metaphysical conclusions. As Robert Batterman puts it regarding emergence: "Instead of focusing on questions of correct ontology, I suggest a more fruitful approach is to focus on questions of proper modeling technique" (2015, p. 133). But the concern for someone primarily interested in studying the ultimate nature of reality is that there is no point in focusing on science if it does not aid this project. For instance, it would no doubt be disappointing if the only conclusions that one could draw about emergence based on science is that on multiple scales, novel processes and patterns come into being that are not predictable or derivable from, nor intuitively understandable in terms of, underlying smaller-scale physical processes. This lesson is trivially made with any number of cases ranging from identical geometric shapes (such as cubes made of different materials) that all fit in the same hole, to natural selection on any planet, no matter how alien the physics or the biological organisms involved. We share this metaphysical concern and will make the case that collectively the sciences have profound things to tell us about the nature of emergence and reality more generally.

We are sympathetic to all the preceding concerns from both scientists and metaphysicians. We aim to speak to both audiences. This book is an exercise in twenty-first-century natural philosophy. That is, we want to use a range of cases from physics to biology and beyond, as a basis to make some strong inferences about the nature of reality—about the nature of emergence in *the actual world*. Thus, unlike many other accounts of emergence in philosophy, we begin not with logic and metaphysics but with science, and then move to metaphysics on that basis. Therefore, after introducing contextual emergence in Chapter 2, we will proceed with the scientific cases (Chapters 3–6) and move on to grander scientific and ontological implications in Chapter 7 for the nature of laws, mechanisms, dispositions, causation, and a host of other metaphysical topics, thereby drawing connections and summarizing what we learned from the preceding chapters.

Our optimism in reaching both audiences is based in part on the fact that, when it comes to emergence, there is a growing community of scientists from different disciplines, philosophers of science, and naturalistically inclined metaphysicians who are working together to get the science right, the conceptual analysis right, and the ontological inferences right. The recent Routledge *Handbook of Emergence* (2019) is an example which grew out of the widely praised multidisciplinary Durham Emergence Project.

We think it is safe to say that most scientists see themselves as engaging in natural philosophy, and are not ultimately satisfied with metaphysical quietism, mere practical problem solving, or instrumentalism. As astrophysicist and writer Adam Becker puts it:

> Science is about more than mathematics and predictions—It's about building a picture of the way nature works. And that picture, that story about the world, informs both the day-to-day practice of science and the future development of scientific theories, not to mention the wider world of human activity outside of science
>
> (2018, p. 7).

Closer to home, take the following claim from Nobel Prize winner Steven Weinberg, "It seems to me that the trouble with this approach [instrumentalism] is not only that it gives up on an ancient aim of science: to say what is really going on out there. It is a surrender of a particularly unfortunate kind" (2017, p.4).

Weinberg is regarded as an arch-advocate of ontological reductionism. As he says himself: "But phenomena like mind and life do *emerge*. The rules they obey are not independent truths but follow [logically] from scientific principles at a deeper level" (2001, p. 115). He is persuaded that scientific inquiry has already demonstrated convincingly that reality is an ordered hierarchical structure, wherein, whatever the fundamental theory of physics is (e.g., quantum gravity), the laws and phenomena of all the other sciences will be entailed by the laws and phenomena of the fundamental physical theory. That fundamental physical theory will be "based on a few simple principles, laws of nature, from which all other regularities can be deduced" (p. 114), and that there are no autonomous special science laws that are logically independent of the principles of physics (p. 112).

Weinberg is defending foundationalism. Foundationalism is, first, the idea that only basic physics contains brute or fundamental laws and entities, that is, only basic physical facts are fully ontologically autonomous in that they depend on nothing else for their existence and are determined by nothing else. Second, the world exhibits a well-ordered objective hierarchy ranging from basic physics on up the spatiotemporal scales, wherein the facts from basic physics unidirectionally determine all the other facts in the world. Foundationalism thus implies that, at least in principle, the complete explanation for phenomena at a particular scale or "level" must come from a theory about smaller scales or "lower-level" phenomena. These grand claims about a particular hierarchical structure of reality and the unity of science are natural philosophy *par excellence*. Contextual emergence challenges this "foundationalist" or hierarchical picture of reality. It is a rejection of the very idea of ontological autonomy at any scale, and is thus a rejection of the idea that determination and hence explanation must strictly be "bottom-up," at least in principle (Silberstein 2017).

Epistemologically speaking, Anderson and Laughlin would no doubt disagree with Weinberg about the logical autonomy of special science laws. For instance, natural selection cannot be derived from any fundamental physics and could be true in worlds with radically different underlying physics. Indeed, natural selection will be true in any world, no matter what the physics, where we have organisms with differential properties conferring differential reproductive advantage in competing for resources in a particular environment. Here is how Laughlin makes the point: "We also know that while a simple and absolute law, such as hydrodynamics, can evolve from the deeper laws underneath, it is at the same time independent of them, in that it would be the same even if the deeper laws were changed" (2005, p. 207). Everyone can agree that special science laws have this kind of epistemic autonomy, but this is also consistent with ontological reduction in the sense of bottom-up *determination*. We will argue there are empirical and scientific reasons to believe in ontological contextual emergence for the so-called special sciences. That is, without resorting to radical emergence, we will push back on foundationalism.

Thus, broadly speaking, when it comes to relatively recent interest in emergence, we can note that, as depicted in the Weinberg quotes above, the widespread assumption in naturalistic metaphysics and in science is that matter alone is fundamental and brute, and thus life, mind, and culture must somehow *emerge* from fundamental physical processes. But what does "emerge" mean here? Laughlin claims: "I prefer the more physical view that politics, and human society generally, grow out of nature and are really sophisticated high-level versions of primitive physical phenomena" (2005, p. 210). Take a popular analogy people often provide for ontological reductionism—that the universe is like a finite cellular automaton, such as John Conway's Game of Life. In this analogy, the cells in the game are like the fundamental physical entities, the rules of the cell's temporal evolution are like the fundamental laws of physics, and robust patterns in the game such as Gliders and Eaters emerge "for free," just as Weinberg says that entities and processes in CMT and biology emerge for free from more fundamental physics. Echoing Weinberg, this is exactly what Laughlin seems to be claiming about psychological, cultural, and sociological phenomena. The preceding quotation from Laughlin and early quotations from Anderson make it clear that for most physicists, even condensed-matter theorists, we can perfectly substitute the claim that chemical, biological, and mental phenomena "emerge from fundamental physical processes," with the claim that such phenomena "are fully determined by fundamental physical processes."

This is really an astounding and counterintuitive claim if one dwells on it for any length of time. Indeed, it is at this point that the idea of ontological reductionism and radical emergence being exhaustive options becomes most troubling. We sense that this is why condensed-matter theorists such as Anderson and Laughlin often seem to want a third option, or at least sometimes appear so. Take the fundamental ambiguity in Laughlin's account, for example. Referring to the interplay between basic physical entities and what he calls the emergent organizational principles, he says,

> all of these miraculous behaviors [the emergent organizational principles such as the laws of hydrodynamics] can be accounted for as spontaneous organizational phenomena that descend from underlying law. We also know that while a simple and absolute law, such as hydrodynamics, can evolve from the deeper laws underneath, it is at the same time independent of them, in that it would be the same even if the deeper laws were changed.
>
> (2005, p. 207)

The idea is that the higher ordering principles arise spontaneously out of the underlying laws but are also independent of those laws. They have universality in that a subsequent change in the lower laws would not affect the higher ordering principles, such as the laws of hydrodynamics. The universality and independence of the ordering principles with respect to the most basic underlying laws seems to be merely epistemic emergence for Laughlin. Nonetheless, his rhetoric suggests that he also treats such universality as a violation of ontological reductionism. He even says that the "epistemological barrier" to understanding how the ordering principles emerge from underlying laws as well as understanding whether the latter are more fundamental than the former is *physical* (2005, p. 207). We think that the question one ought to focus on is *why* are such higher organizational principles and concomitant phenomena multiply realizable? We argue that contextual emergence answers this question and clarifies what Laughlin and others are really after with such claims, without being merely epistemic emergence and without invoking the implausible radical emergence (Silberstein 2017; Bishop and Silberstein 2019).

If ontological reductionism is true, then epistemic or weak versions of emergence are the only kinds possible. For example, David Chalmers defines weak emergence as follows:

> To capture this, one might suggest that weak emergence is the phenomenon wherein [non-obvious] complex, interesting high-level function is produced as a result of combining simple low-level mechanisms in simple ways ... This conclusion captures the feeling that weak emergence is a "something for nothing" phenomenon ... The game of Life and connectionist networks are clear cases: interesting high-level behavior emerges as the consequence of simple dynamic rules for low-level cell dynamics.
>
> (2006, p. 252)

Much like Anderson and Laughlin, Chalmers talks about weak emergence in terms of "*ease of understanding* one level in terms of another. Emergent properties are usually properties that are more easily understood in their own right than in terms of properties at a lower level" (2006, p. 251).[1] However, is weak emergence in principle sufficient to explain everything? We will argue in the negative.

There are certainly others who think that while everything emerges from and depends on fundamental physical processes, they doubt that weak emergence is sufficient to explain everything. These folks thus tend to champion what we called strong or radical emergence, which historically has many different definitions (see, for example, Morgan 1912; Alexander 1923; Broad 1925; O'Connor 1994; Humphreys 1997; Silberstein 1999, 2012; Chalmers 2006). Once again, the basic idea is that if X is strongly emergent with respect to Y, then Y does not determine X. Or, if Y does determine or cause X, it is a brute fact that has no further explanation. Thus, a strongly emergent property is one that, in principle, cannot be derived from, predicted from, or fully explained by some more fundamental (physical) theory, because the emergence of such properties are brute facts. Another common but distinct way to define a strongly emergent property is in terms of its autonomous causal efficacy, what some unfortunately call "downward causation" with

[1] Bear in mind that "weak emergence" is defined differently by different people (e.g., Bedau 2008; Huneman 2008; Wilson 2015).

its suggestion of the brute emergence of some *essentially new type* of phenomena or reified "higher-level" of organization, something force-like that then magically exerts a new causal force on that from which it emerged. Of course, some prefer to conceive of "downward causation" as the parts being synchronically constrained by the wholes that they make up, because they believe parts and wholes, being synchronic, cannot have *causal* relations as such. Either way, this is the radical ontological emergence that threatens a disunified picture of the world.

Historically, mental phenomena have been a driver of the emergence debate, and that has not changed. Witness the hard problem of consciousness and the explanatory gap. Chalmers, for instance, posits universal, brute psychophysical bridge laws, which determine the distribution of conscious properties given physical, functional, or informational properties in this world (1996). Tim O'Connor and Hong Yu Wong are modern-day defenders of strong emergence who hold that given the instantiation of particular complex microstructural properties, basic mental properties will be *caused diachronically* to come into being (2005). These emergent mental properties will have causal powers that are distinct from any microphysical or microstructural properties. Furthermore, these emergent mental properties will, in turn, cause the instantiation of both other microphysical properties and other emergent mental properties. According to this view, physical states play a causal role in the rise of conscious states, *and* these conscious states affect physical states in turn (so-called downward causation). On their view, mental properties would be diachronically or dynamically emergent.

It must be said, however, that these days, with some exceptions (e.g., Symons 2018; Vintiadis 2018; Wyss 2018; Hendry 2019), outside of intentionality and conscious experience, one no longer sees much advocacy for strong emergence. As suggested by Jeremy Butterfield, another notable modern-day exception is Nobel Prize-winning condensed-matter theorist Anthony Leggett. According to Butterfield, "Leggett's work was motivated by his expecting that the phenomena could not be explained by orthodox quantum mechanics—he believed it would need to be modified by admitting new forces, e.g., a quantum analogue of Broad's configurational forces" (2011b, p. 936). Leggett, with his macrorealism interpretation of quantum mechanics, has subsequently attempted and failed to show that unadorned non-relativistic quantum mechanics is incomplete with respect to the emergence of certain classical phenomena (Allen, Maroney, and Gogioso 2017). If it turned out that intentionality and conscious experience were the only possible or plausible candidates for ontological emergence, that would be troubling and puzzling—which is to say that the hard problem of consciousness and the explanatory gap should not drive the larger scientific debate about the structure of reality.

Thus, all these defenders of radical ontological emergence are the exception that proves the rule. When it comes to ontological emergence and epistemic emergence, the consensus is that with the possible exception of conscious experience, the former does not exist at all, and the latter, while ubiquitous (indeed the norm) in science, has no obvious Earth-shattering ontological implications (e.g., it does not tell against ontological reductionism). Here, we have the contours of a false forced choice of either ontological reductionism or radical forms of ontological emergence. But this false forced choice is hardly ontologically innocent. It presupposes the hierarchical and foundationalist picture of reality discussed previously. It must be noted that proponents of both epistemic and radical ontological emergence differ only in that the latter think new (physical or metaphysical) laws, causal powers, or entities must be added to the set of fundamental or brute facts to explain the existence of novel emergent phenomena. Contextual emergence, contrary to both these camps, suggests that relations between smaller and larger scales need not be antisymmetric, transitive, or antireflexive.

1.3 Introducing contextual emergence

What is needed to get beyond this impasse in the debates is a form of ontological emergence grounded in scientific explanation that: 1) does not violate the inherent unity of the world; 2) does not assume that if no reductive explanation exists then there is no scientific explanation to be had other than positing a brute new law or causal power; and 3) does not presuppose foundationalism. The aim of this book is to describe and argue for just such an account of emergence. Contextual emergence is both an explanatory and ontological account of emergence that makes good on all three desiderata, demonstrating that there is a viable, empirically well-grounded alternative between ontological reductionism coupled with explanatory antireductionism (weak emergence) and radical strong emergence.

Contextual emergence, introduced in Chapter 2, focuses on the conditions that make the existence and persistence of emergent systems and their states and observables possible, termed *stability conditions*. Very often these conditions either are or imply inherently and irreducibly multiscale relations, so it is not surprising that scientific explanation is often multiscale (Chapter 3). Contextual emergence highlights the inherently interactive, multiscale interdependence of phenomena at all scales. These stability conditions act as gatekeepers for systems to access modal possibilities (e.g., reducing or enhancing a system's degrees of freedom).

We can begin to illustrate the basic idea of contextual emergence through a vivid and simple example (Shyer, et al. 2017): The ordered pattern of hair and feather follicles on animal bodies is rather remarkable, yet the genes giving rise to follicles do not contribute any location information for individual follicles. The developing skin has an epithelial layer forming the epidermis with the epithelial cells that produce the follicles lying on top of the dermis. The underlying dermis mechanically contracts locally, causing the epithelial cells to bend, forming slight dome shapes where follicles eventually form. These dermal contractions cause compressive stress in the overlying epithelial cells, leading to two critical developments that *would not happen otherwise*. First, the dermal contractions break the symmetry of the random distribution of epithelial cells producing spatial order. Second, these larger-scale mechanical forces activate the smaller-scale genetic machinery producing follicles. Otherwise, the genetic machinery for follicle production *never turns on*.

These larger-scale mechanical forces provide a stability condition sufficient to trigger follicle formation in an ordered array. While the underlying genetic machinery provides some of the conditions necessary for patterned follicle formation, the larger-scale mechanical forces due to dermal contraction provide the crucial condition for patterned follicle formation. The genetics controlling follicle generation is shaped by larger-scale mechanical forces determining typical distance between neighboring follicles. The mechanical contraction of the dermal provides the context for the formation and spatial ordering of follicles—a relationship between the larger-scale mechanical force and the smaller-scale genetic machinery that involves symmetry-breaking.

In terms of our three desiderata, this explanatory pattern does not invoke any mysterious brute forces even though it eschews the ordered hierarchy of levels of reality, where smaller-scale factors unidirectionally determine outcomes at larger scales. It is not a reductive explanation, but it is still scientific. Indeed, it shows us how the smaller-scale genetic and the larger-scale mechanical forces work together to produce astounding phenomena such as the striking pattern of feathering on bald eagles. The emergence of such phenomena do not arise from some set of underlying "governing" or determining dynamical laws, such as in the Game of Life. It is possible for the genes involved in the production of follicles to produce them or not, but this modality is directed one way or the other by a larger-scale non-genetic condition.

More generally, as we will make clear in Chapter 6, the relationship between genes, proteins, and phenotypic characteristics is many–many. Thus, we must now think in terms of gene networks (genomics), RNA networks, protein networks (proteome), and the complex non-linear interactions among them. Furthermore, these relationships are also affected by several global constraints and multiscale contextual features, including cellular environment, the wider organismic environment, and various features of the external environment in which the organism is situated. These interactions are obviously multiscale, multilevel, inextricably interrelated, and interdependent. In the developmental process, what any key biological player does, such as genes and proteins, and what it results in as output, is a function of multiscale contexts and stability conditions.

C. D. Broad wrote that "[A]n emergent quality is roughly a quality which belongs to a complex as a whole and not to its parts" (Broad 1925, p. 23). According to him, if the properties of an irreducible whole are not given by the properties of the basic parts in isolation, they are emergent (see Humphreys 2016 for more details). That is what we have in the case of spatially ordered follicle production, for example, yet without positing any extra brute laws or causal glue. We understand a lot about genetics, we understand a lot about mechanical forces, and we are coming to understand how sets of genes behave under larger-scale mechanical forces, as well as how this involves multiple scales without explanatory gaps. Contextual emergence as a framework captures the pervasive pattern of explanation in the sciences as relational and never devoid of context, though crucial features of the relevant context are often missing or ignored in specific articulations of scientific explanations. As a framework, contextual emergence makes explicit what is too often left implicit.

If one insists on a hierarchical and foundationalist conception of reality as exemplified by the Game of Life analogy, then the only way to produce any kind of robust ontological emergence out of the system—any kind of emergence other than weak emergence—is to add something metaphysical, something *extra* to the game, such as the brute bridge laws of the sort we find in the strong emergence of C. D. Broad (1925). Even so, in this case we are simply adding laws or what have you to the explanatory base. But imagine instead that reality is more like multiscale complex networks or structured graphs of extrinsic dispositions; now robust ontological emergence becomes perfectly natural. Order comes not from anything extra-added, not from any causal or nomological glue, not from any metaphysical grounding whatsoever. What the sciences show us in case after case is that the arrow of explanation and determination is not strictly bottom-up, not unidirectionally from smaller length and time scales to larger scales. It is for these reasons that contextual emergence is common, universal, non-spooky, and does not defy scientific explanation. Nor does contextual emergence imply any kind of discontinuity or disunity in nature. Contextual emergence emphasizes the ontological and explanatory fundamentality of multiscale *contextual stability conditions*, often operating globally over interconnected, interdependent, and interacting entities, and their relations across multiple scales.

Overcoming the false forced choice between ontological reductionism and radical emergence is not just about resolving debates among scientists and philosophers. Providing a viable ontological alternative through contextual emergence has actual-world implications. As one example, recent evidence indicates that while tumor genesis may be more common than previously thought, the tissue microenvironment surrounding tumors determines whether they are suppressed or develop into malignancy (e.g., cancer). This is because normal homeostasis provided by tissue microenvironments surrounding tumor cells controls the behavior of the latter; change the tissue microenvironment, and tumors are no longer inhibited but can change to malignant (e.g., Birbriar 2020; Bissell and Hines 2011; Chung, et al. 2005; Kraman, et al. 2010; Gerasimova-Chechkina, et al. 2021).

Soto and Sonnenschein, for example, argue that cancer is not primarily or solely the result of mutations, but rather, "alterations in the communication among cells and tissues that affect tissue architecture" (2013, p. 90). Given such an account, mutations are primarily caused by such disruptions. They argue that there is much evidence for this claim:

> There are features of the tissue microenvironment that affect either the induction of or progression of cancer, and such features are not reducible to, or explainable in terms of properties or processes at or below the cell and molecular level. Further, Soto and Sonnenschein claim that the somatic mutation theory cannot account for these observations, and that they can only be accommodated on a view that takes that tissue architecture itself as causally relevant to the progression of disease.
>
> (Anya Plutynski 2019)

Different authors use different names for such multiscale cancer-related effects. Soto and Sonnenschein call such multiscale effects "diachronic emergence" (Soto, et al. 2012, p. 271), and Bertolaso (2016, p. 95) calls it "synchronic reflexive emergence." We of course would call it contextual emergence, and we do not doubt that it has both diachronic and synchronic elements and various multiscale stability conditions. The pragmatic point is that reductionistic bottom-up strategies for studying and treating cancer run the risk of missing detection and treatment targets provided by the stability conditions and their breakdown in tissue microenvironments controlling tumor behavior.

As another example, the transmissibility of a SARS-CoV-2 variant, such as B.1.1.7, provides necessary conditions for its spread in a population. Yet stability conditions such as local ecology where the variant first arises, human travel patterns among population centers, presence (or absence) of vaccine uptake, among other social–cultural factors, provide the remaining necessary and sufficient conditions for variant spread in a nation or across the world (Kraemer, et al. 2021), indicating the need for both large-scale public surveillance and social-level public health interventions.

Or consider psychology, where the reductionistic program of biologizing behavior leaves out crucial social and contextual factors, including how one interprets what is pathological and restricts how psychological and social challenges are treated (e.g., Bishop 2007, 2022; Boone and Richardson 2010; Harrist and Richardson 2014; Williams 2001; Chapter 6).

Or consider the problem of the fragility of democracy and democratic backsliding that we are witnessing around the world. Given the right conditions, the forms of discontent and despair in a national population can be triggered by a few influencers and a media ecosystem, including propaganda. Thus, an identity can be formed across disparate socioeconomic lines, making the rise and rule of authoritarian leaders possible (e.g., Applebaum 2020; Hett 2018; Stern 1974; Uscinski, et al. 2021). Reductionistic analyses using only bottom-up approaches leave out elements in the dynamics shaping the identities towards authoritarian rule, making it harder to address contemporary social and political problems with clarity and wisdom. In other words, bottom-up methodological individualist approaches to individual actors can only capture some of the necessary conditions for the political transformations which we are witnessing.

In a recent article in *Harpers* (September 2021) entitled "Bad News: Selling the Story of Disinformation," Joseph Bernstein calls the stability conditions for propaganda, "pre-propaganda." He argues that propaganda, social media, and so on, have no magic intrinsic power to produce social change unless specific stability conditions are in place:

> Another way of thinking about pre-propaganda is as the entire social, cultural, political, and historical context. In the United States, that context includes an idiosyncratic electoral process and a two-party system that has asymmetrically polarized toward a nativist, rhetorically anti-elite right wing. It also includes a libertarian social ethic, a "paranoid style," a deeply irresponsible national broadcast media, disappearing local news, an entertainment industry that glorifies violence, a bloated military, massive income inequality, a history of brutal and intractable racism that has time and again shattered class consciousness, conspiratorial habits of mind, and themes of world-historical declension and redemption. The specific American situation was creating specific kinds of people long before the advent of tech platforms.
>
> (p. 30)

Similarly, think of the social construction of race. Racism is not fundamentally rooted in biology or genetics; rather, the social/political/cultural history of the West supplies the stability conditions for the formation of the concept of race and the construction of racial hierarchies (e.g., Kahn 2017; Kendi 2016; Pustilnik, A. 2012; Van Vechten 1926). Understanding the context of the media ecosystem and how it creates stability conditions for the perpetuation and exasperation of racism and social inequality seems particularly important (Bernstein 2021).

1.4 Where we go from here

A scientifically well-supported path forward beyond ontological reductionism and radical emergence exists. To be sure, there is much more to be said about contextual emergence, and we will say much more throughout this book. After describing contextual emergence in more detail, in Chapter 2 we will explicate and illustrate it further through several scientific examples and case studies ranging from quantum mechanics to classical physics to biology and beyond (Chapters 3–6). We will see that neither foundationalism nor an ordered hierarchical conception of reality are well supported by the sciences, nor are they necessary or sufficient for order, stability, and novelty in our world. Finally, we will explore the implications for metaphysics and the unity of the sciences more broadly in Chapter 7, wherein we will also address possible objections.

2

Contextual Emergence

Something that is not dependently arisen, such a thing does not exist.

Nāgārjuna

2.1 Introduction

As we said in the Chapter 1, this book is an exercise in natural philosophy. This means we will be focusing on contextual emergence and exemplars from across the sciences involving intertheoretic relations, relationships among microscopic and macroscopic scales, relationships among levels of organization, relationships among dynamical and causal processes at various scales, and relationships between parts and wholes. We will start this chapter with some very preliminary issues that sets up defining and explicating contextual emergence (Section 2.2). We then describe the key features of contextual emergence (Section 2.3), followed by a detailed illustration of how it works in a concrete scientific case: Rayleigh–Bénard convection (Section 2.4).

2.2 Preliminaries

For scientists to describe or investigate systems and components, or talk about their properties, they must presuppose a system/environment distinction. Although this distinction is often problematic (e.g., Bishop 2011), no scientific activity of any kind is possible without some such rough and ready distinction.

This is easiest to appreciate when we think of observational or experimental activity. Consider a field biologist observing a particular type of animal. The biologist, even if only implicitly, makes a system/environment distinction between the animal to be observed (the system) and the surrounding elements which are not the direct object of observation (the environment).

Emergence in Context. Robert Bishop, Michael Silberstein, and Mark Pexton, Oxford University Press.
 DOI: 10.1093/oso/9780192849786.003.0002

Similarly, for an experiment in a chemistry laboratory. The chemist always makes a system/environment distinction between the chemical compound under study and the environment of the laboratory where the experiment is taking place. In experimental activity, scientists pay close attention to both the system and the environment, controlling the environment as much as possible to screen off influences from the environment that would interfere with the experiment, or to see how the system responds to different environmental conditions.

Although the environment sometimes gets treated as a kind of boundary condition for the system of interest, this approach sweeps a lot under the rug. Suppose we have a damped driven pendulum at which we are firing photons to investigate their effects on pendulum motion. We can choose to consider the pendulum as the system and the photons as part of the environment, where photons are a perturbing effect of the environment. Or we could choose to consider the pendulum plus the prepared photon as the system we are studying, and everything else as the environment. Which choice we make has implications for how the system is conceived and analyzed, and what we can say about the system under investigation (Bishop 2008a). Returning to the animal/environment example, it makes a big difference to our conception and explanation of cognition as to whether we treat the environment and body as merely input to the brain (the system), as opposed to conceiving of the cognitive system dynamically as being the environment, brain, and body coupled in some complex non-linear fashion. This is currently a major debate in cognitive science.[1] This is not how boundary conditions are typically conceived,[2] so determining the relevant environment needs to be treated with some care. As we will see over the course of this book, contextual emergence begins by drawing attention to subtleties about "environmental" factors that play crucial roles in the behavior of phenomena under investigation.

Even theoreticians must construct a set of variables to be explored against a set of background conditions. This means that theoretical vocabularies *do not exist in isolation*. A particular "higher-level" theory may have a vocabulary in part defined by a choice of system/environment distinction, and in part by the wider conceptual embedding of a theory. The "higher-level" theory's vocabulary may also include states and observables that are robust because of *stability conditions*, where these conditions are neither found in nor expressible by the "lower-level" theory (Section 2.3.2). This will turn out to be

[1] For discussion, see Chemero and Silberstein (2008), Silberstein (2016, 2020), and Silberstein and Chemero (2012, 2013).

[2] We will have more to say about the subtleties of boundary conditions throughout the book.

a key difference between reduction as standardly conceived and contextual emergence.

To elaborate one important often underappreciated aspect in reduction/emergence debates, note the conceptual embedding of vocabularies. The conceptual embedding of a theory is the network of other theoretical resources it connects with to utilize and is implicitly supported by. For instance, the conceptual embedding of a theory might relate to other mathematical resources requiring the "higher-level" vocabulary for those resources to be applied. Take, for example, Leonhard Euler's famous "Seven Bridges of Königsberg" problem. The mathematician was challenged to devise a walk through the city of Königsberg that would cross each of its seven bridges exactly and only once, yet not backtrack on any of them. Euler was able to negatively solve the problem by realizing that the specifics of the route walked was irrelevant. He represented the bridges as connections and the islands they connected as nodes, and was thereby able to specify general topological conditions for whether a walk is possible. In the process, he invented graph theory ("geometry of position," in his terms).

Euler represented the problem abstractly in terms of lines and letters—upper case representing landmasses, and lower case representing bridges. The upper-case letters can be recognized as nodes or vertexes, the objects in a graph that have edges connected to them. The lower-case letters represent the connections or edges. An even vertex has an even number of edges leading to it; otherwise, a vertex is odd. An odd vertex needs to be either the starting point for a path or the ending point; otherwise a person will get stuck at that node and unable to leave except by crossing an edge twice. For Königsberg there are four odd vertices, hence no possible paths where the city could be traversed without crossing a bridge more than once. One more bridge, appropriately located, creating a situation where there were two odd and two even vertices, would enable a path traversing the city and crossing each bridge only once. In general, for there to be a path that traverses each edge of a graph only once—a so-called Eulerian path—the graph can only have at most two odd vertices. Everything depends on the structural details of the graph, the network nodes, and connections.

The significance, here, in part is that it is only specific system/environment distinctions that legitimize particular vocabularies and explanations (e.g., bridges and rivers/islands). Therefore, it is *only those vocabularies that can be connected to relevant wider conceptual resources*. In Euler's case, a vocabulary that can be abstractly mathematized into nodes and connections, and the topological facts about the domain, are only accessible by using that mathematical apparatus. Even if one could reduce the vocabulary of bridges and rivers to some "lower-level" vocabulary, that vocabulary would not be able to connect to the surplus conceptual apparatus outside the vocabulary itself to derive nomic regularities. Only by importing a structuring relevant to the "higher-level" vocabulary (such as a topological constraint) can this be done. We need not only think of mathematical resources. Sometimes the nomic regularities captured by a theory are implicitly buttressed or given applicability only by other theories (Duhem 1954). Contextual emergence starts with making these relations and commitments explicit and tracing their import for the sciences, but it does not end there.

In summary, we often need formal system/environment distinctions from the "higher level" to derive relevant features of interest; hence, we must import this into the "lower-level" theory. Similarly, theories exist in a network of wider conceptual frameworks. To derive relevant features, we may well need wider conceptual apparatus that exists outside of either theory but that can only connect to the "higher-level" theory (or to the "lower-level" theory if it is enriched by contextual resources from the "higher-level" theory; see Chapter 3).

At this point it is important to note, however, that even though our scientific questions and answers are tied to where we draw the system/environment distinction, we are herein not primarily focusing on the perspectival nature of our epistemic access and analysis. That is, we are not merely pointing toward some form of explanatory pluralism in scientific practice. As the book proceeds, we shall see how contextual emergence deals with ontological and epistemological implications of system/environment distinctions and theoretical embedding. Our argument will be that the plethora of explanatory pluralism in the sciences turns out to be good evidence for contextual emergence. But the general point for the moment is that P. W. Anderson and Laughlin are right when they insist that *ab initio* intertheoretic reductions (as well as single-level or single-scale mechanistic reductions) are largely a fiction in actual scientific work.

As a final preliminary comment, "levels" talk needs to be treated with some care, as we have indicated previously with scare quotes. In reduction/emergence debates, there is a tendency to treat levels as reified or pre-given in some universal, timeless sense. That is a problematic oversimplification for many reasons, including the fact that the term "level" has multiple meanings ranging from spatial/temporal scale to some functional description (Thalos 2013). For instance, there are cases of engineered systems, such as computers, where there are pre-given levels because they have been constructed to be so logically, functionally, and physically (e.g., Ellis 2016, chapter 2). By contrast, in natural systems, whatever levels might exist arise over time through dynamical evolution and multiscale interactions. Sometimes such levels are easily identifiable (e.g., elementary particles, molecules, simple crystals, cells, organs, bodies). Sometimes levels can only be identified, if at all, by the length and time scales of the dynamics and dynamical structures or organization of matter (e.g., convection). Sometimes there is an erasure of levels or scales such as nonseparable quantum systems, including even "classical systems" such as Schrödinger's cat or large-scale entanglement, where the joint system now possesses a quantum state or property, which the components taken individually lack. Finally, there are many cases in which there is no ordered hierarchy of levels (e.g., O'Malley et al. 2014; Primas 1977; Silberstein 2016, 2020; Silberstein and Chemero 2013; Thalos 2013) and multiscale relations among different domains are what is relevant for scientific explanation. As this chapter unfolds, we shall see how contextual emergence makes sense of the relationships among different levels, scales, and domains in the absence of the ordered hierarchy that P. W. Anderson (1972) assumed. We again stress that this is not merely or primarily a claim about the perspectival pragmatics of human explanatory practices.

2.3 Contextual emergence

Contextual emergence was originally modeled after nonlinear dynamics, statistical mechanics, and quantum mechanics and applied to cases such as the emergence of temperature and molecular structure. Early work was primarily focused on intertheoretic relationships (Bishop and Atmanspacher 2006), yet contextual emergence has important ontological implications and applications as well (Bishop 2005a,b, 2008b, 2019).[3] As we started arguing in Chapter 1, finding an appropriate ontological form of emergence is important to our understanding of the order and unity of the world.

2.3.1 *The basic framework*

Contextual emergence focuses on contexts, but what does this mean? There is a wide variety of contexts, such as a system providing a context for its components, an environment providing a context for a system, or states of affairs making systems possible. Contexts are always relational. Regarding epistemology, contextual emergence characterizes how different levels of scientific description relate, such as the mechanical description of individual molecules interacting to the statistical description of ensembles of molecules. Regarding ontology, contextual emergence characterizes the relationship between domains of reality, such as how the domain of quantum mechanics is related to the macroscopic domain of rocks and trees, and all the multiscale and contextual features involved in those relationships. Such multiscale

[3] One can interpret Silberstein and McGeever (1999) as defending an ontological form of contextual emergence. See also Silberstein (2005) and Chapter 4.

features can range from the purely physical/mechanical to more mathematical such as purely topological constraints, and everything between. Such multiscale features can be more or less mechanistic/dynamical or more or less acausal/adynamical as in global or systemic conditions such as being a small-world network.

Let us discuss this in a little more detail before we move on. As we have said elsewhere (Bishop and Silberstein 2019; Silberstein 2018, 2021), and as others have noted (Batterman and Green 2020; Green and Batterman 2021; Raja and Anderson 2021; Ross 2021), stability conditions (to be defined subsequently) close off and open up degrees of freedom—they both limit and enable the behavior of a target system. Such stability conditions are often said to have the following features (Huneman, 2018a,b):

1. Acausal or adynamical.
2. "External" to the system in question, as in boundary conditions.
3. Often global or at a higher scale than the target system.

And types are often classified as follows:

1. Purely mathematical (e.g., topological, as with the "Seven Bridges of Königsberg" problem).
2. Law-like or nomological (e.g., a power-law such as the inverse-square law).
3. Causal or structural (e.g., the container walls in Rayleigh–Bénard convection).

We think this taxonomy is a decent idealization but in fact represents categories that are neither mutually exclusive nor exhaustive. There are cases where the very same constraint can, at the same time, be more than one of the types listed previously, or at the very least for which the boundary will be fuzzy. For example, with regard to renormalization group methods and universality classes (see Chapters 4 and 5), there are cases of topological constraints in physics that are both mathematical and nomological. Noether's theorem and group theory also comes to mind here.

There is a great deal of discussion in the literature these days about how to distinguish causal from nomological or dynamical explanation. Our view is to be pluralists about both types of explanation. However, suppose, as many do these days (e.g., Ross 2020), one wants to follow Woodward (2003), defining causal factors as those we can manipulate. We do not see the point in stipulating normative necessary and sufficient conditions for something's being exclusively or essentially causal. We also think the manipulation criterion is neither necessary nor sufficient for something to be a causal variable. There are no doubt causal constraints in nature at various scales, such as the cosmological and Planck scale that we (or perhaps any beings) cannot manipulate even in principle. On the other hand, there are nomological constraints we can and do manipulate, such as various "topological knobs" in network theory, universality classes, and order parameters both physically and in simulations, which would make them also causal on the manipulation account. Per the previous taxonomy, this of course would make such constraints both topological and causal.

As we have said, there are important acausal and adynamical stability conditions that one must talk about. For us the key is that their explanatory or difference-making power does not derive from a causal or dynamical process, dynamical law, or causal mechanism. Nor are such explanations about constitution and composition. A good example here would be conservation laws. Nothing in this definition says anything about manipulability or the lack thereof. As stipulated previously, the other key to such acausal difference makers is that they are often "external" to the system or components in question, and are often global or at a higher scale than the target system or its components (e.g., topological networks).

We must also note some potentially important differences between acausal and adynamical stability conditions, on the one hand, and constraints, on the other. First, constraints, such as the walls of a box constraining and enabling the behavior of a gas, can be causal/dynamical. It is tempting to idealize the walls of the box in this case as being a strictly acausal constraint, but the molecules of gas causally and dynamically interact with the walls of the box per dynamical laws of physics. We would say the walls of the box in this case are both structural constraints and dynamical constraints. Stability conditions can be causal or acausal, and under some descriptions they can be both. Second, stability conditions need not merely constrain and enable specific behaviors of a pre-existing system, but help explain why truly novel and new states and observables, entities, processes, even systems, emerge in the first place (see Section 2.3.2).

It is also important to point out that we can model the very same "chunk" of reality as an acausal global constraint on a causal or dynamical process, or we can sometimes also model it via coupled differential equations as one system whose basic elements have possibly non-linear effects on one another. For instance, we could model the bank of the river as an acausal constraint on the dynamical flow of the river, or we could model the bank + river flow as one coupled dynamical system. It goes without saying that the bank and the river in fact do causally and dynamically alter one another over time. It is also equally obvious that we could treat the flow of the river as the boundary condition on the changes of the bank over time. It is perfectly reasonable to think of structural constraints on causal or dynamical processes as relatively fixed, unchanging, or non-manipulable, but of course (sometimes even quickly) such constraints can and do change with time and scale. And as the riverbank illustrates, such constraints not only can change but also be manipulated sometimes.

As we will discuss in Chapter 7, nothing about any of this suggests that all explanation is purely perspectival, nor does it suggest any sort of anti-realism. Such different explanations are commensurable and equally true, and the properties and features to which they avert in both the explanans and explanandum are all equally real. The fact that there are different ways to carve up the world into system and boundary conditions or causes and background conditions is not at all surprising or problematic, especially give contextual emergence. Certainly, none of that is news. Contextual emergence highlights that what we are often tempted to relegate as mere boundary conditions, or background conditions that are outside the set of explanans, often actually is an important feature of what is doing the explaining. Furthermore, given the contextual nature of reality, these different "carvings" need not be merely perspectival or pragmatic. That is, varying multiscale contexts, boundary conditions, and such, really does lead to new and novel things emerging, as this is the very nature of reality.

The general framework can be summarized as follows:

Ontological Reduction Properties and behaviors in a particular (lower "level" or smaller scale) domain (including its laws) offer all conditions necessary and sufficient to determine properties and behaviors in another domain.

Contextual Emergence Properties and behaviors in a particular domain (including its laws) at best offer some necessary but no sufficient conditions to determine properties and behaviors in another domain.

2.3.2 *Contexts and stability conditions*

In this framework, reduction is distinguished from contextual emergence primarily by identifying the important role of *stability conditions*—the contingent conditions characterizing or defining contexts that guarantee the existence and persistence of relevant systems and their states and observables over time. The key feature of stability conditions is that they are whatever environmental features (i.e., contextual features and constraints), however concrete or abstract, that we are treating as being outside the system, complete the full set of conditions for the emergent in question to come into being. That is, stability conditions enable emergence and robustness.

A basic example of stability is the property of robustness of a system against small perturbations such as an equilibrium state being maintained by the dynamics damping out such perturbations. In Rayleigh–Bénard convection discussed subsequently, the dynamics at larger length and time scales damps out small perturbations so that the convective system maintains stability over time. The large-scale dynamics guarantees the stability of the system over time. More abstractly, the Kubo–Martin–Schwinger condition is a very general acausal stability condition guaranteeing the existence of equivalence classes of statistical states that are the bearers of temperature as a property (Chapter 4; Bishop 2019).

Consider the example of coarse-grained descriptions in statistical mechanics. Such descriptions are possible because of a stable partition of the microstate space, where the macroscopic observables determine the stable partition into equivalence classes guaranteeing that the macroscopic observables are well defined. That is, the macroscopic context constrains the possible partitions or orderings of the microscopic state space so that macroscopic observables function as the relevant observables describing the system. Without knowing the relevant macroscopic observables for the ensemble description, it is not possible to determine the relevant partitioning of the microscopic state space.

States and observables, dispositions, properties, and so forth, are part of the ontological furniture of the systems under investigation. Contextual emergence takes its cues from the context of the target domain under investigation rather than imposing a one-size-fits-all notion of states, observables, and systems for all domains of reality. Such distinctions among domains are sometimes lost in reduction/emergence debates, and this may be one reason why stability conditions relevant for such distinctions typically pass unnoticed. Nevertheless, in every case, there are stability conditions making the variety of systems in our world possible, and it is important to make their role explicit. Not only can stability conditions arise dynamically (and locally) constraining the behavior of basic components (e.g., convection), or be causal-mechanical constraints such as mechanical equilibrium, but stability conditions also include global or systemic constraints such as topological constraints, dimensional constraints, network or graphical constraints, and order parameters, among others. One can point to both dynamical and adynamical, causal and acausal, as well as local and global stability conditions that are difference makers. For example, global adynamical constraints might include conservation laws, free energy principles, least action principles, symmetries, and some types of symmetry breaking. More specifically, think of the light postulate and the relativity principle as instances of adynamical global constraints—they delimit the kind of dynamics physically accessible at all spatial and temporal scales.

To show how easy it is to miss the role of stability conditions as well as to illustrate an acausal and adynamical example, consider spatial dimensionality:

1. In spacetimes with one spatial dimension, Newtonian gravity has constant magnitude between two bodies independent of distance, and there can be no gravitational waves. Electrically charged particles in motion cannot radiate energy.
2. In spacetimes with two spatial dimensions, Einstein's equation in vacuum requires spacetime to be completely flat,[4] and gravitational waves still cannot exist. Moving charged particles can radiate energy. In three-dimensional spacetime, gravitational waves can exist.[5]
3. In spacetimes with four spatial dimensions, stable gravitational orbits cannot exist.[6] The hydrogen atom under some conditions would be unstable because its spectrum would be unbounded from below in these cases.
4. In spacetimes with five or higher spatial dimensions, no stable atoms can exist.[7]

The upshot is that spatial dimensionality makes some stable phenomena possible in some dimensions that cannot exist in others, while also constraining the range of possibilities for the "fundamental" laws of physics to come to concrete expression. This is to say that, for example, the laws of electrodynamics by themselves do not specify all the possibilities for the behavior of charged particles. Spatial dimensionality is one of the conditions that must be added to the laws for stable systems and behaviors to arise. It is an acausal, adynamical global stability condition defining a general context for how all forces can act, what dynamics are possible, or even what system can stably exist.

Why is dimensionality a context for stable atoms or planetary orbits rather than simply part of the system itself? Epistemically this is because the dimensional backdrop is not properly part of whatever level we chose to describe atoms or planetary systems in terms of. Dimensionality is not part of the laws of the system or part of the vocabulary of microscale processes; it is a backdrop to those. This can be appreciated by considering that the consequences of spatial dimensional cannot be inferred from other physics alone without thinking in dimensional terms. This is because the dimensional constraints are multiply realizable with regard to specific laws of physics, particular configurations of matter, causal histories, and local facts about spacetime points and the spacetime metric. By considering these alone, we cannot infer the

4 Both the Ricci and Riemann curvature tensors are zero.

5 The Riemann curvature tensor can be nonzero even if the Ricci curvature tensor is zero.

6 Paul Ehrenfest (1917) was the first to notice that stable planetary orbits cannot exist for spatial dimensions greater than three.

7 The Coulomb potential dominates the $1/r^2$ centrifugal potential for small r close to the nucleus yielding an unbounded spectrum from below, so there are no bound states.

dimensional context. This reflects that the property of dimensionality is not reducible to properties internal to the system itself. Dimensionality is not part of the property bundle of an atom or planetary system that make up its essence. For instance, it is not part of the identity conditions of a planetary system that spacetime has a specific dimensionality. One could recognize a planetary system without recognizing the dimensionality of space. Indeed, this is why it was a discovery that, given the way physics works in our universe, the correct dimensionality is a crucial context to allow the possibility of planetary orbits. Similarly for atoms.

Spatial dimensionality has actual-world effects in phenomena such as diffusion-limited binding times for polymer surfaces, protein searches for specific DNA targets, oxygen transport in muscle tissues, and macromolecular self-assembly. In all these kinds of processes, the difference between two and three spatial dimensions affects the ability of particles to find precise target locations in surprising ways, as Antoine and Talbot (2020) illustrate. Weyl and Dirac semimetals, phases in solid systems, have interesting conduction properties but can only exist in three special dimensions (Armitage, Mele, and Vishwanath 2018). And in general, there are important changes to the properties of topological insulators and superconductors in two dimensions versus three (Qi and Zhang 2011).

Usually, spatial dimensionality is an unacknowledged assumption in reduction/emergence discussions. For instance, when analyzing a system with a chosen level of description, spatial dimensionality is assumed and its constraints on systems typically unacknowledged. Yet the behaviors of the microscale processes being discussed are what they are *because* of the spatial dimensionality of our world. At best, spatial dimensionality is treated as part of the environment in which the system exists, and does not function as a boundary condition in physical modeling. Nor does it typically show up in the vocabulary of the description of the analysis. Moreover, the constraints on system behavior are independent of the systems under study. They apply universally to all systems studied by scientists. This is what it means for spatial dimensionality to be a global stability condition.

2.3.3 *Stability conditions and physical possibility space*

Notice that spatial dimensionality as a global acausal stability condition functions as a modal gatekeeper. It constrains the physical possibilities for systems and behaviors (e.g., whether there are atoms and planetary orbits). While it may be *metaphysically possible* to be a stable planetary orbit or atom in any dimensionality, it is not *physically possible*. Spatial dimensionality is a property that limits what other physical process can occur and how systems can behave.

Hence, we see that stability conditions are deeply connected to accessibility in *physical possibility space*, enabling us to see how these conditions function in contextual emergence. Contexts are characterized by stability conditions, whether these are large-scale dynamical constraints, global acausal constraints, symmetry constraints, or their breakings. The most "basic" laws of physics (e.g., conservation laws) define the physically possible (a space of possibility much smaller than the space of metaphysical possibility). Nevertheless, not all these physical possibilities are actualizable through the basic laws and particles of elementary particle physics by themselves.[8] It is the specific contexts that make particular regions of possibility space accessible, and we can think of stability conditions as *accessibility conditions* for specific subspaces or regions where concrete processes and systems are actualized.

The example of spatial dimensionality illustrates the accessibility of subspaces of physical possibility space. Spatial dimensionality as a stability condition determines the subspace of physical possibility for stable atoms and planetary orbits. The most basic laws of physics alone contribute some of the necessary conditions for there to be atoms and planetary orbits, but a spacetime with three spatial dimensions is needed for stable planetary orbits and atoms to exist. Lasers provide another illustration. According to quantum mechanics, the amplification of stimulated emission of light from atoms is physically possible (in three spatial dimensions). Nonetheless, only appropriately engineered conditions of isolation and stability—specific contexts defined by relevant stability conditions—render that portion of physical possibility space accessible.

Stability conditions, then, define the accessibility of modal space of the system under investigation to specific regions of physical possibility through the constraints at relevant scales defining access to those subspaces. Contextual emergence makes explicit the conditions sufficient for systems to exist and behave as they do, and makes clear that what we take to be the most "fundamental" or general laws of physics are not the only relevant modal constraints as spatial dimensionality nicely illustrates. Much of the reduction/emergence debate is hampered by the unquestioned assumption that the most "fundamental" laws and entities are the only relevant irreducible modal constraints in possibility space. The stability conditions defining concrete contexts in any given case make specific regions of physical possibility space accessible and others inaccessible for the system in question.

[8] We do not mean to imply that elementary particles are the "bottom level" of reality. Quantum field theory treats particles as excitations in fields. For ease of exposition, it is often clearer to stick with particles instead of fields. When it makes a difference, we will advert to field talk.

As we have seen, stability conditions can be causal or acausal, and they can clarify how contextually relevant possibilities become concrete actualities. Explicitly invoking stability conditions supplies a good way to make sense of George Ellis's (2012) notion of causal slack. He maintains that in multiscale systems there is some degree of causal slack or looseness between levels, where higher levels have some degree of causal autonomy from lower levels. These are cases of contextual emergence, where stability conditions at a larger length or longer time scale supply the remaining necessary and sufficient conditions for behaviors of properties and processes at smaller length and shorter time scales. The stability conditions defining the concrete context along with its constraints at the latter scales delimit what specific possibilities are actualized in possibility space for the smaller length and shorter time scale properties and processes.

As a final comment, one implication of how stability conditions work—providing accessibility to specific regions of possibility space, establishing existence and identity conditions for systems in these subspaces, and so forth—is a version of property dispositionalism about modality. All modality turns out ultimately to be grounded in properties, their *extrinsic* dispositions, and the network of relations among these (see Chapter 7). Given contextual emergence, all modality is contextual because the work stability conditions do leads to contextual properties forming a network of compatibility and incompatibility relations with other properties defining the accessibility and concrete actualizations of modality. Such dispositional relationality is reflected in modal relations among properties and their dispositions.

Yet the implication is broader than just philosophy of science. Although we will have more to say about this in Chapter 7, this pattern of dispositional relationality holds for all of reality. This is to say that all modal claims find their grounding in contextually manifested properties. Multiscale contexts are not only physically as fundamental as the most "fundamental" (for us this just means most general) physical laws; they are also *metaphysically fundamental* to producing a world such as ours. As we will see, given contextual emergence, the existence of order, stability and novelty in our world requires no modal glue of any kind, nothing second-order with respect to the natural world. Although contextual emergence starts as a framework based on pervasive patterns in physics and other sciences, it does not include or presuppose any metaphysical grounding, however veiled. Instead, it highlights how an evolving network of multiscale extrinsic dispositions gives rise to an ordered, coherent world. At the end of the day, the Game of Life analogy will fail in every way. Not only are there no autonomous "cells" that compose or otherwise determine everything else—cells whose properties realize all other properties—but there is no "fundamental" physical dynamical law from which everything emerges for free.

2.3.4 *Summarizing and situating contextual emergence*

Contextual emergence makes explicit the conditions and transitions rendering particular regions of possibility space accessible or inaccessible.[9] By way of summary:

1. Contextual emergence is a pattern of scientific explanation that emphasizes the equal fundamentality of what are often multiscale *stability conditions* and interdependent relations at multiple interacting scales. Stability conditions define contexts.
2. Such conditions can be global or systemic, dynamical or adynamical, topological, dimensional, network or graphical, conservation laws, free energy principles, least action principles, symmetry breaking, and order parameters, among other possibilities. Key contextual conditions, therefore, need not involve anything like local or direct causal–mechanical or dynamical interactions, though they often do.
3. Stability conditions can also include global organizing principles such as plasticity, robustness, and autonomy in complex biological systems, or even behavioral, social, and normative in social systems.

[9] Our view contrasts sharply with Jessica Wilson (2010). She defines a version of what she calls *weak ontological emergence*, where there is any restriction, reduction, or elimination of the degrees of freedom (e.g., walls of a box confining a gas). One weakness of this scheme is that it misses cases where systems may gain access to new degrees of freedom (e.g., convection), as well as not identifying the appearance of new states and observables that characterize most cases of emergence in physical sciences. A third weakness is that this account does not clarify whether or not the change in degrees of freedom are due to the "lower level." For example, what is at issue in debates between Batterman and Kadanoff on the one hand, and Landsman, Callender, and Menon on the other, is not *if* there is a reduction in degrees of freedom, but *whether* that reduction is due to lower-level laws and properties alone (see Chapter 4). If changes in the degrees of freedom are brought about by features of the lower level, then this would be a case of ontological reduction, and at best epistemic emergence. Indeed, Wilson contentiously takes it that the theory describing a special science entity *E* is a "fragment" or "extraction" from an underlying and more fundamental theory via heterogeneous bridge principles drawing solely on the resources of the more fundamental theory (2010, 294–295;). In Chapter 3 we shall see why this is the wrong picture for understanding intertheoretic relations. Moreover, Wilson (2013, p. 225) clarifies that the constraints and determination relations of interest to her are all found at the lower level. Such a theory relationship exemplifies reduction as typically understood in the intertheoretic relations literature, and depends on a hierarchical ordering of levels; *hence, the cases of interest to Wilson are not cases of ontological emergence*. Can the entity *E* be considered ontologically emergent even if there is theory reduction? A further contrast between contextual emergence and Wilson's reduction of degrees of freedom approach is that the latter looks to emergence as primarily being a function of the configuration of micro-constituents—following the British Emergentist tradition—whereas contextual emergence reveals that emergence depends on more than the configurations of micro-constituents.

4. Stability conditions can be symmetric, such that X and Y can simultaneously act as contextual constraints for one another.
5. Stability conditions represent both the screening off and opening up of new regions of possibility space (e.g., degrees of freedom), and thereby new patterns of emergence can become robust.

To unpack the last point, contextual emergence provides a framework to understand two things: A) how novel properties arise, and B) why those novel properties matter. This account differs from deductive nomological (D-N) explanations that have received much of the focus in discussions of scientific explanation and reduction/emergence debates.[10] We agree with James Woodward (2003) that there is no need to determine whether something is a genuine, D-N-style law or merely a robust invariance to determine whether it is a viable explanation—that is, not all good scientific explanations refer to laws, even indirectly (e.g., the constraints of spatial dimensionality). Woodward claims that a good explanation allows one to answer a wide range of *what-if questions*: the more what-if questions answered, the more complete the explanation. His goal is to sort causal interactions, so construed, from mere correlations, and he stresses that robust causal explanations exhibit invariance in that the specified relation between the cause variable and the effect variable holds under a wide range of conditions. There are diverse types of invariance, including insensitivity to micro-details, background conditions, interventions, and so forth. The kinds of explanation in contextual emergence illustrated in this book exhibit a high degree and a wide variety of invariances, though the invariances might not primarily be explained by the mechanistic details but by stability conditions, such as dynamical and topological features and the network of other relations involved in the defining of the target system.

Finally, let us compare contextual emergence with typical intuitions about emergence. In surveying the literature on emergence, four typical intuitions or "marks" of emergence are often discussed (e.g., Kim 1999, 19–22)[11]:

Arise Emergents at a higher level arise out of properties and relations characterizing the entities and properties at a lower level.
Unpredictable Emergents are unpredictable, even given exhaustive information concerning the lower level.
Inexplicable Emergents are inexplicable in terms of lower-level features.
Novel Emergents have novel features not found at the lower levels.

[10] More on this in Chapter 3.
[11] Notice the dependence on an ordered hierarchy of levels.

Regarding **Arise**, first, as discussed in Chapter 1, contextual emergence calls into question the fundamentalist and hierarchical assumptions built into this intuition. That is, contextual emergence calls into question the existence of some fully autonomous or fully independent microscale causally/dynamically closed basic physical process sufficient to determine all other phenomena at larger scales. With contextual emergence, emergents only arise from a "lower-level" or smaller scale provided relevant stability conditions, often found at larger scales, are present. While some necessary conditions for the emergents may exist at the "lower level" or smaller scale, nonetheless, for contextual emergence, the sufficient conditions are represented by all the relevant stability conditions at various scales. Second, emergents can also come into being at smaller scales or levels of organization as the result of how these constraints at larger scales or levels of organization are implemented (e.g., the behavior of quantum systems being in part determined by the classical experimental setup). This is modal accessibility in physical possibility space at work.

Emergents are often **Unpredictable** given exhaustive information at the "lower level" or smaller scale alone. However, given contextual emergence, the emergent is often predictable given the "lower-level" information plus the relevant contextual features at other "levels" or larger scales (e.g., the relevant stability conditions and physical states and observables). Similarly for the **Inexplicable** intuition. As we will see throughout this book, the contextually emergent explanation is not going to be fully reductive in either the intertheoretic sense of derivation as discussed by Carl Hempel and company (the D-N model), nor in any synchronic notion of reduction involving the properties of parts determining the properties of wholes (what analytic metaphysics calls realization), and not in any causal–mechanical sense of reduction as with localization and decomposition in biological systems. Thus, in the case of contextual emergence, being unpredictable or inexplicable need not be hallmarks of emergence.

The term **Novel** is certainly loaded. We have already said that an emergent can be predicted and explained, so given contextual emergence, novel means unexpected and irreducibly different in kind from features and concepts connected to the "lower level" or smaller scales. However, it can mean more than this. For example, Jaegwon Kim, among others, has argued that if new "causal powers" emerge at a "higher level" not reducible to or realized by "lower-level" "causal powers," then we face a mystery as to where such "powers" come from (e.g., Kim 1998, 1999). Given contextual emergence, "causal powers" are just extrinsic dispositions that typically require interdependent multiscale conditions. Kim's worries about causal closure and exclusion have no purchase given contextual emergence, as we will discuss in Chapter 7 but will illustrate throughout this book. Finally, contextual emergence explains why novel emergents arise and are robust.

Contextual emergence is multiscale in that "higher-level" or target domain information is required to enrich and constrain the laws and properties of the "lower level" or underlying domain to produce the set of contingent necessary and sufficient conditions for explanation of the emergent. Thus, contextual emergence focuses on making explicit the essential features absent in the "fundamental" level or underlying domain. Scientific explanations do not float free in their own "level" or domain alone. Instead, they implicitly rely on multiscale contextual features not contained in or implied by the "lower level" or smaller scale. Nevertheless, as we will see going forward, the absence of explanatory reduction does not imply explanatory or ontological disunity—pluralism yes, disunity no. Contextual emergence enables explanatory unification across multiple "levels," scales, and domains. While contextual emergence does not suggest the hierarchical structure implied by foundationalism, it also does not suggest a world of reified and explanatorily closed levels of organization. Nor does it does suggest the "gappy" world of C. D. Broad and others.

Lee Smolin argues that "the world around us is nothing but a network of evolving relationships" (2000, p. 96). As we have sketched, contextual emergence implies a contingent multiscale web of inextricably interconnected and interdependent extrinsic dispositions, most of which are in constant flux. Some laws, constraints, principles, and so forth, are more general and subsume more phenomena than others, but such constraints and laws, while not violated by emergents, need not *determine* all the other phenomena at every scale. It is also important to note that the most general constraints need not always come from or be recoverable from the domain of "fundamental" physics. For instance, in physics we use the principle that there are no preferred reference frames. The term "reference frame" has many meanings in physics related to microscopic and macroscopic phenomena, Galilean versus Lorentzian transformations, relatively moving observers, and more. Think again of the light postulate and the relativity principle in special relativity as specific examples of the more general no preferred reference frame stability condition that constrains the possibilities for motion. While we use this general principle in physics more than other sciences do, there is nothing exclusively macrophysical or microphysical about it. Think again of natural selection. This principle is certainly not recoverable from physics, and is every bit as basic to how the world is as any law of physics.

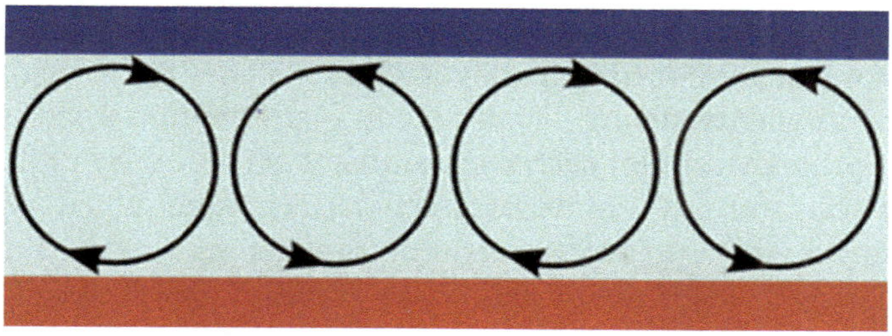

Figure 2.1 Basic Rayleigh–Bénard convection system. The lower plate (red) is held at a warmer temperature than the top plate (blue), creating a temperature difference, ΔT, between the plates leading to convective cells rotating in the directions indicated by the arrows. (Courtesy of the Max Planck Institute.)

2.4 Case study: Rayleigh–Bénard convection

As a detailed case study to illustrate contextual emergence, consider Rayleigh–Bénard convection (Fig. 2.1).[12]

2.4.1 *Basic setup*

A standard setup for Rayleigh–Bénard convection is to sandwich a layer of fluid between two thermally conducting plates. The lower plate is heated while the upper one is maintained at a fixed temperature establishing a temperature gradient, ΔT, in the vertical direction. The fluid near the lower plate undergoes thermal expansion due to heating, becoming less dense than the fluid above. In the presence of gravity, this density difference creates an instability resulting in a buoyancy force tending to lift the whole mass of fluid from the lower plate upwards.

Fluids behave differently from gases. For the latter, individual molecules move independently of their neighbors except for occasional collisions. By contrast, the long-range cohesive forces in fluids cause molecules to pack as closely together as quantum repulsion forces will allow. The result is that fluid parcels roughly are collections of an Avogadro's number of individual molecules (e.g., H_2O molecules) acting as a unit. It is the fluid parcels at the lower plate as a whole unit that tend to start moving vertically under the buoyancy force, not individual fluid molecules.

System-wide global constraints on the motion of parcels due to container boundaries, such as the upper plate blocking upward fluid motion, and symmetries as well as conservation laws, all contribute to a subtle balance in the fluid (Busse 1978). This is to say that the whole system plays a crucial role in

12 For more details, see Bishop (2008b) and Bishop and Silberstein (2019).

determining the allowable states of motion accessible to fluid parcels in the initial uniform state (no rising fluid, yet). This includes the body or volume forces, such as gravity, acting equally on all matter within a given volume of fluid. In principle, absent such constraints, the fluid in the uniform state could flow in an arbitrary number of directions, like water spilled on the floor going "everywhere." Instead, conservation of mass restricts the fluid velocity (Bachelor 1967, 73–75), while fluid flows and allowable pattern formation are constrained by system geometries and symmetries (Cross and Hohenberg 1993). In addition, long-range interactions and the symmetry properties of the state vectors of the fluid parcel positions lead to different values of the relative positions between parcels becoming correlated. The presence of such correlations, and coupling due to correlations, leads to collective behavior restricting the spatial distribution of parcels and contributing to the possible formation of coherent structures. Fluid density variations are relatively small for low ΔT so are rapidly damped out in the initial globally stable uniform state.

2.4.2 *Transition to convection*

Increasing the temperature of the lower plate increases the buoyancy force. The warmer, less dense fluid parcels begin to rise relative to the cooler, denser parcels. As ΔT reaches a critical value, ΔT_c, the correlation lengths among fluid parcels become the size of the container.[13] The fluid now exhibits a property called *generalized rigidity*. Individual constituents are so correlated with all other constituents that no constituent of the system can be changed except by applying some change to the system as a whole (Cross and Hohenberg 1993). Such global changes are applied through change of an order parameter—in this case, ΔT.

As ΔT passes a critical value ΔT_c, the fluid system becomes globally sensitive to small perturbations in density. The homogeneity symmetry of the initial uniform state of the fluid is broken along with some of the spatial symmetries induced by the container. The growth rate of perturbations depends only on the wave number (Cross and Hohenberg 1993, secs. VI and VIII). This means that although the density fluctuations in combination with the system geometry and remaining symmetries lead to a selection of one growing mode that will become the large-scale convection pattern, the mode's growth rate is determined *solely* by global system properties. This is an example of a larger-scale stability condition determining property behavior at a smaller length scale. The most unstable mode scales as the distance of separation

[13] In fluid dynamics models, the correlation lengths become infinite at the critical value.

between the lower and upper plates. The eventual result of the growing mode is the large-scale structure of Bénard cells.

A discernible hierarchy distinguished by dynamical spatial and temporal scales (individual molecules, fluid parcels, Bénard cells) emerges with complex interactions taking place among the different scales. The cells themselves are a large-scale dynamical constraint on the individual motions of fluid parcels. This is due to a balancing among effects owing to the dynamics, the structural relations of each fluid parcel to all other fluid parcels, bulk motion of the fluid, system boundaries and remaining symmetries, and conservation laws. The system displays integrity as the constituents of various levels of organization exhibit highly correlated cohesive behavior. Bénard cells act as a control hierarchy, meaning they constrain the possible motions of fluid parcels in a downward fashion. These cells modify the allowable states of motion for the fluid parcels as well as determine which fluid parcels will stay with their current cell and which parcels will migrate to a different cell. This is remarkable behavior given that Bénard cells emerge out of the motion of fluid parcels after ΔT exceeds ΔT_c. Yet these large-scale structures determine modifications of the configurational degrees of freedom of fluid parcels such that some motions possible in the initial uniform state are no longer accessible, while other motions not available in the uniform state are now accessible. For instance, while $\Delta T < \Delta T_c$, fluid parcels cannot access rotational states of motion characteristic of convection, whereas when $\Delta T > \Delta T_c$, the states of motion characteristic of the uniform state are no longer accessible (e.g., fluid parcels cannot sit motionless).

The stability of the new patterns of constraint that arise is due to the collective effects and steady, large-scale, shear fluid motion suppressing local deviations within the system geometry determined by the container.[14] Body forces, such as gravity and container boundaries, are playing a role as well. Nonetheless, the key stability conditions arise dynamically to shape constituent behavior within the system geometry. It should also be noted that the bulk fluid flow also plays an important role in the formation and dynamics of patterns by contributing nonlocal effects (e.g., acting over many convective roll widths in the fluid; see Paul et al. 2003).

2.4.3 *Rayleigh–Bénard convection and contextual emergence*

Although the fluid parcels and their properties provide some necessary conditions for the existence and dynamics of Bénard cells, they are neither

[14] Shearing can be thought of as adjacent layers of fluid sliding over one another at different rates. For instance, different layers of Bénard cells rotate differentially—inner layers rotate faster than outer layers—about the cell center in a steady bulk shear flow acting on fluid parcels.

sufficient to determine that dynamics, nor sufficient to fully determine their own motions. This is the pattern of contextual emergence.

This contrasts with reductive atomism or "mereological supervenience" which maintains that "the only law-like regularities needed for the determination of macro-features by micro-features are those that govern the interactions of those micro features in all contexts, systemic or otherwise" (Van Gulick 2001, p. 18). The Bénard cells operating at a larger length scale are crucial to determining the behavior of individual fluid parcels. The behavior of such parcels is conditioned by their being situated in a larger-scale dynamical structure that is organizing all matter at smaller length scales. Hence, fluid parcel behavior is not merely the result of the "context-free," law-like regularities envisioned in reductive atomism.[15] The dynamical constraints of Bénard cells in combination with the constraints associated with the whole system enable some possibilities for fluid parcel behavior while restricting other possibilities relative to what would be possible for the parcels in the absence of such constraints. This is the power of stability conditions as gatekeepers of accessibility to specific regions of physical possibility space.

A way to understand the role of stability conditions in contextual emergence is through Mackie's (1965) INUS condition. There are some causal conditions that are Insufficient though Necessary contributors to a condition that is Unnecessary but Sufficient for the effect. By itself, for instance, a short-circuit is neither necessary nor sufficient for causing a fire but can be a necessary contributor to a contingent sufficient condition that started the fire in an old office building. In the convection case, suppose we think of the domain of fluid parcels as the putative realizers of the Bénard cells. For the parcels to be a necessary contributor means that properties and behaviors of the Bénard cells may imply the behaviors of fluid parcels. However, the converse is not true, since the properties and behaviors of the fluid parcels alone do not imply the properties and behaviors of convective cells. Contingent stability conditions specifying the context for the transition from the molecular domain to the properties and behaviors of fluid parcels and from those parcels to the convective cells are required to provide such sufficient conditions (Bishop 2012). It is precisely these stability conditions that are not given by the laws, properties, and behaviors of the smaller-scale fluid parcels domain alone. Hence, the latter are insufficient though necessary conditions for the behavior of emergent phenomena.

[15] As we will see throughout this book, there is nothing that genuinely is context free, hence the scare quotes.

The Bénard cells emerge out of the local dynamics of fluid parcels as a large-scale, nonlocal dynamical process; in turn, these cells constrain or shape the states of motion accessible to fluid parcels. Instant by instant during the pattern formation process, a corresponding large-scale, evolving nonlocal constraint modifies the accessible states of fluid parcel motion. As noted previously, prior to the establishment of this dynamically evolving constraint the fluid parcel trajectories had the property of accessing various states of motion in the initial uniform state. But due to the large-scale, nonlocal constraining effects of the forming Bénard cells, fluid parcels successively lose access to the states of motion of the initial uniform state. This illustrates synchronic emergence, as the self-regulating large-scale pattern is simultaneous with the instant-by-instant modifications of the accessible states of motion for fluid parcels. In other words, the contribution of the parcels at an instant t to the large-scale pattern is conditioned by that pattern at t. If there were no synchronic relationship between the large-scale constraints and the fluid parcels, there would be no modification or selection of which states of motion are accessible to those parcels and, hence, no large-scale structures. This is stability conditions as gatekeepers at work defining the context for fluid parcel behavior.

Note that there are no "new forces" coming out of nowhere, so no radical emergence. Rather, the fluid is governing itself in a complex set of multi-scale interactions among parts and wholes, where the collective effects of the action of the Bénard cells on the fluid parcels is the stability condition delimiting fluid parcel behavior. The jointly necessary and sufficient conditions for the behavior of the convection system and its constituents are provided by the fluid parcels, the dynamics, the large-scale structures and the body forces, system-wide geometry, and symmetries. Individual fluid parcels can only execute motions allowed by all other fluid parcels through the large-scale dynamics and symmetries.

Also note that the fluid parcels and Bénard cells are not wholly distinct entities. This is an example of nonseparability in nonlinear systems (e.g., Bishop 2011). It is the case that fluid parcels can multiply realize Bénard cells similar to how gas molecules in a room can multiply realize temperature. The system can rearrange the fluid parcels without changing the macroscopic properties of the fluid such as temperature, density, and pressure. These different arrangements of the fluid parcels correspond to the same macroscopic fluid properties, hence, form equivalence or universality classes with respect to the bulk fluid properties. Such multiple realizability is typical of emergence in complex systems.

It is also clear that Rayleigh–Bénard convection violates Kim's causal-power actuality principle:

> For an object, *x*, to exercise, at time *t*, the causal/determinative powers it has in virtue of having property *P*, *x* must *already* possess *P* at *t*. When *x* is caused to acquire *P* at *t*, it does not already possess *P* at *t* and is not capable of exercising the causal/determinative powers inherent in *P*.
>
> (1999, 29)

This is a putative metaphysical principle offered to block the possibility of downward synchronous exercises of causal power to defend reductionist intuitions. But there are actual-world systems, such as Rayleigh–Bénard convection, that disobey the principle, indicating that armchair metaphysics is not sufficient to understand nature nor to defend reductionism.

Someone might worry that this situation "involves the controversial idea of simultaneous causation (where a cause and its effect occur at the same time)" (1999, p. 36, n. 34). Yet there are physically coherent expressions of synchronic effects in physics as in relativistically consistent instantaneous action-at-a-distance theories (Dirac 1949; Hill 1967). There are empirical questions whether such synchronous effects exist, but they are physically intelligible.[16] What we observe in Rayleigh–Bénard convection is the work of both adynamical constraints, such as system boundaries and symmetries, and a dynamically growing transformation of the space of possibilities for fluid parcel motion. This is the work of larger-scale stability conditions restricting some states of motion of fluid parcels while enabling others corresponding to the forming Bénard cells. There is nothing spooky or unintelligible about the dynamic changes in accessible states of motion in possibility space synchronic with changed behavior of fluid parcels.

Rayleigh–Bénard convection illustrates how contextual emergence plays an ineliminable role in the determination of the behavior of complex systems violating typical reductionist intuitions. Yet some typical emergentist intuitions (**Unpredictable**, **Inexplicable**) are also revised. The existence and behavior of Bénard cells can be predicted, and they can be explained. What contextual emergence highlights is that relevant information about the properties and behaviors of the fluid parcels plus the specification of relevant contingent stability conditions of the target system not given by the underlying fluid parcels domain allow for the in-principle prediction or explanation of properties and behaviors of the Bénard cells. Stability conditions provided by the large-scale fluid dynamics and system symmetries allow the existence of relevant states and observables for these cells. This is a concrete case of ontological emergence; nevertheless, we have unity, order, and novelty with no forces or bridge laws "coming out of nowhere."

[16] See Brand (1980), Huemer and Kovitz (2003), and Taylor (1963, 1966) for defenses of the metaphysical coherence of synchronic causal relations.

2.5 Summary and moving forward

Contextual emergence is a framework for making explicit the stability conditions enabling the actualization or particular modalities of physical possibility. As we have seen, many of these conditions involve relationships across multiple length and time scales, whether it is physical features of such contexts, more mathematical features such as topological constraints, or the intertheoretic relations involved in the descriptions of said multiscale interactions. Fluid convection offers an illustrative example. The laws of fluid mechanics and dynamics define all the possible states of motion of fluid molecules locating fluid systems in a particular region of physical possibility space. Nevertheless, in the initial uniform state of a Rayleigh–Bénard convection system, fluid molecules *cannot access any of the states of motion associated with convection cells*. After the temperature difference between the upper and lower plates constraining the fluid passes the critical value, many different convective states of motion are accessible. *But now the states of motion associated with the initial uniform state are inaccessible.* Stability conditions in the form of constraints (e.g., geometry, symmetry, symmetry breaking, conservation laws, large-scale dynamical constraints) characterizing the concrete context determines which of these modal subspaces for fluid motion are accessible (e.g., will the cells be cylindrical, hexagonal, or some other shape?).

To further clarify contextual emergence, Chapter 3 will turn to a bread-and-butter philosophy of science topic in reduction/emergence debates: intertheoretic relations. And since multiscale modeling has become crucial to so much scientific work (e.g., Hoekstra, Chopard, and Coveney 2014; Weinan and Jianfeng 2011; Wilson 2017), we also examine how such modeling exemplifies contextual emergence. We will argue that multiscale modeling is a much more interesting and realistic way to illuminate these issues than the standard philosophical debates about intertheoretic relations as such.

3

Intertheoretic Relations and Multiscale Modeling

The ordinary-sized stuff which is our lives, the things people write poetry about—clouds—daffodils—waterfalls—what happens in a cup of coffee when the cream goes in—these things are full of mystery, as mysterious to us as the heavens were to the Greeks.

Tom Stoppard, *Arcadia*

3.1 Introduction

In Chapter 2 we defined contextual emergence and illustrated it with a detailed case study of Rayleigh–Bénard convection. In that example we met the key elements of contextual emergence: fluid parcels contributing some of the necessary conditions for bulk fluid motion, stability conditions in the form of the convection cells defining the accessible states of motion of fluid parcels, an ordered unified account of emergent phenomena with no spooky forces or bridge laws appearing out of nowhere and so forth.

Now, we turn to a popular area of discussion in philosophy of science: scientific descriptions. Scientists use theories, models, and hypotheses to describe and explain the phenomena they seek to understand. Such descriptions have been the subject of much discussion in the philosophy of science. Hence, this chapter has two main focuses. First, we will discuss intertheoretic relations and see that contextual emergence helps us make sense of several moves and tensions present in the literature. Second, we will turn to multiscale modeling in the sciences. Such modeling does not aim at intertheoretic relations as philosophers of science have traditionally conceived, but is central to scientific practice. We will argue that multiscale modeling is best understood within the framework of contextual emergence.

Before we begin these tasks, we should say something more about scientific descriptions and their multiscale nature. Think of convection as an

Emergence in Context. Robert Bishop, Michael Silberstein, and Mark Pexton, Oxford University Press.
 DOI: 10.1093/oso/9780192849786.003.0003

example. A convective system spans length and time scale ranges from individual H_2O molecules to roughly an Avogadro's number of molecules in a fluid parcel to the large-scale Bénard cells. How do scientists develop descriptions for such a complex multiscale system? In this case, we can start with the state space for fluid parcels. An instantaneous state of a system is characterized by the instantaneous values of the variables considered crucial for a complete description of the state. Roughly speaking, any variable that is measurable is an observable. Properties such as position, velocity, linear or angular momentum, energy, mass, charge, and so forth, are examples. The state of motion of a particle in Newtonian mechanics, for instance, will be characterized by its position and momentum at a particular time.

Developing a description for fluids is a multistep process. First, start with the fluid parcel state space—mesoscale between individual chemical molecules and the parcels composed of them—and identify a proper partition of the space into equivalence classes of individual states that are indistinguishable with respect to some ensemble property such as the mean kinetic energy, a momentum distribution of an ensemble of fluid parcels (Fig. 3.1). These equivalence classes are universality classes in that individual

Figure 3.1 A partition of the state space where each color stands for an ensemble of fluid parcels having equivalent mean kinetic energy. The whole circle represents the large-scale bulk fluid that is coextensional with the ensembles.

fluid parcel states multiply realize these statistical states in each partition cell. Furthermore, each cell is the support or range of the distribution representing the statistical state for the value of mean kinetic energy in that cell.

Next, identify individual states in the large-scale fluid dynamics state space (where Bénard cells and other large-scale structures exist) that are co-extensional with these statistical states in the partition over the fluid parcel state space. This second step requires a particular choice of the context of the bulk fluid dynamics state space. The choice determines the set of emergent observables that are constructed from the statistical states defined in the partition over the fluid molecule state space yielding the Navier–Stokes equations.

Note that the context of the bulk fluid does a lot of work in this procedure. This context serves as a stability condition implemented in the partition over the fluid molecule state space ensuring that the emergent states and observables both exist and are stable under temporal evolution in the fluid parcel state space partition.[1] Improper partitions will not guarantee such stability; hence, the large-scale bulk fluid context is crucial to constraining the relevant partition of the fluid parcel state space. Bear in mind that these bulk states and observables do not exist in the original fluid parcels state space. One must already know what the relevant large-scale states and observables are; these cannot be derived from statistical averaging over the fluid parcels. This is because there are structural relations among the bulk fluid properties not reflected by the properties at the scales of chemical molecules and fluid parcels. The choice of larger-scale context along with its states and observables endows the fluid parcel state space with a contextual topology (we shall say more about this subsequently). As we saw in Chapter 2, the context for the fluid dynamics domain description specifies stability conditions found only at the length and time scale of the bulk fluid.

Hence, the procedure that physicists use to construct a multiscale description of a fluid system involves contextual emergence for levels of description, even though textbooks grinding out the gory details do not usually make the contextual emergence explicit. Nevertheless, the same pattern described in Chapter 2 shows up in the construction of fluid dynamics descriptions, indicating that contextual emergence is a fruitful framework for thinking about interlevel descriptions. Furthermore, one sees the same pattern show up in statistical mechanics and thermodynamic descriptions (Chapter 4; Bishop 2019).

The kinds of derivations just outlined involve taking statistical moments that must be closed using large-scale contextual information, continuity

[1] In turn, each fluid parcel represents a stability condition for the dynamics of the chemical molecules composing the parcel.

constraints that must be formulated at the large scale, and so forth (e.g., Batchelor 1967; Pathria 1972). A moment of a function is a qualitative measure of its graph. Suppose a function represents mass. Its first moment is the center of the mass, while the second moment is the rotational inertia. In physics, a moment is an expression involving the product of a distance and physical quantity, such as the torque, the product of a force on an object, and the distance from the reference point to the object. In statistical mechanics and fluid dynamics, moments are the expectation values for powers of the observables thought of as random variables in the statistical sense in a probability density function:

$$\int_r \rho\,(r)\,r^n dr.$$

Here, $\rho(r)$ is the probability density function for mean kinetic energy, say, r is the distance, and n is special dimension. The integral sums the contributions over the whole space. Such moments are "closed" (think computed) by either placing conditions on the larger-scale space (e.g., its boundaries) or by invoking experimentally determined values of large-scale parameters. Either way, these closure procedures make irremovable reference to large-scale properties of the bulk fluid and its container/constraints (e.g., the pipe through which a fluid is flowing). The texts discussing these derivations typically fail to explicitly note the multiscale information invoked in the construction of these descriptions as well as the implications of this information.[2]

Something else to notice about the procedure for constructing fluid dynamics descriptions is that it is not a matter of formulating scientific theories in logical languages as the logical positivists sought or as most of the intertheoretic relationship literature in philosophy of science has concentrated on. As we will see subsequently, the point of such procedures is aimed at producing relevant descriptions of physical phenomena to better understand how the actual world works. *Scientists do not do logic*—they do astronomy, physics, chemistry, geology, biology, and such. And, as we shall also see, although a lot of sophisticated mathematics is involved in the construction of multiscale scientific descriptions, philosophers' concerns about how different theories are mathematically related is not the main point of scientists' work either. Understanding how different theories or levels of description are related is

[2] These construction procedures often produce partial differential equations such as the Navier–Stokes equations. While such equations are powerful descriptions of how physical phenomena change in space and time (e.g., fluid flowing through a pipe), they are computationally difficult to solve and usually involve supercomputers. Recent progress has been made in devising machine-learning algorithms to solve such equations (e.g., Ranade et al. 2021), but solving such equations by these or other means does not reduce the equations to something else (e.g., fluid parcel dynamics or neural networks). They simply solve the equations (i.e., machine learning techniques basically guess and fit functions to equations to achieve solutions).

relevant to grasping the unity, order, and coherence of the natural world, so we cannot fault philosophers for being interested in that. Nonetheless, the focus on logical or mathematical relations tends to obscure the actual unity, order, and coherence that scientific investigation reveals.

3.2 Scientific theories, observables, and contextual emergence

Discussions about theory reduction and intertheoretic relations take place almost exclusively in the context of the physical sciences. Moreover, these discussions presuppose that different levels of description are valid (e.g., theories formulated at different descriptive levels such as the microscale for fluid parcels and the macroscale for bulk fluids). Contextual emergence was originally articulated for understanding the relationship of scientific descriptions at different levels (Bishop and Atmanspacher 2006). In this section we will outline how theories and theory relations are understood in contextual emergence.[3]

3.2.1 *States, observables, and topologies*

A well-defined concept of states and observables of systems is needed to discuss relationships among levels of description in the physical sciences. The algebraic approach in physics offers just such well-defined concepts and provides some background for what follows (see Bishop 2019 for details). This is not to imply that biological or other special science descriptions must be put into the same descriptive apparatus as that of physics or chemistry. Indeed, although some areas of mathematical biology can be cast in this apparatus, evolutionary theory generally cannot. Rather, the point is that each of the sciences—sometimes even subfields within a particular science—uses its own form of descriptions because they focus on states and observables relevant to their domains of inquiry and draw on concepts and vocabulary distinctive of their objects of inquiry. Contextual emergence takes its cues from the practices of individual sciences rather than being dependent on any specific descriptive framework at one level of description in a particular science. We adhere to the algebraic framework because it has wide applicability to the philosophy of science literature and allows us to display some of the unifying features of patterns of contextual emergence in the physical sciences.

[3] We will not engage realist/antirealist arguments about scientific theories, and nothing discussed in this chapter will depend on adjudicating these debates.

For any lower-level theory $\mathbf{T_L}$, one can speak of its algebra of observables A_L and how this algebra can be related to that of any higher-level theory $\mathbf{T_H}$.[4] With the definition of state space given previously, the "most fundamental" or lowest-level theory in physics is chosen to be as context-independent as possible, meaning that only the most basic assumptions are required for its definition (recall the discussion of possibility space in Chapter 2). This is equivalent to specifying the algebra of observables with the most basic topology, $\mathcal{T}_L$.

The simplest way to understand the relationship between an algebra of observables and its topology is by considering sets and spaces. A set is a mathematical representation of a group of objects, where the objects in question could be anything: numbers, language terms, physical objects, or what have you. Indeed, one can think of the fluid parcels state space where the observables relevant for fluid parcels is a set of objects. Suppose we have two simple sets, h and f, whose objects or elements are numbers: $h = \{3,7,9,14\}$, $f = \{9,14,28\}$. Some important properties of sets can be summarized and illustrated as follows:

The intersection of h and f, $h \cap f$, is the set of elements that are in both h and f: $h \cap f = \{9,14\}$.
The union of h and f, $h \cup f$, is the set of elements that are in h or f: $h \cup f = \{3,7,9,14,28\}$.
The proper subset h of f, $h \subset f$, is $\{9,14\} \subset \{9,14,28\}$.
The proper superset f of h, $f \supset h$, is $\{9,14,28\} \supset \{9,14\}$.

There are various ways a set can be partitioned. Each partition defines a sense of "nearness" of points within the set. Yet nearness is not intrinsic; instead, it is defined by specifying a topological structure for the set. Let S be a set, where its elements, x, are points. Let N be a function assigning to each x in S a collection $N(x)$ of subsets of S. The elements of $N(x)$ are neighborhoods of x with respect to N. Together, S and N define the topological space V. Often we say that N is a closure function, as it assigns to each subset of S a defined negation set, a set of all the elements not in each subset of S. The topology, $\mathcal{T}_S$, is representation dependent because we decide which subsets to pick out from S.

When the topological properties are added to mathematical spaces, the spaces are often referred to as *topological vector spaces* independent of whether the spaces are composed of functions, vectors, or more abstract elements. A norm function induces a natural *metric* or measure of distance $d(a,b)$ between two points a and b with the properties

[4] By using symbols, such as **T**, to denote theories, we do not mean to imply that these are logical systems. Physical theories are not logical systems *passé* logical positivist analyses. Scientific theories are systematic bodies of knowledge, assumptions, and concepts about the domain or scale of the physical world on which they focus, but they are not deductive logic systems.

$$\begin{aligned} d(a,b) &= 0 \text{ iff a} = \text{b}, \\ d(a,b) &= d(b,a), \\ d(a,b) &\leqq d(a,c) + d(c,b), \end{aligned}$$

for all points a, b, and c in a suitably defined mathematical space.[5] The norm function is an example of a metric for vectors in a Hilbert space, which has played an important role in the mathematical formulation of quantum mechanics (von Neumann 1955/1932). For our fluid model example, the fluid parcel state space forming the microscale description has a norm topology induced by the norm function defining distance between the points in the space.

The concept of metric is one way of defining the topology of an algebra of observables, where the arguments for the function d are observables from the algebra A. Any space where $\mathcal{T}$ can be defined in terms of a metric d is said to be *metrizable* and is known as a *metric space*. Defining different topologies leads to different mathematical spaces.

Finally, let Ω be a set of points, and suppose there is a closure function assigning to each set $S \subset \Omega$ a set $\bar{S} \subset \Omega$ (called the closure of S) satisfying the properties $S \subset \bar{S}$, $\overline{S \cup M} = \bar{S} \cup \bar{M}$, for some set M, $\bar{\varnothing} = \emptyset$, $\emptyset$ being the empty set, and $\bar{\bar{S}} = \bar{M}$. A topology with the closure function $S \rightarrow \bar{S}$ is said to be *finer* than that given by the closure function, if $S \subseteq \overline{S'}$, for each $S \subset \Omega$. The topology is said to be *coarser* if $S \supseteq \overline{S'}$ for each $S \subset \Omega$.

The important point to take away is that the algebra of observables A for a theory **T** is always associated with a topology $\mathcal{T}$ defined for a metric space in the physical sciences. One can compare two different theories by comparing their associated algebras of observables and topologies.

3.2.2 *Contextual topologies*

In physics, the algebras of observables characterizing different theories have different topologies and are usually related via singular limits in some parameter (or other nontrivial mathematical relations, such as renormalization). When the limiting procedure is singular, this means that the equations at the limit are not smooth approximations of the equations when the parameter is approaching (but has not reached) the limit value. Singular limits can be characterized as those in which the *final set of equations has a qualitatively distinctive character compared to the starting set*. The former set of equations will have a different algebra of observables compared to the starting set.

Roughly, one can think of the difference between sets of equations connected by regular (non-singular) limits and singular limits this way. Imagine

[5] The metric function can be generalized to vectors.

there is a country path that runs through a forest to a meadow. If, as you walk along the path, the number of trees diminishes bit by bit until finally you reach the meadow, then the limit is regular. On the other hand, imagine, as you walk the along the path, that the density of the trees stays the same until you reach a stream that cuts across the path. You must jump across the stream to continue the path, and on the meadow side there are suddenly no trees. In this case the limit is singular: mathematically we have an obstacle to jump, and on either side of that obstacle the system is qualitatively different.

To regularize these limits (i.e., to get convergent limits in an appropriate metrizable space) requires specifying new contextual topologies. A *contextual topology* is a topology of a mathematical space of elements for a domain or scale of physics (e.g., statistical mechanics) that has been suitably enriched to correspond to a mathematical space that contains the relevant algebra of observables for a target system (e.g., a Carnot engine in thermodynamics). This enrichment is carried out by incorporating information from the target domain and its context (e.g., states and symmetries of the target domain). The contextual information necessary to produce a suitably enriched topology is not given by the finer topology of the lower-level description (i.e., statistical mechanics simpliciter) conceived as being as context-free as possible based on first principles (Bishop 2019; Primas 1998). An example would be how the fluid dynamics scale of the bulk fluid provides the relevant information to produce a contextual topology for the partitioned fluid parcels state space.

The stability conditions (Chapter 2, Section 2.3.2) associated with the higher-level theory $\mathbf{T_H}$ allow a contextual topology to be introduced by picking out relevant reference states and defining the appropriate observables for these states (e.g., for the fluid dynamics context). Topologies define the convergence properties for a sequence of elements in a space, so we can characterize topologies as strong/fine or weak/coarse to sharpen the definitions given in the previous section. An example of a strong topology would be the $\|\cdot\|$-norm in a vector space (e.g., Banach or Hilbert space) inducing a topology $\mathcal{T}$. An example of a weak topology would be the dual, the set of all continuous linear functions, inducing a topology σ on the vector space. The latter topology is coarse, while the former is fine; that is to say, $\sigma \subseteq \mathcal{T}$.

Convergence in strong and weak topologies can be illustrated by series expansions. Uniform convergence of a Taylor series of a function within its convergence radius is an example of convergence in a strong topology. A Fourier series of a function, which converges only in quadratic norm L^2, would be an example of convergence in a weak topology. Hence, even if two theories, such as statistical mechanics and thermodynamics, are related by singular limits (Bishop 2019), we can compare the algebra of observables and their respective topologies. Thermodynamics as the "higher-level theory" has a coarser topology characterized by an algebra containing observables not found in the finer topology of the algebra associated with statistical mechanics simpliciter as the lower-level theory.[6]

Contextual emergence shares some similarities and differences with Robert Batterman's (2002) account of asymptotic theory emergence. Like him, we consider asymptotic singularities often to be important for marking out crucial instances of emergence. But whereas Batterman focuses on the "no man's land" in the asymptotic region between theories (2002, pp. 115–120), contextual emergence focuses on stability conditions defining the contexts for theories and their relations. A more substantial point of distinction is that Batterman explicitly restricts his account of emergent properties to the role those properties play in asymptotic explanations (2002, pp. 126–129). By contrast, contextual emergence addresses *how* properties arise and why they are robust, how descriptive levels are related, the relationships between parts and wholes, and so forth. In more recent work he has emphasized, similar to us, how scientific theories are not related via reductive relationships (e.g., Batterman 2018; Batterman and Green 2020; Green and Batterman forthcoming). However, Batterman does not draw any ontological conclusions, whereas we shall discuss ontological implications of contextual emergence particularly in Chapter 7.

As we saw in the contextual emergence pattern discussed in Chapter 2, cases of emergence are those where neither the stability conditions nor the complete set of states and observables of the higher-level description are given by or derivable from the lower-level description simpliciter (and its state space). The contexts characterized by stability conditions are contingent on those conditions and are related to the new states and observables, or new structures and other aspects of target systems. These conditions, states, and observables represent the salient features of the description of a target system at a relevant level of description. To develop any description requires "abstracting from" or ignoring those details of a given system (and its environment) that are irrelevant, as is illustrated by constructing the Navier–Stokes equations.[7] The stability conditions along with their associated states and observables provide the foundation for all descriptions in the case of physics. Cases of universality are particularly interesting here, as they illustrate how descriptions ignoring microconstituent details yield accurate descriptions of mesoscopic and macroscopic properties and behaviors, but are also multiscale (e.g., thermodynamics and fluid dynamics).

The implication is that some features irrelevant in one context may be relevant in another. Temperature is an example of such a feature; it is relevant in

[6] For details, see Bishop (2019) and Bishop and Atmanspacher (2006).

[7] The typical gloss on these constructed higher-level descriptions is that they involve approximations and idealizations that deviate from the lowest-level description. We will see why this gloss underestimates what is going on in these "approximations" and "idealizations."

thermodynamics but irrelevant in Newtonian mechanics. A second example would be the shape of molecules, which is relevant in physical chemistry but irrelevant in quantum-mechanical descriptions using Schrödinger's equation focused on energy levels. Nevertheless, it is possible to implement contexts where temperature is relevant at the level of statistical mechanics, or contexts where molecular shape is relevant at the level of quantum mechanics (Chapter 4 and Bishop 2019). The natural way to represent contexts of these kinds mathematically is the enrichment of the original topology of the lower-level state space illustrated in the construction of fluid dynamics descriptions, hence the name contextual topology. The finest topology corresponds to the most context-free, lowest-level "first-principles" descriptions, while coarser topologies represent an increasing amount of contextual information not encoded in first principles. Specifying the difference in topologies of the corresponding state spaces is a precise means for relating properties at different levels of description to each other in the physical sciences.

The mathematics involved in carrying out these procedures is quite challenging. However, these ideas can be illustrated through some simple toy examples involving sets. Let the set $S = \{A, B, C\}$, then the trivial topology is $\mathcal{T} = \{\varnothing, \{A, B, C\}\}$ (i.e., the empty set and the whole rest of the set). $\mathcal{T}$ can be represented pictorially as follows.

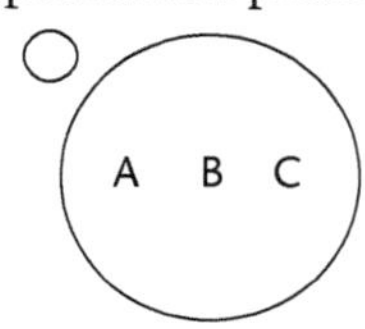

We can structure S as long as we preserve every relation. Yet doing so introduces a new, contextual, topology. For example:

$$\mathcal{T} = \{\varnothing, \{A\}, \{A, B, C\}\}$$

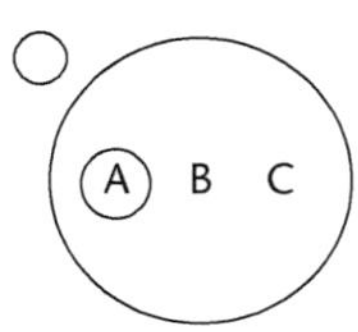

Now let us write this as $\mathcal{T} = \{\varnothing, \alpha, \beta\}$, where $\alpha = \{A\}, \beta = \{A, B, C\}, \alpha \cap \beta = \{A\} = \alpha$, $\alpha \cup \beta = \{A, B, C\} = \beta$. So, we have covered every relation between α, β.

Similarly, we could have $\mathcal{T} = \{\varnothing, \alpha, \beta, \gamma, \delta\}$:

$$\mathcal{T} = \{\varnothing, \{A\}, \{B\}, \{A, B\}, \{A, B, C\}\}$$

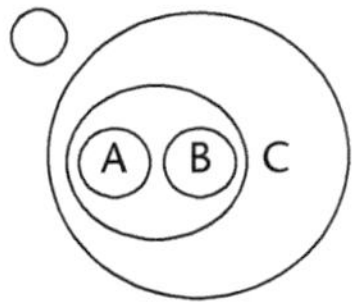

Or:

$$\mathcal{T} = \{\varnothing, \{B\}, \{A, B\}, \{B, C\}, \{A, B, C\}\}$$

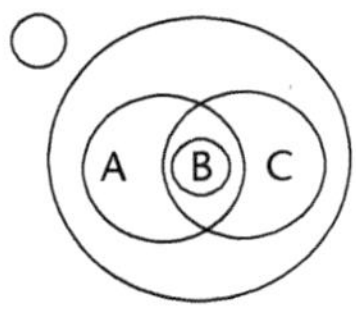

But not every ordering is allowed. For instance, $\mathcal{T} = \{\varnothing, \alpha, \beta, \gamma\}$:

$$\mathcal{T} = \{\varnothing, \{A\}, \{B\}, \{A, B, C\}\}$$

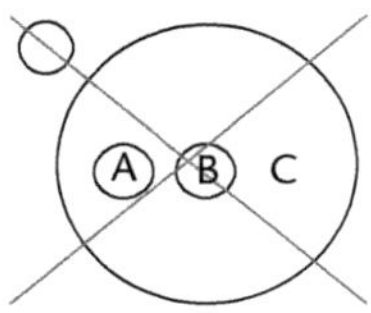

This topology is not allowed, since $\{A\}$ and $\{B\}$ are in the topology but their union is not in the topology (i.e., is missing a subset defined by the topology, which is the objects in α or β, that is $\alpha \cup \beta = \{A, B\}$, and this is not specified).

Similarly, for $\mathcal{T} = \{\varnothing, \alpha, \beta, \gamma, \delta\}$:

$$\mathcal{T} = \{\varnothing, \{A\}, \{A, B\}, \{B, C\}, \{A, B, C\}\}$$

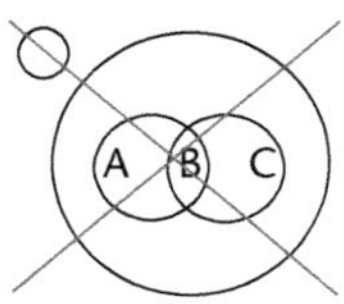

This is not an acceptable topology, because $\{A, B\}$ and $\{B, C\}$ are in the topology but their intersection $\{B\}$ is not.

Therefore, many different topologies are possible, each of them compatible with the same starting set. This illustrates how higher-level descriptions are not implied by lower-level descriptions because contextual topologies are underdetermined by the norm topology of the lower-level description. For instance, starting with the fluid parcel state space different for the microscopic description, different macroscopic descriptions can be constructed by adopting different contexts not implied by the microscopic description.

The purpose of theories and models is to reflect law-like and other robust relations and patterns in the world. Specifying a topology is picking out the important relations in the set so that these relations can be captured by relevant descriptions. By imposing a topology, an abstraction, to group together elements of the set, we create a hierarchy of structure within the set that orders the set, but that ordering is not fixed by the lower-level description.

Moreover, we can compare topologies using the earlier definitions.[8] When every set in a topology $\mathcal{T}_1$ is also in the topology $\mathcal{T}_2$ and $\mathcal{T}_1$ is a subset of $\mathcal{T}_2$, we say that $\mathcal{T}_2$ is finer than $\mathcal{T}_1$ and $\mathcal{T}_1$ is coarser than $\mathcal{T}_2$ as described previously.

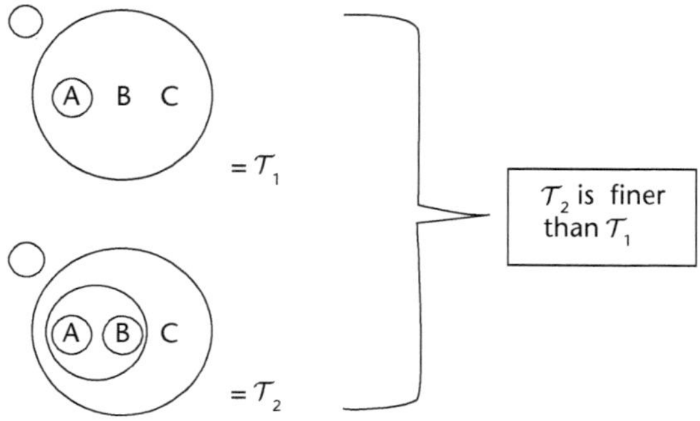

Consider the topology of $\mathbf{T_L}$, presumably more fundamental, and the contextual topology of the $\mathbf{T_H}$ description. This contextual topology depends on the abstractions scientists make that are relevant to the higher-level context. Contextual topologies encode new context-dependent features which are not already present in the fundamental theory. Nonetheless, like some contextual topologies of our toy example were not allowed, the norm topology of the lower-level description restricts what contextual topologies are possible in the sense of setting limits on consistency (recall our discussion of possibility space in Chapter 2). This is all summarized by Hans Primas (1998)—one of the first to articulate the importance of contextual topologies in physical theories:

> In a mathematically formalized theory, a context can be introduced by choosing a new coarser *contextual topology* which is compatible with the finer topology of the more fundamental theory. This new topology is never given a priori but depends in a crucial way on the abstractions made by the cognitive focus or the pattern recognition devices used by the experimentalist. The closure of such higher-level descriptions in the chosen contextual topology captures new context-dependent features which are not already present in the fundamental description. In this mathematically precise sense, one can speak of the *emergence of novelty* in descriptions of a higher level.
>
> (1998, p. 86, original emphasis)

[8] Although some topologies cannot be compared in principle meaning, it is not possible to produce a total ordering for all topologies.

3.2.3 *Contextual topologies and intertheoretic relations*

"Higher-level" theories, then, can be rigorously related to a basic "lower-level" theory through introducing relevant contextual topologies as described in the previous section. This variety represents the fact that there are numerous possibilities for choosing coarser contextual topologies compatible with the topology of the underlying "fundamental" theory as illustrated by our toy examples in the previous section. For instance, beginning with the norm topology for Hilbert space quantum mechanics, one can complete different singular limits in different orders to produce a variety of different macroscopic descriptions other than Newtonian classical mechanics. This is equivalent to saying that there are several different contextual topologies one can equip the Hilbert space with compatible with the norm topology generating coherent macroscopic descriptions. Nonetheless, none of these macroscopic descriptions is implied by the Hilbert space norm with its associated algebra of observables.

This contextual emergence pattern implies that contextual information is at least as important as the "first principles" of the lower-level theory. This pattern also has another implication first noted by Primas:

> If the derived theory would not be closed in the new contextual topology, the validity domain of the derived theory would be contained in the validity domain of the basic theory. To get a mathematically complete and consistent theory, one has to close the derived theory in the new [contextual] topology. The result is that the validity domain of the derived theory and of the basic theory intersects nontrivially. Neither domain is contained in the other. Although higher-level theories can be derived rigorously from a more fundamental theory, they are not sub theories of the basic theory. Therefore, higher-level theories cannot be totally ordered, and theory reduction is not transitive.
>
> (Primas 1998, p. 86)

Contextual emergence makes clear how a lower-level description $\mathbf{T_L}$ contributes some necessary conditions, while the context for the higher-level description $\mathbf{T_H}$ contributes the remaining necessary and sufficient conditions for the derivation of the target phenomena in question. The target system and its context (e.g., the dynamics of fluid flow) provide the added necessary and sufficient conditions for the higher-level description. The implication is that the $\mathbf{T_H}$ descriptions "are context-dependent, they cannot be deduced from context-independent first principles without taking into consideration the abstractions evoked by the pattern recognition devices necessary for the observation of phenomena" (1998, p. 85).

Moreover, in physics, as described previously, a typical procedure for doing this is implementing a concrete context as a contextual topology. But this is a nontrivial process! A typical tool used for such purposes is an asymptotic expansion.[9] To formulate such an expansion, a reference state must be specified in the $\mathbf{T_L}$ state space representing essential features of the target context. Electronic ground states of a molecule in quantum chemistry and Kubo–Martin–Schwinger states in statistical mechanics would be examples of such reference states. If the expansion is singular, as an appropriate parameter tends to some limit in the fine topology of the $\mathbf{T_L}$ description simpliciter, then it is not uniformly convergent in the topology intrinsic to $\mathbf{T_L}$. Such discontinuous limiting behavior indicates the need for a context-driven change of description. The number of degrees of freedom for thermodynamics (continuum limit), or the electron mass divided by nuclear mass for physical chemistry, are examples of such parameters (Chapter 4 and Bishop 2019).

The key idea is in identifying a new topology that regularizes the asymptotic expansions that are singular with respect to the old, finer topology; hence, these expansions converge with respect to the new, coarser topology.[10] In many cases, regularization is possible by introducing a hierarchy of length and time scales. One can then consider the motion of the system on the "fast" time scales relative to almost fixed reference states corresponding to the "slow" time scales (e.g., Batterman 2002; Bishop 2019; Primas 1998). The separation of time scales yields a coarser contextual topology of the state space often leading to a partition into equivalence or universality classes of states. The new, coarser topology is *compatible* with the original, finer topology if they and their dynamics are *topologically equivalent* with each other (e.g., Atmanspacher and beim Graben 2007; Bishop 2019). Nevertheless, the contextual topology is contingent in that it is not given by the original finer topology nor any other elements of $\mathbf{T_L}$. The $\mathbf{T_H}$ description associated with the coarser contextual topology and the target system allow new context-dependent features of a higher-level description to be defined that are not definable in the original state space under the finer topology (e.g., novel properties such as molecular shape).

We have outlined a formal procedure for the contextual emergence of higher-level properties based on contexts *in addition to* the features of the $\mathbf{T_L}$ description. This is the sense in which contextual emergence describes how qualitatively new properties arise which are unavailable in $\mathbf{T_L}$ simpliciter.[11] A new contextual topology accommodates novel properties as such within a higher-level description *rather than approximating them* as a limiting case in the topology of a lower-level description.[12] After all, new stability conditions,

[9] Asymptotic expansions are widely used in physics (e.g., Batterman 2002; Dingle 1973; Friedrichs 1955; Primas 1998).

[10] To the unsuspecting eye these can appear to be cases of theory reductions!

[11] In this way, contextual emergence can make sense of why the fundamental context-free descriptions of physics might appear to "not describe true facts about reality" (Cartwright 1983) without endorsing Nancy Cartwright's radical conclusions that the fundamental laws "lie."

[12] This is a point of contrast with Jeremy Butterfield (2011b), where he misleadingly speaks of approximating emergent properties before reaching the asymptotic limit. Such an approach can

states, and observables not derivable from the lower level simpliciter appear. The main "*task of higher-level descriptions is not to approximate the fundamental theory but to represent new patterns*" (Primas 1998, p. 87, original emphasis) in contrast with how so much of the philosophy of science literature has framed intertheoretic relations. These patterns typically are not reducible to a lower level in the strict sense of reductionism.[13] Reducibility in this sense would mean that only the principles and properties of the lower-level description (e.g., first principles of a "fundamental" description, elementary constituents, norm topology, and so on) are needed to describe new patterns exhaustively. When higher-level contexts in addition to the lower level must be considered to rigorously derive descriptions of these new patterns, then intertheoretic reduction fails. An example would be asymptotic series relationships between the quantum-mechanics description ($\mathbf{T_L}$) and molecular descriptions used by chemists ($\mathbf{T_H}$) via adiabatic procedures. The latter is not a simplification or approximation of the former; rather, it is a qualitatively different description (Chapter 4; Bishop 2019; Primas 1998).

To recap, contextual emergence describes cases where contexts—their stability conditions, properties, structures, and so forth—are at least as important as the first principles and properties of any $\mathbf{T_L}$. Given such a theory *and* a suitable contextual topology, novel properties can be rigorously derived using asymptotic expansions, for instance, that are regularized in the new contextual topology. These $\mathbf{T_H}$ descriptions cannot be considered as sub-descriptions of $\mathbf{T_L}$. Nor are $\mathbf{T_H}$ descriptions extractions from or fragments of the $\mathbf{T_L}$, in contrast to Jessica Wilson (2010), *because that theory simpliciter does not imply the new contexts or contextual topologies*, and hence does not imply any of the possible higher-level descriptions. Expressed another way, since the contextual topologies cannot be totally ordered with respect to $\mathcal{T}_L$, the corresponding higher-level descriptions cannot be sub-descriptions, extractions from or fragments of $\mathbf{T_L}$.

The typical case in physics is that we have some description, $\mathbf{T_L}$, and we want to construct a new description, $\mathbf{T_H}$, for the new context. This was illustrated by construction of the Navier–Stokes description for bulk fluid flow from the fluid parcel state space and description. It is important to note that the $\mathbf{T_L}$ also has a context defined by stability conditions guaranteeing the states and observables relevant to the lower-level description exist and are persistent. For example, the context for fluid parcels specifies a set of stability conditions for the existence of the states and observables relevant for the description of fluid parcels. These stability conditions, states, and observables

make it appear that there is theory reduction when in actuality a robust form of emergence is present.

[13] Hence, one should not expect there to be many cases of what Ernst Nagel called "approximative reduction" (2008/1979, pp. 311–363, 371–373).

are not given by the $\mathbf{T}'_{\mathrm{L}}$ description for individual H_2O molecules. In other words, the contextual emergence pattern shows up across multiple length and time scales in the descriptions relevant for different levels of organization. Even the most "fundamental" or "lowest"-level description depends on stability conditions guaranteeing the existence and identity through time of its states and observables (i.e., there is no genuine context-free level of description).

As a point of clarification, note that we do not take the various singularities arising in asymptotic expansions in physics to necessarily be representative of actual infinities in nature. Rarely is it possible for physical quantities to become infinite in actual-world systems. For instance, when a correlation length becomes mathematically infinite, as in the transition to convecting states in a fluid, this indicates that the interactions span the length scale of the system.[14] Rather, the emphasis in contextual emergence is on the contextual topology—connected to relevant stability conditions—regularizing the asymptotic series and the states and observables associated with that new topology. In this sense, arguments that getting rid of the infinities gets rid of the case for emergence are red herrings (e.g., Menon and Calendar 2013).

Finally, we note that the notion of contextual topologies nicely expresses the importance of the system/environment distinction discussed in Chapter 2. All observation requires a distinction between observable and environment, and this can be represented set-theoretically by separating out states and observables from an environment based on the relevant abstractions for the level of description. For example, when the research focus is on the context for a fluid parcel, the larger-scale bulk fluid and its dynamics form the environment while the fluid parcel is the system. This means that the contextual emergence pattern we see repeated in the sciences tracks with the system/environment distinctions that scientists make, as these distinctions are always contextual.

3.2.4 *Abstractions, idealizations, and contextual topologies*

Since abstraction plays such a significant role in scientific enquiry and is intimately related to developing scientific descriptions and models, it is important to clarify the difference between idealization and abstraction. Idealizations are common in models, where scientists leave out details to simplify analysis *making it more tractable*: planes are regarded as frictionless, planets as point masses, and so on. But if the details are put back in, the model would be more accurate. Ernan McMullin's (1985) now classic analysis

[14] See Chapters 4 and 5 for more discussion of infinities and phase transitions.

suggests that we should adopt a sanguine attitude towards such idealizations because they are merely pragmatically useful approximations and can be removed in principle through a so-called Galilean strategy systematically removing the idealizations until desired accuracy is achieved. Idealizations make up for our limited analytic or computational power. When we idealize, we simplify in a harmless manner allowing problems to be solved more easily. These "white lies" of science work so well precisely because they can be removed, sometimes in practice, but always in principle.

In contrast, abstraction removes unnecessary details *to reveal patterns and constraints*. Abstractions are not idealizations (although the two are often conflated in the philosophy literature and are used interchangeably as folk terms among scientists). There are many case studies in which the modeling assumptions cannot be removed and, as well, where modeling abstractions are necessary for the definition of quantities in terms of which we wish to parametrize the world. Abstractions help define new categories of variables as described previously. Hence, an abstraction is not the same as a pragmatic convenience.

Sometimes, abstractions are required to capture specific modal facts. If we want to know modal facts about generalities that apply across radically different microphysical systems, then abstracting away from those details is not just convenient. It is essential. This is because the abstract model can help identify why a range of systems have specific features in common (e.g., universality discussed in Chapter 5). Abstraction allows scientists to identify the modal information concerning the circumstances under which diverse systems share relevant features, and therefore answer counterfactuals about when they would not share those features. This is much more powerful than simply listing the properties of systems and noting that some have commonalities. Abstract models are necessary to map the modal landscape of those commonalities in a principled way and capture larger-scale or global constraints across systems sharing common features/behaviors. Think of the commonalities at the large-scale fluid dynamics that hold independent of details at the smaller scale of individual molecules.

The Ising model is an illustrative example. In his 1924 thesis, Ernst Ising was aiming to understand ferromagnetism. To do so he imagined that the up/down polarizations in the atomic dipoles of a metal could be abstractly thought of as a regular array of points, each assigned a value +1 or –1. If S_i(a) is the value of site i in lattice arrangement a, then $\Sigma_i\, S_i$(a) is the difference between +1 and –1 sites in arrangement a. The quantity M, which is the sum of S_i(a) divided by the total number of sites, describes how ordered the system is. If M is +1 or –1, then the lattice is maximally ordered, since all the sites are the same (either all +1 or all –1). If M is zero, then the lattice is maximally disordered, since just as many sites are +1 as –1.[15] With each adjacent pair of sites <*j,k*>, there is an associated interaction energy, JS_jS_k, and the total energy, E_a, is the sum over all adjacent sites in a. E_a is at a minimum in a maximally ordered arrangement. As temperature decreases, if E_a is positive, the probability, P_a, of disordered arrangements falls to zero, whereas as temperature increases disordered states become more likely. The order parameter is the sum of all P_aM_a, so each arrangement is weighted by its probability. By counting the number of ways the lattice could be arranged, and using statistical methods, Ising was able to calculate the resultant values of bulk properties such as the magnetization.

Ising's model is highly abstract, but it is not simply an idealization of a ferromagnetic. It is not the case that by adding in more details the Ising model would become more accurate. In fact, it *would lose all its power if it were more accurate*. This is not a point about practical utility. In the idealization of a frictionless plane, we may not be able to include an exact representation of the friction, and still calculate in practice; but, if we could, we would use a maximally detailed model for maximum accuracy.[16] For Ising, this is not the case. Ising's abstraction is ruined by more detail. We would not be able to define a new variable, the order parameter, if the model did not abstract in the way that it does.

Another example of the necessity for higher-level abstract structural explanation to capture modal facts is dimensional explanation, discussed at length by Pexton (2014). In dimensional explanation, even the precise form of the laws of physics is irrelevant. This irrelevance could never be accessed by considering those laws themselves but only by exploring the constraints on the laws.

Dimensional analysis has been around a long time (Newton called it the "Great principle of Similitude"), but the modern form can be traced back to James Clerk Maxwell. It was Maxwell who distinguished mass [*M*], length [*L*], and time [*T*] as the independent dimensions from which others could be derived. (Notationally, from now on a [ ] will be used to denote the dimensions of a variable, e.g., speed has dimensions [*L/T*].) These three dimensions are not the only possible choice as a variable base (e.g., in mechanics one might choose to categorize a system in terms of force [*F*] as well as length and time).

Consider a straightforward example: a spring undergoing simple harmonic motion. We have a mass *m* attached to an ideal linear spring of constant

[15] The sites on the lattice can be visualized as arrows that can point in one of two directions, up or down. If the arrows all point up or all point down, then it is ordered; if an equal number point up as point down, then it is maximally disordered.

[16] Friction is far more complicated than the simple coefficient of friction introduced in first-year physics textbooks, but we will set those details aside.

k suspended by gravity with acceleration g. What is the period of oscillation P? The system is categorized by four variables, $\{P,m,k,g\}$ with dimensions $[T,M,M/T,L/T^2]$, where the dimensions of k are determined by Hooke's law. This is a significant assumption and illustrates that dimensional analysis always relies on strong physical intuition and background theories. From these there is only one dimensionless constant that can be formed. Let us call it C, so we have

$$C = \frac{P^2 k}{m}$$

$$[1] = \left[T^2 \left(\frac{M}{T^2}\right)\left(\frac{1}{M}\right)\right]$$

Hence, from considering the dimensional structure of the system alone, we can say that

$$P \propto \sqrt{\frac{m}{k}}$$

Notice that the period does not depend upon the gravitational acceleration, g, at all. This cannot be part of the expression, because g is the only variable with the dimension of length which cannot be cancelled out by another variable. This example is typical of the kind of dimensional analysis needed for a first-year physics examination question. In such cases, reading off the dimensional structure seems straightforward and natural, but even in simple systems dimensional subtleties can come into play.

Consider a baseball thrown at an angle with horizontal velocity V_x and vertical velocity V_y, which moves a horizontal distance R under the effects of a gravitational acceleration g. If we proceed as before we have $\{V_x, V_y, R, g\}$ with dimensions $[L/T,L/T,L,L/T^2]$. Let a, b, and c be unknown exponents:

$$R \propto V_x^a V_y^b g^c$$

$$[L] = \left[(L/T)^{a+b}(L/T)^c\right]$$

Balancing the different dimensional components L and T separately gives:

$$1 = a + b + c, 0 = a + b + 2c$$

We have two equations and three unknowns, so there is no unique solution. However, Huntley (1967) proposed an extension that can allow us to find a solution. We need to artificially add extra dimensions—"directed dimensions." In these, the magnitude and direction of a vector are considered dimensionally distinct, so in our baseball case the horizontal and vertical components of velocity are now categorized as having distinct dimensions $[L_x]$ and $[L_y]$, hence $[L_x/T, L_y/T, L_x, L_y/T^2]$. This leads to

$$[L] = \left[(L_x/T)^a (L_y/T)^b (L_y/T)^c\right]$$

which can be solved to give $a = 1$, $b = 1$, and $c = -1$, and so we can say that

$$R \propto \frac{V_x V_y}{g}$$

The illustrative point from this example is that even in simple cases dimensional explanation requires abstraction, and that abstraction is contextual. The abstraction here reflects other contextual facts about the underlying physics as applied to the system in question; for instance, that gravity acts in the y-axis only. Hence, although speed has the same dimensions strictly speaking, the physics of our target system creates the appearance of the different axes having different dimensions. Note that the distinction between abstraction and idealization or approximation is very important. *We have abstracted the system as having different dimensions in the different axes, so this is not an approximation.* To proceed with the analysis a system must be formalized into having a specific number of dimensions; we cannot deidealize it to having a different dimensionality. *The description of the system with directed dimensions is formally a different conceptual partitioning of the same empirical domain, and it is this different partitioning that allows us to capture the correct modalities.*[17] In physical terms, one dimension is so small that it can be neglected, but in modeling/theoretical terms, to describe the system we must *formalize* that as the system having only the reduced number of dimensions.

Abstraction, then, is what the introduction of a new contextual topology represents. *A higher-level contextual topology does not approximate a more accurate lower-level topology.* Rather, it abstracts, and through those abstractions it allows the definitions of new states and observables relevant to the context defined by the relevant stability conditions. We might be able to deductively connect two theories together, *but not by deidealizing the higher-level theory.* Instead, we must *enrich* the lower-level theory state space with structure derived from the *contextually defined abstractions* of the higher-level theory. These contextually defined abstractions are never given by

[17] This will be a critical point in Chapter 5 when we look in detail at condensed-matter cases. Often, condensed-matter systems are strongly dependent on dimensionality.

first principles of a lower-level or "more fundamental" theory or its domain. We see this in the case of the fluid dynamics scale where the context through the stability conditions establishes the relevant states and observables for the Navier–Stokes equations. It is only by abstracting away from the microphysical details of molecules that we access the multiscale relations, states, and observables of a flowing river. Abstraction is crucial for obtaining the right states and observables for the relevant level of description, and we will see that abstraction is crucial to multiscale modeling discussed in Section 3.4.

We can summarize this section as follows. In algebraic approaches and other approaches, a fundamental theory *formally* is as context-independent as it can be. Call this theory $\mathbf{T_F}$, which has a purely algebraically-only derived ordering. The ordering of its set of observables—the algebra of observables—is its topology. A fundamental theory has what is called a "norm" topology. To describe a target system, we restrict the domain of $\mathbf{T_F}$; that is, we introduce *new system/environment distinctions*, and introduce a new topology relevant for the new context. Topology is linked to *abstractions*—it is not given by first principles. Think of this as taking a nomically open theory and closing it in the new contextual topology. Specific nomic relations are only extractable from an algebra of observables if it has a particular topology. One crucial point is that the validity of the domain of the higher-level description intersects non-trivially with the validity of the domain of $\mathbf{T_F}$. The topology defines the domain of applicability. Higher-level descriptions are not approximations of lower-level theories or mere subsets of them; instead, higher-level descriptions have unique domains of applicability because of a unique contextual topology. All this is illustrated by the different contexts of the molecule, parcel, and bulk scales of a fluid, where the algebras of observables are distinct at each scale and the topology of the relevant algebra of observables at the bulk scale is a contextual topology with respect to topology of the algebra of observables at the parcel scale which, in turn, is a contextual topology with respect to the topology of the molecular scale.

3.3 Intertheoretic relations and contextual emergence

The literature on intertheoretic relations is vast, and we do not pretend to cover it exhaustively. Rather, our aim is to illustrate how contextual emergence makes better sense of the relationship of scientific theories than the accounts that focus on relations in a framework of predicate logic. What we shall see is that the framework explained in the previous section tracks more closely with actual scientific theories, and that the unity, order, and coherence of scientific theories is not a matter of logical relations and domains of definitions that has shaped so much of the literature in the philosophy of science.

One form of intertheoretic relation is reduction. The classic formulations appeared under the influence of positivist thinking in the mid-twentieth century (e.g., Nagel 1961), but the contours are still very much characteristic of reductionist thought (e.g., Papineau 2002, Chapter 2). A typical example of theory reduction discussed in the literature is the relationship between special relativity and Newtonian mechanics, where the latter is a special case of the former in the small velocity limit $v^2/c^2 \to 0$, where c is the speed of light. Or, discussed less often, Newton's mechanics being a special case of general relativity in the weak-field limit.[18] Another example comes from the British emergentists (e.g., see McLaughlin 1982), who distinguished between resultant and emergent properties. Resultant properties were modeled after the vector addition of forces in classical particle mechanics, where the total length and direction of a force vector is the sum of its components. When higher-level properties can be analyzed in such fashion, they are said to be resultant. Outside of these and some other rather trivial cases, there are few viable examples of intertheoretic reductions in the sciences.

As we have emphasized, stability conditions are a key feature of contextual emergence and highlight the important role that contexts play in scientific descriptions. In typical philosophical treatments of reduction and emergence, states and observables are taken for granted, meaning that the stability conditions guaranteeing those states and observables are also taken for granted. Underestimating such conditions, and their role in the existence and identity of states and observables, can lead to misidentifying a case of contextual emergence as a case of reduction (e.g., Nagel 1961, p. 341; Butterfield 2011b).

3.3.1 *Intertheoretic reduction and context enrichment: comparison with Nagel and Nickels*

Here is a typical question in the intertheoretic reduction literature driven by thinking of theories as sets of sentences in logic. Can two different theories involving distinct theoretical vocabulary each describing the same range of phenomena be said to be fully equivalent? If the answer is "yes," this leads to another question. Can one of them be privileged such that one theory is entailed by the other but not the other way around? The standard view of the philosophy of science suggests that the answer to such questions is

[18] We leave aside tricky questions of how the masses in these three different theories are related. Potentially, these mass differences block the kind of straightforward intertheoretic reductions being illustrated.

"yes." On the surface, it appears there are many examples where one theory is reduced to another by the progress of science looking deeper into nature and examining its most basic building blocks. The usual examples given are classical Newtonian mechanics reducing to Einstein's special relativity, thermodynamics reducing to statistical mechanics, and chemistry reducing to quantum mechanics. Comparing contextual emergence's account of intertheoretic reduction illuminates how our account differs from typical discussion in the philosophy of science literature.

In general, there are two dominant accounts of intertheoretic reduction due to Ernst Nagel (1961) and Thomas Nickels (1973), with the former receiving more attention in the literature. Nagel's account is based on the derivability of laws of science (nomic relations): *theory* $\mathbf{T_H}$ *reduces to another theory* $\mathbf{T_L}$ *when the laws of* $\mathbf{T_H}$ *are derivable from* $\mathbf{T_L}$. For Nagel, laws are logical deductions meaning the relation of derivability is itself logical. He discusses two types of intertheoretic reductive relationship. The first is "homogeneous" reduction. Here, there are no terms in the vocabulary of $\mathbf{T_H}$ that are not in $\mathbf{T_L}$. It is the received view that these types of reduction are few and far between in the history of science, to say the least (e.g., see Sklar 1967; Primas 1998).

Nagel's second type, "heterogeneous" reduction, is much more philosophically interesting as an account of scientific practice. Here, not every term in $\mathbf{T_H}$ is in $\mathbf{T_L}$, and those missing terms must be connected to the terms in $\mathbf{T_L}$ and be derivable from them. This process of connection and derivation is achieved with the help of bridge laws which join the dots between the two theories. Nagel's account has influenced many neo-Nagelians who propose modifications of the original scheme mainly concerning the nature of the bridge laws and the specific role they play (e.g., Schaffner 1993; Butterfield 2011a, 2011b).

By contrast, Nickels proposed an account of reduction through a mathematical limit process based on scientific practice as found in physics and other physical sciences, where a fundamental parameter in one theory is taken to a particular limit and the new theory arises out of the old one.[19] His account of reduction is sometimes described as the "physicists' sense of reduction" (juxtaposed with Nagelian reduction being a "philosopher's sense of reduction").

Yet we know that not every theory can be reduced straightforwardly in this way. As described previously, sometimes, as the relevant parameter(s) are taken to the limit, the set of equations radically changes form compared to

[19] Note that Nagel's and Nickels' senses of reduction work in opposite directions. In Nickels' sense we take $\mathbf{T_L}$ to be a more complete theory and produce the reduced theory $\mathbf{T_H}$ out of it. In Nagel's sense of reduction, we start with $\mathbf{T_H}$ and then build definitional extensions to connect it to $\mathbf{T_L}$.

the starting equations. Mathematically, these are cases where the limit is singular. Different algebras of observables—as referred to previously—would be important examples of such qualitative distinctions between theories related by a singular limit. Michael Berry (2002), Batterman (2002), and Primas (1998) argue that when limits produce singular relations there is no genuine entailment of one theory by another. The relationships between statistical mechanics and thermodynamics or between quantum mechanics and chemistry would be examples of such theories related through singular limits.

We are in sympathy with the idea that singular limits imply a lack of reduction, but not so straightforwardly as Berry (2002) and Batterman (2002) might suggest. Clearly, if one accepts Nickels' reduction scheme as the only legitimate sense of intertheoretic reduction, then there are failures of reduction. But there is no particular reason why anyone should accept Nickels' reduction in this way. It is a sense of intertheoretic reduction that can only be applied to cases where theories are almost completely mathematized and related via relevant parameters so limits can be taken. Outside of physics and chemistry, such cases are few and far between. For instance, one may *state* that the limit between, say, classical genetics and molecular genetics is singular, but we are not in a position to *demonstrate* it. For the moment we simply note that *the fact that two sets of equations are linked by a singular limit tells us nothing about what is special about what those equations represent.* Furthermore, the fact that two theories might be related via limits does not imply that any mathematical mapping between the two theories exists. For example, the reduction of classical mechanics to special relativity in the limit of small velocities does not imply any mapping since the notions of mass, space, and time differ significantly in qualitative ways between the two theories.[20]

Regarding Nagel, bridge laws are not the main point of contextual emergence (nor particularly relevant to what physicists and other scientists do in their explanatory practices so near as we can tell). Rather, the emphasis is on what contextual conditions are necessary and sufficient for the relationship between lower- and higher-level descriptions. We can see the contrast with Nagel and Nickels' accounts this way: Consider $\mathbf{T_H}$ and its associated vocabulary $\mathbf{VT_H}$ and $\mathbf{T_L}$ and its vocabulary $\mathbf{VT_L}$. Let these theories have (partially) overlapping empirical domains. As stated, the intertheoretic reduction question has been concerned with whether $\mathbf{VT_H}$ can be expressed in terms of $\mathbf{VT_L} + \mathbf{B}$, where $\mathbf{B}$ is a set of appropriate translation terms (i.e., bridge laws). Contextual emergence implies that discussion of intertheoretic reduction in this way ignores an important and often implicit component: *the respective*

[20] We ignored these issues previously when first introducing presumed examples of intertheoretic reduction.

contexts in which the theories are developed and applied. If we ask why a way of describing the world in terms of vocabulary $\mathbf{VT_H}$ does not reduce to a way of describing it in terms of $\mathbf{VT_L}$, the answer is that the vocabulary $\mathbf{VT_H}$ is only actually sustained by a set of implicit contexts of theory $\mathbf{T_H}$, $\mathbf{CT_H}$—for instance, a particular system/environment distinction, stability conditions that allow observables to be well defined, or distinct theoretical assumptions valid only at a particular level of coarse graining.[21] The notion of sustaining, here, is meant to imply "is made conceptually robust and applicable."

This means that the intertheoretic reduction question is not one of whether $\mathbf{VT_H}$ can be entailed by $\mathbf{VT_L}$ (with suitable translations) but whether $\mathbf{VT_H}$ sustained by $\mathbf{CT_H}$ reduces to $\mathbf{VT_L}$ sustained by $\mathbf{CT_L}$. Our example of the context of fluid dynamics compared to the context of fluid parcels illustrates the issue. These are two distinct contexts with different stability conditions, different states, and observables where the context of fluid dynamics and its states and observables is not reducible to the fluid parcel context and its states and observables. The examples discussed in subsequent chapters will all be cases where reduction cannot be done. So, even if $\mathbf{VT_H}$ can be connected to $\mathbf{VT_L}$ successfully, if $\mathbf{CT_H}$ is not entailed by $\mathbf{CT_L}$ then there has been no reduction in any of the senses usually discussed in the literature.

Note that contextual emergence suggests why it may seem as if many intertheoretic reductions have happened in the Nagelian sense. The contextual aspects of theoretical vocabularies are often overlooked. Crucially, it is often possible to connect vocabularies if we *import* contextual aspects from one vocabulary to the other. We may have the appearance of Nagelian reduction because $\mathbf{VT_H} + \mathbf{CT_H}$ has been entailed by $\mathbf{VT_L} + \mathbf{CT_L}+$ (elements derived from) $\mathbf{CT_H}$.[22] Again, the construction of fluid dynamics from the context for fluid parcels by partitioning the state space and closing the statistical moments using information from the fluid dynamics context illustrates this. The *contextual enrichment* of $\mathbf{VT_L} + \mathbf{CT_L}$ by elements from $\mathbf{VT_H} + \mathbf{CT_H}$ is what we have in the case of fluid dynamics, not reduction in any relevant sense. Fluid dynamics certainly has not been reduced to the resources of the fluid parcel states pace and description. Contextual emergence makes explicit the components that have been imported from $\mathbf{CT_H}$ undercutting the claim that $\mathbf{T_H}$ has been "reduced away," since even if $\mathbf{VT_H}$ has been connected to $\mathbf{VT_L}$, $\mathbf{CT_H}$ has not been connected.

The point is that *vocabularies do not exist in isolation*—a point (made in Chapter 1) that is often overlooked in discussions of intertheoretic relations

[21] Recall the discussion of theoretical vocabulary in Chapter 2, Section 2.2.

[22] This is an important reason why we think Butterfield's (2011a, 2011b) reconciliation fails to recognize cases of contextual emergence and misidentifies them as cases of reduction in a neo-Nagelian sense.

considered only in terms of relating sets of sentences via logical machinery. $\mathbf{T_H}$ may have a vocabulary that is defined, in part, by a choice of system/environment distinction and by the wider conceptual embedding of a theory (e.g., bulk flowing of water in a river or through a pipe). $\mathbf{VT_H}$ may also include states and observables that are robust because of stability conditions, where those stability conditions are not found or expressible in the $\mathbf{VT_L}$, as is the case with fluid dynamics. Discussions of intertheoretic reduction largely ignore contributions of context because they tend to focus on logic, yet these contributions are necessary aspects facilitating the definition of our theoretic vocabularies to begin with.

Now consider the alleged reduction of classical genetics to molecular genetics. Specific nomic relations of classical genetics guarantee the stability and applicability of relations in classical genetics, but arguably it does not provide such conditions for the finer-grained description of the genetic world in terms of molecular genetics. Seeing theory reduction as not just a two-place relation, between $\mathbf{T_H}$ and $\mathbf{T_L}$, but instead between $\mathbf{T_H}$ and similar theories at that level of abstraction and a lower-level theory suggests that alleged reductions often implicitly import some of the wider contextual features at the higher level without acknowledging them. Contextual emergence seeks to make such contextual information importation explicit. Often the discussion compares those two theories of genetics to each other and ignores how they fit into the wider conceptual apparatus of reproduction which involves multiple levels of organization: gene/alleles, chromosomes, and gametes (contribution of one chromosome from each paired set), zygote (contribution of two gametes) for example. Each level of organization contributes to the context for formulating classical genetics and explaining transmission of classical genes, plus the assumption that there is a one-to-one correspondence between genotype and phenotype (expressed characteristic). Molecular genetics focuses at lower scale on molecules and their arrangements in sequences and their structure in the famed double helix. Classical genetics operates in terms of identifications of a gene with its corresponding phenotype. Molecular genetics operates in terms of the synthesis of RNA and polypeptides, but the latter have no one-to-one correspondence with a classical gene or its corresponding phenotype. Any "reductions" of classical genetics to molecular genetics will have to *import key structural relations from classical genetics (e.g., the cytological context of chromosomes) because those are the very ones that allow it to fit together with reproduction.*[23] Hence, purported reductions of classical genetics to molecular genetics will involve stability conditions associated with the context of chromosomes

[23] Of course, there is the larger context of how classical genetics and molecular genetics are each embedded within evolutionary theory.

and gametes implemented at the smaller scale of molecular genetics. In other words, the relationship between classical and molecular genetics is inherently multiscale.

One other aspect of conceptual embedding is the way by which contexts can change the *semantic content* of a theory. Just from ordinary language we know that the context in which something is said often changes its meaning, and this can be true of terms in a theoretical vocabulary as well. The *meaning* of a term can be quite different, depending upon how a theory uses that term to represent aspects of an empirical domain and how a theory uses that term in conjunction with conceptual resources not contained in the theory itself. Sometimes even the same word used in two different theories has a different meaning in each of them, and only in the context of one of the theories does the term adequately capture something about the world.[24]

In summary, we often need formal system/environment distinctions from the higher-level description to derive specific features, so we must import this into the lower-level theory. Similarly, theories exist in a network of wider conceptual frameworks. To derive a target feature it is possible that we need a wider conceptual apparatus that exists outside of either theory but that can only connect to the higher-level theory (or to the lower-level theory if it is enriched by contextual information from the higher-level theory).[25]

A second contrast with Nagelian frameworks is that there is no attempt to establish logical entailments in contextual emergence. Nagel's deductive–nomological framework, in our judgment, is not the best way to reconstruct the explanatory practices of scientists or the relationships among theories and models. For instance, theories and models in the sciences are not logical sentences as the logical positivists conceived, nor are positivist-style reconstructions convincing.

To try and systematically treat scientific theories and their relations as matters of logic as the received view following Nagel does is highly problematic. As Ian Hacking (1983) points out, the practice of physics and other sciences requires that we move beyond the realm of first-order languages and incorporate higher-order quantifiers and indexical expressions. Such extensions raise difficult questions regarding logical notions of completeness, decidability, and other deduction properties (Hacking 1983, 104–108). Simple examples show why it is reasonable to make these extensions if we are to attempt to reconstruct physics theories as sets of logical sentences with entailments. To represent properties of matter such as mass or charge as intrinsic to a classical

[24] See Jammer (1961, 2000) for discussion of how the meaning of mass undergoes qualitative changes in different theories.

[25] The role of system/environment distinctions, stability conditions, and embeddings is one way to see that theories at distinct levels rarely, if ever, are related via some kind of formal mapping or logical relations.

particle requires the use of second-order predicates with identity. Additionally, many properties of classical mechanics and fluid dynamics take on a continuum of possible physical values requiring a formal language powerful enough to describe continuous functions of real numbers.

A more complex example involves the mathematical language of physics itself. Although some of what mathematicians call real analysis appears to be expressible in terms of second-order predicates with identity, physics also makes use of mathematical concepts such as manifolds and the calculus of manifolds, both of which require going beyond second-order languages. The situation is similar in general relativity and quantum mechanics when we shift from using separable Hilbert spaces to nonseparable ones or to even more abstract spaces (Hellman 1989, 108–117; Thirring 1983, 29–31).[26] For instance, the canonical commutation relations are one of the defining features of quantum mechanics and are closely associated with the Heisenberg relations. When operators representing position and momentum are substituted into these fundamental relations, however, problems arise for separable Hilbert spaces in that the domain of definition for such unbounded operators becomes problematic (Reed and Simon 1972, 270–6; Dubin and Hennings 1990, 10–13). For all these reasons, choosing to stick with a logical framework for physical theories appears hopeless (not to mention that it runs completely out of gas for biology and ecology).

A third contrast is between Nagelian bridge laws and stability conditions. The latter are not co-extensional definitions of nonlogical terms from the target domain (e.g., temperature in thermodynamics) in the elementary domain (e.g., statistical mechanics). As we saw in Chapter 2, stability conditions are required for the existence and identification of states and observables at the larger length and longer time scales and even the target systems themselves. It is not definitional work being done here, but existential work, as in the example of the relationship between the fluid dynamics and fluid parcels contexts. Furthermore, as we will see shortly, it is the robust emergent actual entities that need accounting for—a task that Nagelian bridge laws do not accomplish but which scientific descriptions do. Existence is not a matter of logic.

3.3.2 *Comparison with Butterfield's neo-Nagelian intertheoretical reductionism*

In this connection, it is worthwhile comparing contextual emergence with Jeremy Butterfield's (2011a,b) discussion of intertheoretic reduction and emergence. We should say at the outset that unity of the scientific picture

26 Rigged Hilbert spaces are examples of nonseparable spaces.

of the world is a theme that often crops up in Butterfield's (2011a)—a value that we share.

His characterization of emergence is

> properties or behaviour of a system which are novel and robust relative to some appropriate comparison class ... So the idea is that [a system's] properties and behaviour are novel and robust compared to those of its component systems, especially its microscopic or even atomic components. I shall also put the idea in terms of theories, rather than systems: a theory describes properties or behaviour which are novel and robust relative to what is described by some other theory with which it is appropriate to compare—often a theory of the system's component parts.
>
> (2011a p. 921)

Meanwhile, he takes reduction to be:

> Essentially, deduction of one theory from another; though usually the deduction goes through only if the other theory is augmented with appropriate definitions or bridge-principles linking the two theories' vocabularies. This construal will mean that I come close to endorsing the traditional account of Nagel.
>
> (2011a, p. 921)

Butterfield acknowledges that there probably is no "best meaning" for terms such as reduction and emergence. Such an approach may be largely unobjectionable; however, a point of contrast between ourselves and Butterfield is precisely whether derivability of emergent phenomena reasonably counts as reduction. Here, much turns on his invocation of Nagel's framework. Butterfield confines his understanding of theories to that of logic, where a theory is "a set of sentences of an interpreted language closed under logical consequence" (p. 926). We note that in addition to there probably being no working scientific theory that meets Butterfield's characterization (e.g., evolutionary theory is not like this at all), Nagel's framework always appears more plausible in a purely logical construal of theories. Additionally, we note that the logical framework misses out the importance of contexts defined by stability conditions and the differences in topologies between $\mathbf{CT_H}$ and $\mathbf{CT_L}$, as we pointed out earlier regarding Nagel's heterogeneous form of reduction. We conjecture that the reason Butterfield thinks the mathematics relating different scientific descriptions, such as asymptotic series and renormalization techniques, count as Nagelian reductions is his adherence to the abstraction of theories as "a set of sentences of an interpreted language closed under logical consequence."

A key aim for Butterfield is to establish that "emergence is not in all cases failure of reduction, even in this strong [Nagelian] sense of reduction. There are cases combining emergence with reduction—as well as cases of one of these without the other (of course)" (p. 923). In his (2011b) he makes this more precise through the notion of limits, such as thermodynamic limits, or where some parameter approaches a critical value. As we described previously above, the association of novel phenomena with limits is also part of contextual emergence. In our terminology, emergence for Butterfield essentially means that emergents are not explainable, derivable, or predictable based on the appropriate comparison class in $\mathbf{T_L}$.

Cases where there is Nagelian reduction and emergence appear to be in strong tension at first glance. Butterfield argues that such cases involve theory relations through limits (or through renormalization techniques). How are we to understand this? This is the burden of Butterfield's (2011b).

He begins by assuming it is always possible to define a definitional extension D (a move in systems of logic) for $\mathbf{T_L}$ such that the latter is implied by $\mathbf{T_L} \cup D$. Here is where treating theories abstractly as sets of logical sentences closed under logical consequence does a fair amount of work. A great deal of work also is done by the procedure Butterfield follows, by which "one performs the deduction *after* taking a limit of some parameter. So, one main moral will be that in such a limit there can be novelty, compared with what obtains away from the limit" (2011b, p. 1068, emphasis added).

One could complain that invoking limiting procedures represents a significant modification to Nagel's framework (see previous section). After all, the use of limits indicates that more than logical deduction is going on in the derivation of the emergent phenomena, because—as we have indicated and will illustrate in subsequent chapters—contextual information not found in the lower-level theory is involved in dealing with the limits. In our language, $\mathbf{CT_H}$ and $\mathbf{CT_L}$ (and usually $\mathbf{VT_H}$ and $\mathbf{VT_L}$) are not the same and do not stand in a reductive relationship, logical or otherwise.

Another significant modification to Nagel's framework is that Butterfield extends the notion of bridge laws to encompass definitional extensions, D, where the extension is chosen such that any given $\mathbf{T_H}$ supposedly can be logically deduced from some $\mathbf{T_L}$. Clearly, these definitions or bridge laws do much work for this scheme. Butterfield points out that they "must of course be judiciously chosen, with a view to securing the theorems of" $\mathbf{T_H}$ because of the kind of relationship between the two theories that needs to be forged (2011a, p. 931).

This broadening of the Nagelian notions of "deducibility" and "reduction" makes it easier for him to argue that reduction and emergence are consistent with each other, since everything hangs on what is packed into "deducing" one theory from another or "deducing" novel emergent properties.

Contextual emergence is modeled after the practices of scientists, unlike Nagel's logical reconstruction of theory relations or Butterfield's broadening of Nagel's framework. Where Butterfield stretches the notions of "deducibility" and "bridge laws," contextual emergence focuses on what is required to relate two different descriptions to each other (e.g., thermodynamic and statistical mechanics descriptions) via singular limits and their regularization, say. The fact that by drawing on statistical mechanics one might be able to derive, explain, and predict thermodynamic features can only be counted as "reduction" if that term has been stretched far beyond what Nagel had in mind. But it also is worth noting that Butterfield's "reductions" always involve ineliminable references to $\mathbf{CT_H}$ without acknowledging the crucial role played by the concrete context for $\mathbf{T_L} \cup D$. In essence, Butterfield hides the contextual emergence in his reconciliation strategy.

As a last contrast, Butterfield often speaks in terms of approximating new properties, whereas, as described previously in Section 3.2, the real issue in emergence is the transitions leading to new states and observables. As we discussed, it mischaracterizes what is going on to say that $\mathbf{T}_{\mathrm{H}}$ approximates a more elementary $\mathbf{T_L}$, with the latter being "more fundamental" or "truer" in the narrower reductionist's sense. If one abstractly conceptualizes scientific theories as sets of logical sentences, it is easy to misconstrue $\mathbf{T_H}$ as some kind of approximation to $\mathbf{T_L}$. Contextual emergence challenges this bias (e.g., stability conditions are "just as fundamental" for the existence of molecular states and observables as elementary particles and forces). The "secondary conditions" involved in handling the limits between theories, laws, properties, and so forth, are minimized in Butterfield's approach, packed into a definitional extension D, so the work that such conditions are doing is left implicit rather than made explicit.

For example, consider Butterfield's discussion of "meshing dynamics" (2012). He presupposes that the lower-level microdynamics mapping is deterministic. But, since there could be one-to-many and many-to-one microdynamical mappings, microindeterminism is possible (p. 107). Macrostates are related to microstates through coarse graining by adopting an appropriate partition of the microstate space in the simplest cases. Butterfield's idea is that the microdynamics defines a macrodynamics in the following way. The mapping representing the microdynamics—the mapping from microstates to microstates—is restricted to the appropriate partition yielding the coarse-grained macrostates. Roughly, since, under the partition, two distinct microstates s_1 and s_2 can be sent to the same partition cell, macrostates are multiply realizable, and Butterfield characterizes these situations as nonmeshing (i.e., there would be microdeterminism, but macroindeterminism in the macrostates due to the one-to-many mappings). Micro- and macrodynamics mesh otherwise (pp. 107–108).[27]

[27] Compare with our discussion in Section 3.2 and Primas (1998). Given a particular $\mathbf{T_L}$, there are always multiple ways to define a new contextual topology for a specific $\mathbf{CT_H}$ involving distinct

To discuss actual physical systems, Butterfield adds the following "more realistic" conditions to the meshing dynamics:

> (a) the meshing may not last for all times;
> (b) the meshing may apply, not for all microstates *s*, but only for all except a "small" class;
> (c) the coarse-graining may not be so simple as partitioning [the microstate space]; and indeed
> (d) the definition of the microstate space ... may require approximating and/or idealizing, especially by taking a limit parameter: in particular, by letting the number of microscopic constituents tend to infinity, while demanding of course that other quantities, such as mass and density, remain constant or scale appropriately. (p. 108)

These meshing conditions/dynamics are related to the regularization of singular limits, so there is a relationship to contextual emergence here. Nonetheless, Butterfield, like many, underestimates what is involved in what the "approximating and/or idealizing" actually refers to. Indeed, as we indicated previously, approximating the microstates is not what actually happens; rather, new macrostates and observables arise as signaled by the contextual topology of $\mathbf{CT_H}$. The contrast with Butterfield is that contextual emergence clarifies the roles that stability conditions and contexts play in the transitions leading to new states and observables (no approximations of microstates). In this sense, we think the best way to interpret and clarify Butterfield's work is to reformulate and amend it as contextual emergence.[28]

Butterfield has many things to say about the intricacies involved in the emergence of new properties. At these points, he seems comfortable with emergence-talk. Yet he also insists that there is a relationship of derivation between $\mathbf{T_H}$ and $\mathbf{T_L}$; at these points he seems comfortable with reduction-talk. In other words, successful intertheoretic relationship for him comes with the emergence of new properties. His cases of "property emergence" matches well with contextual emergence. However, we would argue that even the relationships which Butterfield thinks are derivations between theories (e.g., between thermodynamic descriptions and statistical mechanics descriptions) are successful *only if* concepts and information in $\mathbf{T_H}$ are used to supplement $\mathbf{T_L}$ in the derivation (i.e., the ineliminability of $\mathbf{CT_H}$).

sets of algebras of observables. But this is not Nagelian reduction at all (e.g., multiscale modeling discussed subsequently).

[28] For example, Butterfield's meshing dynamics can be recast in terms of topological equivalence between lower- and higher-level descriptions' algebras of observables (Atmanspacher 2012; Bishop 2019).

We think such cases of intertheoretic relations find their natural home in contextual emergence.[29]

As a last comment on Butterfield, we suspect that he finds himself in this somewhat strange situation because he focuses on intertheoretic relations as a matter of logical relations rather than the physics and concrete contexts. He is not alone in this focus. Many philosophers of science tend to focus on logic and abstractly on mathematics as well. Focusing on intertheoretic relations exclusively this way leaves much of the interesting physics—the important transitions leading to new states and observables—out! Therefore, we claimed previously that singular limits and emergence are formally separable (something with which Butterfield agrees). Emergence is more about the actual sciences—the physics, chemistry, biology, and so on—than the formal relations among theories.[30]

3.3.3 *Schaffner, interlevel theories in biology, and intertheoretic reduction*

Kenneth Schaffner has suggested other modifications of Nagel's account to deal with theories in biology (e.g., 1969, 1993). Although he acknowledges that there are some instances in biology where theories look like what one typically expects in the physical sciences, biological theories often have a quite different structure. One important difference is that many theories in biology are not amenable to an axiomatic structure and mathematization.[31] For instance, Schaffner argues that most theories in biology and the biomedical sciences are structured "as a *series of overlapping temporal models*" (1993, p. 320, original emphasis). He elaborates:

> The models of such a structure usually constitute a series of prototypical mechanisms and variations (mutants) which bear family or similarity resemblances to each other, and characteristically each has a (relatively) narrow scope of straightforward application to (a few) pure types. The models are typically *interlevel* in the sense of levels of aggregation, containing component parts which are often specified in intermingled organ, cellular, and biochemical terms. Stages of temporal development in the models—or life cycles in some—may represent either deterministic, causally probabilistic, random (Markovian), or even mixed

[29] We are reminded that Primas originally used the term "weak reductions" when additional contextual information must be brought into the explanations or derivations (e.g., 1977, pp. 138–139), but dropped such language due to the considerations we have discussed in this chapter.

[30] For example, Butterfield (2011b) identifies renormalization as a kind of neo-Nagelian reduction, but this is only possible if important details about the role which $\mathbf{CT_H}$ plays in the renormalization process is ignored.

[31] For those theories in biology that can be mathematized and/or given axiomatic structure, Primas (1977) has argued that such theories fit in the framework that eventually has become contextual emergence. He stresses that these theories are non-Boolean, so cannot be given a logical ordering of "higher" or "lower," nor can be fit into a logical ordering of physical theories as more basic or fundamental.

> connections. This probabilistic character of some causal connections (a *failure of determinism*) should be distinguished from the *conceptually distinct failure* of the exact *match* of a model to a non-pure type to which, nonetheless, it is closest.
> (1993, p. 321, original emphasis)

To handle such biology cases, Schaffner modified Nagel's account and developed what he calls the general reduction–replacement model of intertheoretic reduction. On this model one seeks to construct an analogue of $\mathbf{T_H}$ in $\mathbf{T_L}$—call it $\mathbf{T^*_L}$—such that $\mathbf{T^*_L}$ can be logically deduced from $\mathbf{T_L}$. However, many cases in biology do not allow for the construction of such a $\mathbf{T^*_L}$. Instead, one must allow for the replacement of $\mathbf{T_H}$ by a new $\mathbf{T_L}$ that includes the empirical domain of $\mathbf{T_H}$, where the empirical domain is connected to the new $\mathbf{T_L}$ by appropriate bridge laws (p. 329). As an example, Schaffner gives the replacement of the demonic theory of disease by the germ theory of disease, where the empirical range of phenomena formerly "explained" by the demonic theory are now explained through the germ theory. He also allows for instances where a part of $\mathbf{T_H}$ is replaced by the new $\mathbf{T_L}$. The whole empirical domain of the original $\mathbf{T_H}$ is preserved by the new $\mathbf{T_L}$ plus the remaining adequate component $\mathbf{T^*_H}$ of the original $\mathbf{T_H}$.

Notice what now seems to be a recurring theme. Even replacement of $\mathbf{T_H}$ by a new $\mathbf{T_L}$ or $\mathbf{T_L} + \mathbf{T^*_H}$ still must draw upon $\mathbf{CT_H}$ and $\mathbf{VT_H}$, meaning that the stability conditions defining $\mathbf{CT_H}$ and driving $\mathbf{VT_H}$ are not reduced. Schaffner's model simply ignores these as do most abstract formalizations of intertheoretic reduction. What happens in Schaffner's model is that a new theory that is inherently multiscale is produced, replacing two previous theories in part or in whole. Moreover, even though the machinery of states and observables from physical sciences is not applicable, there are still observables—the higher-level phenomena—which biologists deal with that do not get reduced or replaced on Schaffner's account; rather, they get *incorporated into* a new replacement theory. These phenomena must be brought into the explanatory theory or interlevel theories in the case of $\mathbf{T_L} + \mathbf{T^*_H}$ (p. 332). A kind of logical mapping of these observables into lower-level phenomena does not happen, nor can such mappings explain the transitions that must take place from the lower-level phenomena (e.g., DNA) to higher-level phenomena (e.g., phenotypes). Fins and limbs may have been built by similar regulatory DNA during the developmental process, but these DNA neither replace nor explain the behavior of fins and limbs. Indeed, Schaffner's discussion of cases such as learning and memory in *Aplysia* do not read anything like what we would expect from a Nagelian framework; rather, we see broad generalizations operating at a higher level that shape or constrain what kinds of narrower generalizations at the lower level are relevant and implemented (e.g., more general biological principles get implemented at the cellular level of description, while narrower cellular principles get implemented at the molecular level of description).[32]

Indeed, Schaffner, while acknowledging that even the general biological principles may be different for different organisms, fails to account for how the embodiment of the *Aplysia* acts as a stability condition for the phenomena in the successively lower levels of description. Even the causal explanations are interlevel in descriptive vocabularies drawing on multiple levels. We suggest that the "partial model reduction" which Schaffner was pursuing is better understood as a form of contextual emergence.

3.3.4 *Wimsatt and functional reduction*

William Wimsatt (2007) eschews formalistic accounts of reduction such as Nagel's. Instead, Wimsatt's point of departure is taking levels of organization as primary, and theories being about entities at particular levels of organization. Levels of organization "are primarily characterized as local maxima of regularity and predictability in the phase space of different modes of organization of matter" (2007, p. 249). This does not mean that levels are always well defined or ordered in a linear fashion, but that they are distinguishable. Wimsatt does recognize that many of the biases built into the Nagelian framework tend to lead people away from taking contextual information seriously. One effect of such biases is "to assume that lower-level theories correct upper-level theories, but not conversely" (p. 250).

The basic idea of Wimsatt's functional account of reduction is that mechanisms and other causal factors at a lower level of organization explain the phenomena observed at a higher level of organization, but he has three cases for interlevel explanatory relations in mind:

> We wish to be able to explain every phenomenon under every informative description by showing, first if possible, how it is a product of causal interactions at its own level, but barring that, how it is a product of causal interactions at lower levels (a microlevel or reductive explanation), or least probably and desirably in our reductionist conceptual scheme (but absolutely unavoidably in a world of evolution driven by selection processes), how it is a product of causal interactions at higher levels (most commonly, a function explanation).
>
> (2007, p. 258)

This prioritizing of cases reflects both a commitment to levels of organization as local maxima of regularity and predictability and to "acceptance of a weakly but generally reductionistic worldview." These priorities are cashed

[32] In Chapter 6 we will see why reductionist intuition in biology fails. It is this intuition, plus treating theories as abstract logical structures, that leads someone such as Schaffner to try to shoehorn biological theories into a neo-Nagelian reductive framework.

out under a constraint that scientists' search "for explanatory factors is also conducted according to some sort of efficiency optimizing or cost-benefit considerations" (p. 258). What this means in practice is that if scientists find that causal factors at a given level of organization are sufficient to explain a phenomenon observed at that same level, there is very little to be gained by expending the effort to ground the phenomenon and its causal factors in a lower level of organization. In other words, scientists are pragmatic, and while Wimsatt believes that reductionism is the ideal strategy for scientists they do not press for reductions when there is little to be gained by that effort. Whether or not one buys Wimsatt's characterization of scientists, this is all consistent with epistemological emergence and a commitment to ontological reductionism as described in Chapter 1.

Wimsatt mentions the supposed reduction of the laws of thermodynamics to statistical mechanics as an example of a reduction of higher-level laws to lower-level laws (p. 261). But, as we will see in Chapter 4, supposed reductions of thermodynamics to statistical mechanics are examples of contextual emergence. A biological example discussed by Wimsatt is the ultimate identification of Mendel's factors with molecular genes (pp. 261–269). As Wimsatt notes: "Identity claims are often made on the basis of correspondences between or explanations of only two or three properties, often together with some subsidiary background information of a non-correlational nature" (p. 267). There clearly is a sense of replacement of Mendel's factors and the laws (statistical patterns) he derived by observation. The functionality of the factors is replaced by the functionality of molecular genes.

Nevertheless, whether there is a reductive explanation of Mendel's work by a putative lower level of organization is debatable because of a crucial ambiguity regarding background information. The identification of his factors with molecular genes takes place within a substantial amount of background information. If that information truly is located at the higher level or organization, then it must be implemented in the lower level of organization or functions as a constraint on entities and causes in the lower level or organization. This looks more like contextual emergence and stability conditions than a case of reduction. On the other hand, if the background information is at the lower level or organization, it looks like Mendel's factors and patterns also exist at that same level of organization. The identification or replacement of Mendel's factors by molecular genes is no reduction, but an instance of discovering what the factors are and why they functioned according to the statistical patterns which he discovered.

Wimsatt seems to recognize that there clearly is not a standard reduction going on in the Mendel case (and for many other cases in biology). He acknowledges that there is a concept of emergence that is compatible with his understanding of reductionism as spelled out previously: "One can be a reductionist and an emergentist, too, with a proper understanding of these notions" (p. 274). Sharpening his view, Wimsatt writes: "*A reductive explanation of a behavior or a property of a system is one that shows it to be mechanistically explicable in terms of the properties of and interactions among the parts of the system*" (p. 275, original emphasis).[33] On the other hand, emergence is understood in terms of context-sensitivity: "This account analyzes emergence in terms of dependence of a system property on the arrangements of the parts and, ultimately, on the context-sensitivity of relational parts' properties to intrasystemic conditions. Some accounts of emergence suppose extrasystemic context-sensitivity of system properties" (p. 275). Wimsatt draws this out in an important footnote:

> [Earlier] I called an account reductionist if there were a mechanism at any level or combination of them that explained the phenomena—even if it included mechanisms at a higher level or external to the system in question. In drawing these distinction, I would now require explicitly relativizing an account to two (or more) levels: (1) a bottom level—the lowest level at which specific parts of a mechanism must be invoked to explain the phenomena (so an explanation of "position effect" requires characterizing an operon as a genetic control structure and how it works, but not the atomic physics accounting for the relevant electron orbitals that yield the binding which occurs); (2) a top level—drawing the system boundaries broadly enough so that all relevant parts of the mechanisms involved are included. (Still broader things may sometimes be required, but taken for granted, as features of solar and planetary dynamics may be crucial to evolutionary processes on a variety of different time scales.) This "multilevel reductionist analysis" picks out the appropriate levels for objects, processes, and phenomena, and explicates their relations to complete the explanatory task with no further mystery.
>
> (2007, p. 394, n. 6)

Hence, we see that what most counts for Wimsatt as a "reduction" is explanation in terms of mechanism(s) even if that involves multiple levels of organization. He insists: "We must work back and forth between levels to check that features crucial to a phenomenon at an upper level are not simplified out of existence in modeling it at the lower level" (p. 276).

The multilocus genetic system serves as an example for Wimsatt, where genes are organized into chromosomes (pp. 287–296). In sexual reproduction the sperm and egg are gametes, each typically contributing half of the genotype (haploid). Moreover, genes come organized as chromosomes, so there are multiple levels of organization in play: the gene/allele, the chromosome, the gamete, the genotype (haploid/diploid), population of genes,

[33] We will say more about the vagaries involved in such appeals to mechanisms in Chapter 6.

assortative vs. obligate mating, and more. The resulting gamete frequencies are largely determined by the chromosome frequencies (i.e., what chromosomes are available for parents to contribute). But sexual reproduction is fairly complicated, as Wimsatt elaborates:

> One inherits *gametic compliments of chromosomes*, not chromosomes drawn independently at random from their sets of homologues in the population at large. These are assembled into diploid genotypes, not arbitrarily large polyploid assemblages of chromosomes, from which are drawn the haploid gametic complements of chromosomes and genes of the next generation. *Gametes and diploid genotypes are each assemblages structured via a systematic combination of constraints and random elements, and the results reflect both* ... If close together on the same chromosome, genes will show a continuing slowly decaying statistical association, given by the linkage distance r, ranging from 0 to .5 as a function of how frequently recombination events separate them—a property of their relative locations on the chromosome. But if they are either very far apart on a long chromosome, *or are on entirely different chromosomes*, they will still show an identical linkage effect of .5. *The identity of this effect (when r is .5) with the genes so far apart on the same chromosome or on different chromosomes shows that this association is not—or not just—a product of chromosomal organization.* It is a property of the genotypic–gametic life cycle.
>
> (2007, pp. 289–290)

He examines cases to demonstrate how the "structural constraints of the diploid–haploid life cycle" (p. 291, original emphasis) affect recombination.

Nevertheless, there are more complications in the modeling of sexual selection:

> Boveri's classic (1902) experiments showed that dispermic fertilization, where disturbances of mitosis make genome size and composition a random variable, had disastrous consequences to be avoided. *That one has to impose constancy of size, composition, and arrangement as side conditions in population genetic models of genome formation reveals that we are not dealing with aggregative properties.* (These conditions are so taken for granted that they are rarely stated—or studied.) *With these conditions and theories, things may appear aggregative that are not.*
>
> (2007, p. 295, original emphasis)

Non-aggregativity is an emergence proxy for Wimsatt, so here we can see that constancy of size, composition, and arrangement act as stability conditions in population genetics modeling to control the level of emergence that such models might involve! In other words, though biologists often leave out the stability conditions necessary for population genetics models—Wimsatt's side conditions—contextual emergence focuses us on making such conditions explicit. Constancy or conservation of genome size is a stability condition on molecular genetics in population models capturing the larger-scale fact that chromosome size-change often has adverse effects on the fitness of an organism. Again, we see larger-scale constraints acting as a stability condition for the dynamics of population genetics, so no reduction in the molecular scale in any straightforward sense.

Wimsatt (pp. 304–308) summarizes his discussion by noting:

1) Emergence is common in biology.
2) Properties might look like they are aggregative (non-emergent) depending on how the model decomposes properties into components. In other words, "*we will tend to see aggregative or more aggregative decompositions as natural decompositions, and their parts as instances of natural kinds, because these decompositions provide simpler and less context-dependent regularities, theory, and mathematical models for their behavior* (p. 304). The more context left out, the less emergence is apparent.
3) Related: approximations are often made in descriptions at different levels of organization to produce tractable mathematical models that are aggregative, hence emergence can disappear due to these assumptions. These approximations are often aiming for context-independence.
4) As scientific analysis becomes more detailed, higher-order relational properties and more complex interactions among parts are included in the descriptions at different levels of organization.

Yet, as noted previously, Wimsatt takes "a reductive explanation of a system property of behavior" to be "one showing it to be mechanistically explicable in terms of properties of and interactions among the parts of the system" (p. 308). On the other hand, "to say that the behavior of a system is totally explicable in terms of behavior of its parts is *not* to say that it is an aggregative function of the parts" (p. 309). This is to say that for Wimsatt, a property at level of organization n has been successfully reductively explained if the mechanism (or mechanisms) or components of a mechanism may exist at the same level or at both level n–1 and n+1. If the explanation is in terms of mechanism(s), it is reductive.

This is a form of functional explanation in terms of mechanisms, but it does not locate the necessary and sufficient conditions in the mechanisms or components at level n–1 for phenomena at n. Additional contextual information is often needed from level n and even n+1 (or higher levels or organization) to complete the explanation. Irrespective of whether mechanistic explanations exhaust explanations in biology (see Chapter 6), we can see from how Wimsatt describes his account of explanation involving multiple levels of organization that it is better to reserve the term "reductive account" for one that exclusively locates all the resources needed for the explanation of a phenomenon at level n to level n–1 (or lower). Otherwise, we clearly have cases of contextual emergence, where stability conditions at level n or n+1 (or higher) are needed to supply the full set of necessary and sufficient conditions reflecting the context of the phenomenon in need of explanation.

Maybe one reason why Wimsatt characterizes his account as reductive is the alignment of the intuition that scientific explanations are reductive with the general intuition that mechanistic explanation is always reductive. In any event, the context-sensitivity of the multiple levels of organization involved in what Wimsatt counts as successful explanations in biology fits better with the contextual emergence framework. For instance, he acknowledges what has become standard understanding in biology that a given molecular pathway may have different effects in different cellular contexts, meaning that the contribution of molecular resources at level n–1 to higher levels of organization is conditioned by the context represented by the latter levels. Genetic function provides a good example. Given that the function of any gene or group of genes is dependent on the total context (whole genome, cell, organ, body plan, population, ecological niche, climate), the molecular properties come to expression in and are shaped by cellular and organismal contexts. This not only blocks the Nagelian approach to theory reduction but also implies that functional accounts of explanation drawing on multiple levels of organization cannot be cashed out purely from molecular resources available at a lower level.

Therefore, we think it is misleading to use the term "reductive explanation" or "reductionistic framework" for Wimsatt's account of explanation involving multiple levels of description. Rather, contextual emergence is the more fruitful and perspicuous framework for the kinds of multiscale context-sensitive explanations he (and biologists) pursues where making details and constraints explicit matters crucially.

3.4 Multiscale modeling

In the preceding discussion of intertheoretic relations we saw that the interlevel or multiscale nature of levels of description recur often. The fluid dynamics example we have tracked throughout this chapter illustrates this well, as does Wimsatt's multilevel account of explanation in genetics. Multiscale modeling has been a mainstay in the natural sciences for more than a century, which is not surprising, since scientists have largely always realized that they are working with a multiscale world. The computational modeling of soft flowing materials provides a challenging example, where models must span six or more spatial decades in scale and twice as many in time (Montessori et al. 2020). Recently, philosophers have been paying more attention to such modeling (e.g., Batterman and Green 2020; Hoekstra, Chopard and Coveney 2014; Wilson 2017).

One connection between our discussion of levels of description and intertheoretic relations and multiscale modeling is provided by Mark Wilson's observation that the success of such modeling approaches "can be achieved only if the descriptive vocabularies utilized are subjected to a carefully monitored set of contextual restrictions" (2017, p. 2). Each model has its own vocabulary tied to the context relevant to that scale as emphasized in our discussion of intertheoretic relations. For instance, the vocabulary relevant for the molecular scale of H_2O molecules is irrelevant to the bulk scale of fluid flow and vice versa.

Nonetheless, the context of the modeling effort is also important when trying to understand wind flow across a windmill farm, complex materials such as steel, soft condensed matter systems, or the cardiovascular system. Context is crucial to multiscale modeling because the physical processes and properties at one level of description are relevant at a specific length or time scale and have some degree of autonomy from those at a different length or time scale, as we saw in Section 3.2. As well, recall the case of Rayleigh–Bénard convection in Chapter 2, where the fluid parcels and convection cells have some degree of autonomy from each other though clearly are not independent. The context of each length or time scale is tied to the relevant processes and properties for modeling at each level of description. Hence, the need for multiscale models to deal with different properties, processes, vocabularies, and stability conditions at different scales.

These contextual differences in properties, processes, vocabularies, and so forth, for different levels of organization, also lead to quite different methodologies for how different scales are bridged in terms of information transfer among the scales. The variety of bridging methodologies includes coarse/fine graining, homogenization, sampling, projection, lifting/upscaling, refinement, micro–macro coupling, and boundary methods, among many others. This variety reflects the contextual nature of different levels of description in multiscale models, as well as the challenges of working out appropriate information flow between the different computational techniques applied at each scale (e.g., particle methods for a smaller length scale submodel, and finite element methods for the continuum model at a larger length scale).[34]

We demonstrate that multiscale modeling is an application of the contextual emergence framework and is best understood in that light. One thing to keep in mind in what follows is that there is no consensus on the definition of what a multiscale model is or on the appropriate vocabulary (e.g., Hoekstra, Chopard, and Coveney 2014). This is probably because multiscale modeling

[34] Of course, questions of computational efficiency, error propagation, numerical accuracy, and stability, come into play for these interlevel bridgings as well as for the computational methods chosen for each scale.

is such a multidisciplinary affair, with modelers working in diverse fields such as physics, geology, biology, ecology, and economics. Nothing we say will depend on any disciplinary community's definitions or distinct vocabulary. We will work with the rough and ready idea that a multiscale model involves modeling phenomena at distinctly different scales where the information in the models at different scales involve various interlevel dependences and communication among scales.

3.4.1 *Contexts in multiscale modeling*

Multiscale models represent a modeling approach to handling complex phenomena that involve multiple length and time scales. The basic idea is to break the phenomenon of interest into a nested set of interrelated models at distinct levels relevant to distinct contexts. Each descriptive level involves its own length and/or time scale, vocabulary, and stability conditions enabling effective modeling at the different scales. The results of each modeling scale are brought into relationship with those at the nearest neighboring scale, sometimes across multiple scales.

Understanding turbulence in liquids such as flowing rivers has been of interest since at least Leonardo da Vinci. Turbulence remains one of the most challenging phenomena to model on a computer. One of the features of turbulent flow is the formation and persistence of eddies. An eddy is a swirling structure where current flows opposite to the motion of the bulk fluid. An eddy itself also may not flow with the bulk motion, indicating a kind of quasi-detachment from the bulk. In turbulent flows, kinetic energy of the larger-scale bulk cascades downwards from larger scales to smaller scales through eddies. In eddies, the energy continues to pass down to the smallest scales where viscosity eventually prevents any smaller-scale motions dissipating energy as heat.

Physically, there are two mechanisms by which energy passes from larger to smaller scales involving eddies (Fig. 3.2). The first is eddy stretching. Eddies get stretched by the pull of the bulk fluid straining motion along the axis of rotation. This tends to cause the cross-section of eddies to shrink. Conservation of angular momentum constrains eddy rotation rate to increase (think of spinning figure-skaters pulling in their arms, increasing their rotation rate). The region of coherent strain is usually slightly larger than the scale of the eddy. The work involved in eddy stretching transfers energy from larger to smaller scales. The second mechanism is strain self-amplification. As faster-moving fluid overtakes slower-moving fluid in its path, a strong compressive strain rate steepens (analogous to ocean-wave steepening before breaking on shore). Strain self-amplification reduces the size of the region being squeezed, distributing the associated kinetic energy toward smaller scales of motion.

Recent work demonstrates that strain self-amplification typically contributes more to the energy cascade in eddy dynamics relative to stretching (Carbone and Bragg 2020).

This energy dissipation involves strain at multiple scales: the scale of the bulk fluid, the scale of the eddy, and strain at successively smaller scales within eddies. This is challenging to model due to the inability to run computer models with fine-grained resolution from the size of the flowing river to the eddies to the length scales where the energy is being dissipated. To model large-scale eddies in turbulence, subgrid models are incorporated into the larger-scale computer models of turbulent flow (e.g., Sukoriansky et al.

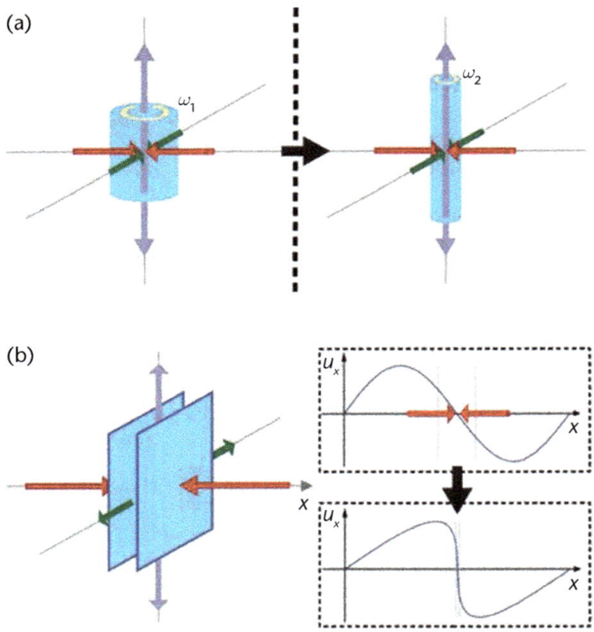

Figure 3.2 (a) Vortex stretching. When under strain (red, green, and violet arrows), a vortex rotating at a rate ω_1 becomes stretched, resulting in a higher rotation rate ω_2 (by conservation of momentum). Energy is transferred to successively smaller and smaller vortices. (b) Strain self-amplification. Regions where the strain rate is strong leads to a sheet-like area of high compression naturally growing thinner as the strain rate becomes steeper. This is because faster-moving fluid (peaks in the two graphs of u_x as a function of position x at two different points in time during the compression) overtakes slower fluid ahead of it and squeezes the fluid parcle. Reproduced from "The squeezes, stretches, and whirls of Turbulence," by Perry L. Johnson, *Physics Today* 74, 4, 46 (2021), https://doi.org/10.1063/PT.3.4725, with the permission of the American Institute of Physics.

1996; Shang 2019). The bulk turbulent flow is typically modeled by discretizing the Navier–Stokes equations using finite difference methods. Similarly, subgrid models are based on finite difference methods applied to the various equations used to model the physical behavior of fluids on smaller length scales than can be resolved by modeling of the large-scale bulk fluid. Subgrid models for eddies take advantage of the fact that eddies are quasi-independent of the bulk flow. Eddies exhibit universality near boundaries such as walls in that they are independent of both the Reynolds number (a dimensionless parameter characterizing the relative importance of inertial force to viscous force in a liquid) and the type of bulk flow in which they arise (e.g., Piomelli 1999; Shang 2019).

A filtering operation is applied to filter out larger-scale variables from those relevant to the size scale of the eddies. This provides the basis for the scale separation for the large-scale bulk fluid model and the eddy subgrid models, as well as for removing the smallest-scale fluctuations too computationally expensive to calculate. The idea is to model the energy dissipated by the eddies to produce an estimate for the energy to be subtracted from the bulk flow model. This mimics the energy cascade to smaller scales taking place through eddies. Basically, the role of the subgrid eddy model is to relate the small-scale stresses in the eddy to the large-scale strain rate tensor of the bulk model. The latter tensor tracks the rate of stretching and shearing of fluid parcels (e.g., the tendency of the bulk fluid to stretch, squeeze, and shear eddies).[35] The aim is to predict the local energy transfer between bulk fluid and subgrid scales for physical systems such as flow through pipes, artificial heart valves, turbulent wakes from wind turbines, and sneeze droplet spreading.

A basic turbulence computational scheme runs like this:

1. Calculate the bulk flow of a fluid at the largest scale computing stress and strain (strain rate tensor) advancing the discretized equations in time until formation of eddies.
2. Scan representative volume elements at smaller scale locating the formation of eddies and switch to a subgrid model appropriate to treating eddy response to the computed bulk strain rate tensor.
3. Calculate an estimate for the energy dissipated by each eddy and subtract energy from the bulk fluid model across the computational grid representing each eddy's contribution.
4. Recompute the bulk fluid model for the next time step advancing the stress and strain.
5. Repeat steps 2–4 for the length of the required simulation time.

[35] There are additional approaches to modeling eddies and relating their energy dissipation to the bulk flow, some of which produce fewer numerical errors when applied to fluid flow near walls, while others produce relatively greater errors (Piomelli 1999). But the approach sketched in the main text is the most intuitive to understand.

Subgrid models for eddies presuppose the bulk fluid modeling for stress and strain (e.g., eddies undergo strain self-amplification and stretching due to bulk stress and strain, and one or more parameter values for the subgrid models may be set by the bulk flow model). In other words, the bulk fluid model computer stress and strain acts as a stability condition for the subgrid eddy models constraining the behavior of eddies.

Clearly, the questions of relating descriptions at different scales in multiscale modeling are different than those of intertheoretic reduction. For instance, in modeling turbulent flows it makes no sense to try to reduce the description of the large-scale bulk flow to the smaller scales of the energy dissipation dynamics in eddies. The latter descriptions do not capture bulk flow behavior; the subgrid models only tell the bulk model code how to subtract energy from the flow. Recall that all the equations are filtered so that each model has a resolution for its own length and time scale (e.g., the bulk flow model has the scale information relevant for the eddies removed).

Moreover, as with Rayleigh–Bénard convection in Chapter 2, the properties and behaviors of the smallest length scales of turbulent fluids are insufficient to yield the larger-scale bulk flow (e.g., the bulk flow places constraints on the smaller scales that are not generated by those smaller scales). This is to say, that there is crucial context captured in the bulk flow model that is missing from the subgrid models. As well, there is information generated by the subgrid models that informs energy dissipation for the bulk fluid model. Indeed, there is no attempt to send the full information available at the subgrid scale to the bulk model. Rather, only the energy to be subtracted due to each eddy's dissipation is communicated to the bulk model. The bulk flow stress and strain are updated for the next time step, and the new constraints these represent are communicated to the subgrid model for the eddies.

Briefly consider another example: the multiscale geophysical modeling of a slab "of subterranean granite subject to extreme geothermal pressures" described by Wilson (2017, pp. 13–16). To learn about the Earth, geologists are interested in how sound waves move through different materials. When stresses on granite are moderate, it behaves much like an elastic solid. Such solids can be characterized by three elastic moduli tracking pressure on the surfaces of a representative block of the solid: Young's modulus, shear, and the bulk modulus. The first two moduli are the most important for the geologist's modeling. Young's modulus (the slope of the stress–strain curve of the granite) tracks the stress normal to opposite faces that tend to stretch the block longer and thinner or shorter and fatter. Shear tracks stress tangential to opposite faces that tend to deform rectangles into parallelograms (i.e., where material layers slide over one another). The elastic moduli are measures of bulk response of the granite block to external pressures and are not definable at the atomic or even granular level. This is to say, that such moduli do not represent parameters coarse-grained from smaller length scales such as the atomic lattice. Rather, the moduli represent whole block response to external pressures.

Under sustained higher pressures, granite is recrystallized into layered gneiss. This is important, because isotropic elastic solids like granite are not sensitive to the direction of the applied pressure, but gneiss is, making it an anisotropic solid. To track the behavior of the latter solids, models need five to six more elastic parameters. The bottom line is that seismic waves will travel through granite differently than through gneiss, so geologists need models that will estimate the portion of underground granite slabs that will convert to gneiss, and locate where this metamorphosis occurs.

A molecule-by-molecule model is a hopeless task for this purpose, so geologists turn to multiscale models employing what is known as finite-element modeling, where the submerged block is computationally decomposed into a set of smaller cells or volume elements following this scheme:

1. Start at the highest level computing internal stresses across the entire granite block as an elastic solid (tracking Young's and shear moduli). Only reassess if a representative volume element at a lower level indicates that this is needed.
2. Scan representative volume elements at smaller scale, locating which volume elements are exhibiting the highest local stresses, and switch to an anisotropic subgrid model appropriate to treating the volume element "as a laminate comprised of different minerals in a manner appropriate to granite grain" (2017, p. 14).
3. Check for shear (consistent with elastic solid modeling) or for recrystallization (consistent with anisotropic modeling). If the volume element shears, accept the higher-level stress modeling. If the volume element recrystallizes, replace the higher-level modeling equations with anisotropic equations consistent with gneiss.
4. Recompute the stress distribution on the entire block, incorporating information from the subgrid models.
5. Repeat steps 2–4 to check for consistency until the model achieves self-consistent assignments for stresses and responses for all representative volume elements.

Again, the point of such multiscale modeling is not to reduce higher-level descriptions to lower-level ones. Rather, the models pick out the relevant description for the relevant scale to obtain relevant information. Also notice that multiscale modeling is neither bottom-up nor top-down. It is not even accurate to say that the subgrid models are "more correct" or the higher-level equations are "approximations" to the lower level. In the granite example, the anisotropic subgrid model presupposes the stresses on the volume element in question as calculated by the bulk elastic model for the granite block. In other words, in the multiscale model scheme, the subgrid model is asking the question: "Given the stresses on the representative volume element calculated by the higher level, will a statistical array of minerals shift their relative positions consistent with an elastic or an anisotropic solid?" In other words, the stresses on representative volume elements calculated by the elastic bulk modeling act as a stability condition constraining and shaping the behavior of the volume elements described by the subgrid model.

3.4.2 *Multiscale data analysis*

One sees similar lessons in the modeling of complex data from nonlinear dynamical physical, biological, ecological, or social systems, where the dynamics are multiscale. The dynamics of such systems are characterized by the time-evolution of the patterns corresponding to scalar and vector fields by the systems. Capturing these patterns experimentally or numerically results in multiscale high-dimensional datasets. When features of interest in the data are multiscale effects, they cannot be reduced to the smallest-scale dynamics in the dataset. Mathematicians have developed topological tools to analyze data in such situations.

Persistent homology is one of these tools (Edelsbrunner and Harer 2008; Weinberger 2021). It provides a way to characterize nearness of "distance" between two functions on homeomorphic topological spaces, nearness of shapes (e.g., dog faces, plants), and nearness between positive and negative simplices, among other mathematical objects or structures (see Appendix). Similar to what we have seen in multiscale models (and in mathematical procedures such as asymptotic series and the renormalization group), this is a means of ignoring irrelevant information to focus on what is crucial at the scale or scales of interest. Basically, the technique works by assigning collections of points in a dataset to a scalar or vector field in what is known as a persistence diagram (see Appendix). Persistence diagrams encode topological features of subsets in the data at each scale, capturing pattern information. Such topological features might be the geometry of a container providing constraints on a fluid, for example. One interesting feature of persistent homology techniques is that one does not have to start with any assumptions about which scale, or scales, are important to find persistent patterns in the data.

The type of persistence defined for persistent homology (see Appendix) is stable under perturbations. A stability theorem (Cohen-Steiner, Edelsbrunner, and Harer 2007) functions as a stability condition for when persistence is stable in the sense of the stability of persistence diagrams to small perturbations. This means that there is a stability condition guaranteeing when this technique works on complex datasets. This is equivalent to showing that the mapping producing the persistence diagrams is a Lipschitz condition, so small changes to the data contributing to patterns in a persistence diagram yield small changes in the diagram itself.

The technique can be used to draw inferences about homology between two persistence diagrams, among other uses (Edelsbrunner and Harer 2008, pp. 275–277). This has immediate consequences for analysis of time series, bringing us to persistent homology's use in complex multiscale datasets (pp. 277–279). The mapping of data under persistent homology into a persistence diagram leads to a space of persistence diagrams that can be endowed with a metric (hence the ability to compare the "nearness" or similarity of two persistence diagrams). Analyzing a time series of patterns leads to a time series in the space of persistence diagrams that can be used to quantify the average rate of change in the geometry and topological properties of the patterns in the diagrams; in other words, the dynamics of how patterns in the data change over time.

As mentioned previously, obtaining these patterns in complex multiscale and multidimensional datasets is a challenge. Persistent homology provides a way to reduce the dimensionality of a complex dataset to a time series of persistence diagrams. The technique depends only on the transformation of the data, not on any assumptions about the system or equations that generated the data as long as conditions of the stability theorem hold.

To illustrate, consider Mengsen Zhang and colleagues' "human firefly" experiments revealing simultaneous coordination at larger ensembles, subgroups, and dyad scales. (Zhang, Kelso, and Tognoli 2018; Zhang et al. 2020). Ensembles of eight research subjects were instructed to tap rhythmically on a touchpad, and each member of the ensemble started a trial with a pace set by a metronome. After the pacing period, subjects were allowed to see each other's taps as flashes ("human fireflies"). Each subject was instructed to keep tapping at their own metronome frequency (tempo) throughout the entire trial. No instructions were given to coordinate with other subjects.

To study how patterns of coordination among participants formed or dissolved, investigators introduced different levels of diversity through manipulating the assignment of metronome frequencies to each participant. The experimental setting was to vary the distribution of frequency predispositions. Frequency difference was the chosen observable due to the ease of control and systematic measurability. Hence, frequency difference was varied among group members.

This experiment produced a complex multiscale high-dimensional dataset. The multiscale topological features in the data are related to the dynamic metastable patterns in the following way. Taking the time derivative of phase yields a relation between phase and frequency. This relationship connects the dynamics of relative phases of the observed frequencies to the dynamics of instantaneous frequencies, so that metastable phase coordination in the data is characterized by specific patterns in the collection of instantaneous frequency trajectories. This meets the conditions to apply persistent homology to analyze the patterns of metastable coordination and their transitions at multiple scales in the data—information that remains obscure in other data analysis techniques (Zhang et al. 2020). Investigators detected a major transition in coordinated patterns among subgroups of ensembles initiating around 30 seconds, and a minor transition around 10 seconds. The major transition was associated with two pairs suddenly breaking up and switching partners with others at 28 seconds, then breaking off and returning to their original partners at 33 seconds. Such collective transitions are only revealed through topological analysis, not any other data analysis methods (p. 9). Meanwhile, the minor signal around 10 seconds represented a lone wolf breaking off from a coordinated pair—a transition at a lower scale than the 30-second transition.

Using persistent homology, global, intermediate (mesoscale), and local scale dynamics are identifiable in the data. The structural patterns detected in the data are irreducible to the dynamics of individual points in the dataset, revealing collective transitions taking place on a larger scale than point dynamics (p. 10). There is a correspondence revealed in the data between larger-scale collective behavior and persistent patterns in the data reflective of a stability condition at larger scales for metastable collective behavior of individual "fireflies."

Metastable systems arise in dynamical systems where the system's stability is characterized by how quickly it returns to a stable state following a perturbation. When a system has multiple coexisting attractors and fluctuations are sufficiently strong to cause switching among different accessible stable states, it is multistable. In the "human fireflies" experiment it is the intermediate-scale groups, their coexisting tendencies for integration (more within-group coordination leads to more between-group coordination) and segregation (more within-group coordination leads to less between-group coordination) that represent multistability across space and time scales. The value of the frequency diversity δf turned out to have a critical point that served as a stability condition for integration/segregation: values above critical value δf^* led to segregation. As well, dyadic pairs were affected by higher-scale groupings, not the other way around. It turns out that the presence of intergroup difference dramatically elevates metastability in the system that one sees in the persistent homology analysis—a structural relationship revealed under this form of topological analysis that is very difficult to discern otherwise.

As with multiscale modeling, we see contextual emergence illustrated in both the persistent homology data analysis technique as well as in the dataset generated by the behavior of human subjects in the "firefly" experiment.

3.4.3 *Stability conditions and multiscale models*

Let us take a step back and see how contextual emergence is exemplified by multiscale modeling. Our illustrative examples highlight the role that higher-level stability conditions play for subgrid models. In the turbulence case, the stress and strain generated by the bulk fluid model provides the stability conditions constraining eddies and activating the energy dissipation mechanisms which their subgrid models track. In the granite case, it is the stresses generated by the elastic bulk model that provide the stability condition constraining and activating the crystallization which the anisotropic subgrid model tracks. In turn, the subgrid models provide some necessary conditions for the higher-level models. The eddy subgrid model tells the bulk fluid model how much heat to remove and to where, even though the heat removal information is not necessary and sufficient for the behavior of the bulk fluid model or even the eddy subgrid model. The anisotropic subgrid model tells the elastic model what subvolumes within the granite have undergone crystallization, though the crystallization information is not necessary and sufficient for the behavior of the elastic bulk model or even the anisotropic subgrid model.

In the "human fireflies" experiment, persistent homology techniques identify topological features in the dataset indicating where larger metastable behaviors and their transitions provide stability conditions constraining or shaping dyadic pairs. Although there would not be any data or patterns unless individual humans were tapping the touchpads, their individual tapping is not necessary and sufficient for the behavior of the larger-scale metastable behaviors, nor even for the individual tapping behavior.

Another point to note is well illustrated by turbulence modeling. The stress and strain calculated in the bulk fluid model provides a stability condition for the behavior of the smaller-scale fluid eddies. But the bulk stress and strain are different from the stress and strain calculated in the eddy subgrid model (the latter stress and strain are due to the dynamics of the eddy swirls on the fluid parcels composing the eddy). The subgrid model calculates the local stress and strain down to the length scale allowed by its filtering, and then estimates energy loss due to heat as dissipation at that smallest-scale limit. In other words, the local stress and strain defined by the eddy as a unit acts as a stability condition for the behavior of the smaller-scale fluid parcels in a way similar to the Bénard cells discussed in Chapter 2. Hence, in multiscale modeling we see a layering of stability conditions across scales defining contexts relevant for physical modeling at each of those scales.

A third point to note about multiscale modeling follows from the last one. As illustrated by our examples, models at different scales are focused on different regions of physical possibility space. For instance, the bulk fluid model calculates the possibilities for bulk fluid flow, while eddy subgrid models explore a more restricted region of possibility space where the bulk stress and strain stability conditions act as a gateway to that restricted region of possibility. Similarly, the elastic bulk model calculates the possibilities for large-scale stress and shear on the granite block, while the anisotropic subgrid models explore a more restricted region of possibility space where the elastic stress and shear stability conditions act as a gateway to that restricted region of possibility space. Likewise, the multiscale analysis of the "human fireflies" data reveals similar kinds of restrictions and explorations of regions and subregions of possibility space.

3.4.4 *Multiscale modeling and contextual emergence*

This is the contextual emergence framework described in Chapter 2 and illustrated previously for intertheoretic relations. Recall the definition of contextual emergence:

> **Contextual Emergence** Properties and behaviors in a particular ("lower-level" or smaller-scale) domain (including its laws) at best offer some necessary but no sufficient conditions to determine properties and behaviors in another domain.

The pattern of a subgrid model at a lower level of description contributing some of the necessary conditions needed for the bulk model at a higher level of description fits this framework well (compare the computational schemes outlined in the previous section). The eddies forming in turbulent bulk flow provide some necessary conditions for the bulk flow (e.g., serving as energy sinks dissipating heat) but no sufficient conditions for the latter. It is the bulk flow where the remaining necessary and sufficient conditions for both bulk flow and eddy behavior are found. Hence, eddy subgrid models take in information representing constraints from the bulk model to estimate the energy loss due to heat in the downward cascade, and communicate that back to the bulk flow model.

Similarly, the gneiss forming in granite blocks provide some necessary conditions for the elastic bulk model (e.g., representing sites where crystallization occurs) but no sufficient conditions for the latter. It is the elastic bulk stress and sheer where the remaining necessary and sufficient conditions for both the granite bulk and the crystallization behavior are found. Hence, anisotropic subgrid models take information representing constraints from the bulk model to estimate sites where crystallization is likely to take place, and communicate that back to the elastic bulk model.

Indeed, though he does not use the terminology, Batterman's discussions of multiscale modeling articulate the contextual emergence nature of such models (e.g., 2012, 2018, pp. 11–14; Batterman and Green 2020; Green and Batterman, forthcoming). Consider what Batterman says about the multiscale modeling of steel:

> Somehow between the lowest scale of crystals on a lattice and the scale of meters or millimeters, the low-level ordered structures must disappear. But that suggests that properties of the steel at its most basic, atomic level cannot, by themselves, determine what is responsible for the properties of the steel at macroscales.
>
> (2012, p. 258)

What does he mean by this regarding modeling phenomena such as the bending of steel? He uses renormalization group methods as an example of upscaling information from microscopic to macroscopic models/levels.[36] In essence, the bulk behavior of bending steel is remarkably independent of the atomic-level description. What information is relevant becomes encoded in elasticity parameters such as Young's modulus characterizing "universal behavior despite potentially great differences in microscale physics" (p. 265).

For modeling macroscopic behavior of steel, we must be mindful that

> microstructural features include mesoscale dislocations, defects of various kinds, and martensitic transformations. If we engaged in a purely bottom-up lattice view about steel, paying attention only to the structures for the pure crystal lattice, then we would get completely wrong estimates for its total energy, for its average density, and for its elastic properties. The relevant Hamiltonians require terms that simply do not appear at the smallest scales.
>
> (p. 268)

The mesoscale dislocations and defects, for instance, disrupt the symmetries of the atomic lattice of steel viewed on the atomic scale (the pure crystal lattice). Renormalization group methods provide a way of accounting for physics at mesoscopic to larger scales. As Batterman puts it,

> to extract a continuum phenomenology [behavior of the bulk steel], we need a methodology that enables us to upscale models of materials that are heterogeneous at small scales to those that are homogeneous at macroscales, as is evidenced by the fact that only a very small number of phenomenological parameters are required to characterize their continuum level behaviors.
>
> (p. 269)

The kind of information available at each scale (atomic, various mesoscopic scales, bulk scale) each vary, and each has a role to play in the behavior of

[36] We will have more to say about renormalization in Chapter 5.

steel. The forces applied to a steel bar are not given by the arrangements of the atoms in the lattice. The symmetries of the atomic lattice are broken by larger-scale impurities, dislocations, and grain boundaries. The elastic parameters characterizing the bulk steel are not derivable from the atomic lattice (indeed, starting with the atomic lattice and attempting to calculate these bulk parameters bottom-up yield values disagreeing with empirical measurement) so must be input as experimentally measured values or estimated when measurements are not possible.

> Bottom-up modeling of systems that exist across a large range of scales [such as steel] is not sufficient to yield observed properties of those systems at higher scales. Neither is complete top-down modeling. After all, we know that the parameters appearing in continuum models must depend upon details at lower-scale levels. The interplay between the two strategies—a kind of mutual adjustment in which lower-scale physics informs upper-scale models and upper-scale physics corrects lower-scale models—is complex, fascinating, and unavoidable.
>
> (p. 283)

Or consider Batterman and Green's (2020) discussion of multiscale modeling of steel:

> A model at the lowest scale (just as an upper-scale continuum model) would not be able to see the microstructures—the voids, cracks, grain boundaries, etc—that exist at intermediate scales ... How should we understand material parameters such as Young's modulus? Typically one determines its values for different materials via tabletop experiments. However, one must also recognize that material parameters reflect lower-scale structures in the material—structures that are neither observable at continuum scales nor at atomic/lattice scales. The important features responsible for the values for the material parameters are mesoscale features ... In the context of steel we should think of material parameters as reflecting mesoscale structures significantly above the atomic scale and significantly below the continuum scale.
>
> (pp. 8, 10)

Upscaling from the mesoscopic modeling of voids, cracks, grain boundaries, and so forth, to derive estimates for parameters such as Young's modulus for the bulk steel model, allows for modeling how a steel bar behaves as a whole when it may be difficult to experimentally measure such parameters. But this means that the properties and processes at the atomic scale do not determine the bulk model behavior; they contribute some of the necessary conditions through the different length and time scales for the bulk model. In turn, "finite element methods ... are often used to discretize macrolevel parameter values into boundary conditions at the mesoscale. The boundary conditions can in such cases be considered as macrolevel inputs to boundary value problems at lower scales (p. 12). In this way, the bulk properties calculated by the bulk model supply constraints for subgrid models at smaller length and faster time scales (e.g., to adjust the mesoscopic details to align with the response of the steel bar to external forces).

Commenting on a variety of multiscale examples, Green and Batterman (2020) summarize: "Thus, the idea of a scale dependent *relative autonomy* offers an alternative to more extreme positions (reductionism or antireductionism), while capturing aspects of how scientists deal with multiscale system."[37] This is the contextual emergence framework that we have seen for multiscale modeling: larger-scale bulk model properties that cannot be generated from smaller-scale subgrid models alone are crucial input for subgrid models as stability conditions characterizing the largest-scale context. In turn, information from the subgrid models needs to be communicated back to the bulk model as part of the necessary and sufficient set of conditions for modeling the object in question. While each scale of modeling has some relative autonomy from other scales, there is neither full independence, nor bottom-up-only or top-down-only dependence.

We could provide a similar analysis to demonstrate that Wilson (2017) discusses multiscale modeling in the same terms as contextual emergence, though not explicitly adopting the framework or making explicit how many of the constraints he discusses serve as stability conditions.

3.5 Summary and moving forward

As we saw in Chapter 2, contextual emergence makes explicit what is implicit. In this chapter we see it brings to the fore details normally missing from intertheoretic reduction discussions. Our diagnosis of the allure of reductive accounts such as Nagel's and Butterfield's is that they implicitly mislead one into thinking that reduction is going on when relationally something much more interesting is the case. Moreover, we have seen that multiscale modeling illustrates how complicated the interaction and relationship among different contexts is, where models embody important features of their specific contexts. One upshot of taking contextual emergence seriously is that the world—and the sciences exploring it—is more complicated and more interesting than our metaphysical and philosophy of science analyses typically can encompass. Old philosophical intuitions (e.g., that theories are set of sentences of an interpreted language closed under logical consequence) are not only outmoded but also are inconsistent with the world we live in and the scientific practices exploring it.

[37] Unfortunately, Green and Batterman (2000, n. 10) refers to Bishop (2012) as an example of higher-level variables derived by coarse grainings from lower-level variables where the former constrain the latter, missing the contextual emergence in fluid convection and temperature.

This chapter has been focused mainly on descriptions, modeling, and intertheoretic relations—what can be called epistemological concerns and how the contextual emergence framework handles such concerns. Going forward, we continue our case-study approach begun in this chapter. We will look at several examples in physics (Chapter 4), condensed-matter systems (Chapter 5), biological, neural, and more extended cognitive systems (Chapter 6), where the emphasis will be on how contextual emergence addresses both explanatory and ontological issues raised by various sciences.

Appendix

Following Edelsbrunner and Harer (2008), as a simple example of persistent homology, consider real-valued functions of a single variable $f : \mathbb{R} \to \mathbb{R}$, with the usual continuity conditions, where f has non-degenerate critical points with distinct critical values (i.e., each critical point is either a local minimum or local maximum). The critical points can be paired in the following way. Let a local minimum, say, represent a component. Proceeding along the axis, the function passes through a local maximum, then through a new local minimum representing another component. Pair the maximum with the "younger" of the two local minima representing the two components, and let the other minimum now represent the component resulting from the merger.[38] When critical points x and y are paired in this fashion, the persistence can be defined as $f(x) - f(y)$. One can then create a persistence diagram by mapping each point $(f(x), f(y))$. This amounts to plotting the critical points as coordinates in the diagram.

This procedure generates a set of connected points in the persistence diagram. Next, let $s < t$ and consider the subsets $\mathbb{R}_s \subseteq \mathbb{R}_t$. In going from s to t, components of $\mathbb{R}_s$ may merge, or new components may be born and merge with other components in $\mathbb{R}_s$, or even merge with components of $\mathbb{R}_t$. We can define $\beta_0^{s,t}$ as the number of components born at a finite time or before s belonging to distinct components in $\mathbb{R}_t$. It turns out that $\beta_0^{s,t}$ for the previous pairing procedure has the property that it is equal to the number of pairs (x, y), where $f(x) \leq s < t \leq f(y)$. No pairs of points other than paired critical points have this property.

Building on these properties, one can define a simplicial complex K with p-dimensional simplices and a p-th simplicial homology group (pp. 259–260). Two simplicial complexes K and L can be related by a continuous simplicial mapping $f: K \to L$. By the simplicial approximation theorem, a continuous map of simplicial complexes f can be approximated by a simplicial map g, implying that such continuous maps induce a homomorphism on the homology $f_p : H_p(K) \to H_p(L)$. Persistence can be defined for simplicial complexes and other more complicated structures (e.g., pp. 261–262), and an algorithm exists to compute persistence (pp. 264–266).

The persistence diagrams represent a mapping assigning every scalar or vector field to a collection of points $\mathbb{R}^2$. A feature in the data encoded by the point $(s, t) \in \boldsymbol{f}_p$ represents a feature at the P-th homology level. One can speak of s and t as the birth and death coordinates, respectively, of this feature. The lifespan s–t characterizes the prominence of the feature. Features with long lifespans are considered important, while features with short lifespans are often associated with noise (though one must keep in mind that one person's noise is another person's data of interest!). In this way, persistence diagrams are lower-dimensional representations of the field generating the pattern in the higher-dimensional dataset, and track the persistence of patterns.

[38] Paired critical points need not be adjacent.

4

Classical and Quantum Physics

There are more [and fewer] things in heaven and earth, Horatio, Than are dreamt of in your philosophy.

Shakespeare, *Hamlet* (with a little rewrite by us)

4.1 Introduction

We introduced contextual emergence as an ontological alternative to reductionism and radical emergence. Rayleigh–Bénard convection provided a case study illustrating the pattern contextual emergence takes in the physical world. Chapter 3 discussed how this pattern shows up in intertheoretic relations and multiscale modeling.

In contrast to analytic metaphysics, which starts with intuitions to find bedrock truths, we take the sciences to be the best guide to the ontological furniture of the world—the states, observables, stability conditions, systems, levels of organization, and transitions between domains of reality and their multi-length and time scales. Hence, going forward we will focus on examples of how the contextual emergence pattern shows up across sciences such as physics and biology, and draw out some of the consequences.

This chapter focuses on examples from physics because it is often thought of as the home of reductionism and as the reductionistic model for the rest of the sciences (e.g., Atkins 1995). Elementary particle physicists are particularly wont to proclaim the reductionistic character of physics (e.g., Weinberg 1995), but as we have seen, condensed-matter physicists also endorse ontological reductionism (e.g., Anderson 2011; Laughlin 2005). Many philosophers also take physics to be the archetype for reductionism (e.g., Nagel 1961; Menon and Callender 2013). Physics, then, is an appropriate place to start for demonstrating how contextual emergence upends these reductionist views of physics (as we have already seen in the case of Rayleigh–Bénard convection in Chapter 2).

Emergence in Context. Robert Bishop, Michael Silberstein, and Mark Pexton, Oxford University Press.
 DOI: 10.1093/oso/9780192849786.003.0004

What follows is a series of case studies, some in more detail, some in less, illustrating the pattern of contextual emergence across subdisciplines in physics. We shall begin with some further remarks on contexts, then dive into the cases, some of which have been taken to be workhorse examples of successful ontological reductions, such as temperature and chemical potential (Section 4.3), the macroscopic world's relationship to the microscopic (Section 4.5), and molecular structure (Section 4.6), while others illustrate particular aspects of contextual emergence nicely, such as white dwarf stars (Section 4.4), quantum entanglement (Section 4.7), and renormalization group explanations (Section 4.8). We round out with some briefer miscellaneous examples of contextual emergence.

4.2 Contexts and interactions

Contexts are not just aspects of the environment that are part of the contingent necessary and sufficient conditions to produce emergent properties. Contexts are also the environment with which those properties, once produced, interact. This is illustrated well in complex systems such as Rayleigh–Bénard convection. Such systems exhibit interleaving multiscale systems of dynamic feedback, among elements of the system, the system, and the environment. The formation of Bénard cells illustrates how emergent structures and properties, in turn, provide a stability condition constraining and shaping the allowed motions of the fluid parcels from which the convection cells arose. Similarly, in an ecological system, the environment serves as a necessary context to produce a specific set of ecological interactions. Yet once that web of ecological interactions is manifested it, in turn, can change the environment, altering the context from which it was produced. It is the constant dialectic between context and manifestation that produces the complexity of the system, and is one of the differences between a complex and a simple system.[1]

As we have stressed, contextual emergence makes explicit how a system provides a new context for the components of the system through the internal contexts created by multiscale relations among parts. A simple example of this would be a mean field system, where a particular charged particle experiences and interacts with a field which is the average of the rest of the charges in the system. The collective field produced by the charges creates a new context for each charge in that system. Similarly, the Bénard cells provide a context for the behavior of each fluid parcel in the system.

[1] This is not to imply that simple systems lack such a dialectic, but only to point to the most striking illustrations. After all, simple systems also have their necessary contexts to screen off or open modalities to allow for the production of properties.

It should be clear that contextual emergence is not (w)holism or priority monism (Schaffer 2010) in the sense that there is not a whole apart from or prior to all the contextually driven multiscale interactions. Again, Rayleigh–Bénard convection illustrates that the whole (Bénard cells) *emerges* from the parts (fluid parcels), while the smaller-scale parts become subject to the larger-scale whole that eventually arises. Therefore, contextual emergence not only enables us to clarify how emergents arise over time, but also how those emergents continue to interact with their context as well as with "lower-level" system constituents.

4.3 Property emergence and global stability conditions: temperature, chemical potential, and pressure

The thermodynamics property temperature is the usual example of a successful reduction, where temperature is thought to be merely the result of mechanical properties of the mean motion of molecules in a gas or liquid, say (e.g., Nagel 1961, p. 341). But such a rough textbook picture is a gross mischaracterization due to underestimating some important details (Bishop 2019, Section 4.4.2).[2] Temperature has a much more complicated relationship to molecules.

Individual particle states do not have a temperature, so temperature is not a mechanical property (hence, temperature is an irrelevant property for the classical mechanics of particles). We first must look to the larger-scale distributions of statistical mechanics (SM). In theoretical derivations these distributions are derived from statistical moments for the kinetic energy, say, of the particles.[3] Essentially, the SM domain is the domain of large-scale ensembles, where these ensembles represent equivalence classes of single-particle states having the same properties. Yet the property of temperature is not found at the scale of ensemble properties simpliciter. In textbooks, properties such as temperature, chemical potential, and pressure are recovered through continuum limits (e.g., Bishop 2019; Compagner 1989; and Pathria 1972), though these discussions rarely point out that these limits involve information from the target domain and systems of thermodynamics.

[2] Batterman's discussion of thermodynamics' relationship to statistical mechanics emphasizes the singular nature of the limiting relations at criticality (e.g., 2002, 121–126). However, as we will see, he underestimates the complexity of the relationship of thermodynamic observables such as temperature to the scale of statistical mechanical ensembles.

[3] As outlined in Chapter 3, the statistical moments must be closed by using information from the larger-scale contexts of particle distributions—information not found at the small scale of individual particle states. The contextual emergence pattern is crucial to these derivations.

On the other hand, at the scale of thermodynamic systems (e.g., steam engines and refrigerators), temperature presupposes thermodynamic equilibrium—a particularly special stable state. Thermodynamic equilibrium is based on the zeroth law of thermodynamics:

Zeroth Law If two systems are in thermodynamic equilibrium with a third system at time *t*, they are in thermodynamic equilibrium with each other at *t*.

In such a state of equilibrium, thermodynamic properties, such as temperature and chemical potential, are uniform and unchanging throughout the system (e.g., the gas in the room having no variations anywhere in space at *t*). Hence, if two systems are in thermodynamic equilibrium, they are at the same temperature. Indeed, temperature is an indicator of an equilibrium state where heat is no longer flowing between the systems. In this sense, the definition of temperature is relational and indicates a particular kind of stability for thermodynamic systems.

The problem is that thermodynamic equilibrium is neither formally nor conceptually available at the scale of the ensembles of SM. Such ensembles are always subject to fluctuations, meaning there are no uniform, unchanging properties exhibiting the appropriate stability of thermodynamic equilibrium. Temperature is usually introduced into SM through a phenomenological argument based on the equivalence relation provided by the zeroth law. This, of course, is not a rigorous relationship between the properties of the ensembles of SM and thermodynamics, nor does it provide insight into how temperature arises.

4.3.1 *The ontological origin of temperature*

Thermodynamic equilibrium[4] serves as a context providing stability conditions for the existence of temperature (e.g., no heat flow, stable states bearing properties such as temperature and chemical potential). The emergence of temperature amounts to how the stability conditions of thermodynamic equilibrium are implemented at the scale of ensembles in SM. This is accomplished through the Kubo–Martin–Schwinger conditions guaranteeing the existence of a class of statistical states and observables with very special properties: the KMS states. Physically, the KMS conditions represent a stability condition guaranteeing the existence and persistence of KMS states that are stable against local perturbations and having temperature as a well-defined

[4] Technically, what follows is based on so-called quantum statistical mechanics, but since the result is always a mixed quantum–classical algebra of observables (Bishop 2019), it makes no difference to the case for emergence whether one discusses classical or quantum statistical mechanics.

observable. Some of the stability properties of the KMS conditions are (Emch and Liu 2002, p. 355)[5]:

1. Local thermodynamic stability. Fluctuations of KMS states confined to bounded regions of space cannot increase the system's free energy.
2. Global thermodynamic stability. KMS states that are translation invariant do not raise the spatial density of the free energy.
3. Local dynamical stability. Arbitrarily small local perturbations of the system dynamics admit stationary states that are arbitrarily close to KMS states.
4. Passivity. If a system is initially in a KMS state, no local perturbations applied over any finite time interval to the dynamics can result in energy being extracted from the system.
5. Reservoir stability. Any system in a KMS state can serve as a thermal reservoir when weakly coupled to a finite system.

These conditions guarantee the stability of states against both dynamical and thermodynamically localized perturbations.

Originally, the KMS conditions were derived for infinite systems (Kubo 1957; Martin and Schwinger 1959; Haag, Hugenholtz, and Winnink 1967). Later, Rudolf Haag and collaborators (1974) produced a rigorous derivation of the KMS conditions for finite systems based on three physically reasonable stability conditions for equilibrium states:

1. Stationarity. Expectation values of observables do not change over time.
2. Structural stability. The dynamics in which such states participate is stable against perturbations.
3. "Asymptotic abelianness." Correlations between local quantities at different times vanish sufficiently fast as the time differences go to infinity.[6]

The first condition is a standard expectation for an equilibrium state. The second condition amounts to placing an extra condition on stationary states that are stable under arbitrarily small perturbations; namely, such states are always given by density matrices that are functions of the system Hamiltonian.[7] The third condition ensures that there are no fluctuations in the intensive variables of a system—key information from the domain of thermodynamics via the zeroth law.

[5] For more discussion, see Sewell (1980).

[6] This condition ensures "that any two local observables measured at different times become more and more compatible when the time difference goes to infinity (asymptotic Abelianness in time)" (Haag, Kastler, and Trych-Pohlmeyer 1974, p. 176).

[7] A technical way to look at the second condition is as a replacement for the assumption that the system satisfies ergodicity.

The KMS conditions guarantee the contextual contingent stability condition of the zeroth law of thermodynamics for the scale of statistical ensembles relevant for SM. The second law of thermodynamics as a stability condition is equivalent to a maximization of entropy (Emch and Liu 2002, pp. 335–337).[8] The KMS conditions guarantee that if an ensemble system is in a KMS state, then it is in the canonical Gibbs state. This uniquely defines a parameter whose natural interpretation is (inverse) temperature. Moreover, every KMS state ψ can be decomposed into extremal KMS states φ. Such decompositions are often accompanied by spontaneous symmetry breaking (SSB) so that the states φ do not necessarily share the full set of symmetries of ψ. Takesaki (1970) has shown how temperature must be a classical observable, rather than an element in a pure quantum-mechanical algebra of observables. Temperature is an element of an algebra M of contextual observables, where the context is defined by the KMS conditions as a stability condition guaranteeing the Gibbs state. The algebra of observables for the smaller-scale of single-particle states is of type I, while the contextual algebra of observables M is of type III (Appendix 1)—hence the previous claim that temperature cannot be a mechanical property. This implies that the larger-scale statistical ensembles of SM do not have the property of temperature unless they conform to the KMS conditions. Therefore, temperature, and thermodynamics in general does not reduce to SM in any straightforward sense. In contrast to some textbook claims, this fact implies that thermodynamics is not simply a coarse-grained description of SM. Furthermore, the KMS states of different temperatures are disjoint (Takesaki 1970)—a property inconsistent with mechanical states underlying the SM ensembles. Finally, KMS states of different temperatures fall into different superselection sectors, meaning that temperature is dispersion-free; that is, a classical observable (Takesaki 1970).

This analysis of temperature helps us to understand that it is not merely a property of any mechanical force interactions. Indeed, temperature is a much broader property than the simplified average kinetic energy of molecules in a container. For instance, the kinetic energy in a paramagnet is small, but the temperature can be very high because it is related to the change of internal energy with respect to the change in entropy.[9] As such, the absolute temperature of systems can be negative when the kinetic energy of their constituents can never be negative.

[8] Equivalently, the free energy of the system is minimized in thermal equilibrium.

[9] Technically, the temperature in such situations is defined as $\left(\frac{\partial S}{\partial U}\right)^{-1}_{f.e.p.}$, where S is the entropy, U is the internal energy, and the subscript *f.e.p.* indicates fixed external parameters. For instance, for the microcanonical ensemble this would be a fixed number of particles (or spins) and constant volume.

Recall that observables are the measurable quantities of systems. Hence, we identify them with physical reality and take them to be ontological features of the world as revealed by empirical sciences. Temperature, then, is an example of a contextually emergent property neither contained in nor derivable from the laws and properties of the mechanical properties of single-particle states or their ensembles. The KMS conditions form the stability condition for the existence of thermodynamic states and properties. None of the laws and properties of the smaller-scale single-particle states nor the larger-scale ensembles of SM imply the KMS conditions; rather, SM depends on this crucial stability condition. Suppose that a quark–gluon plasma was the first object satisfying the KMS conditions in the early universe. The plasma would have acquired both quantum and classical observables once the stability condition was met. Temperature is one of the earliest examples of the contextual emergence pattern in the universe, and the KMS conditions form a global, acausal stability condition not dependent on space or time.

The contextual emergence of temperature does not exclude or contradict its being multiply realizable through equivalence classes. This is multiple realizability as sufficient conditions for different realizations of temperature. Different statistical mixtures (distributions) of particles can serve as multiple realizations of a thermodynamic state with the same temperature.

When crucial contextual details are made explicit, we can see that thermodynamic properties act as constraints for the mechanical properties of particles. Although the mechanical single-particle states and larger-scale ensemble states of SM contribute some necessary conditions for thermodynamics, the context of thermodynamics (crucially defined by the KMS conditions) contribute the rest of the jointly necessary and sufficient conditions for concrete systems and their behaviors. For instance, the behavior of individual particles may be constrained by the collective behavior of many-particle systems as a whole similar to Rayleigh–Bénard convection (e.g., Bishop 2004).

Therefore, we see that the textbook story of reduction of temperature to mean kinetic energy of single-particle states is misleading. There must first be KMS states with temperature as a relevant observable which is guaranteed by the KMS stability condition.

4.3.2 *Chemical potential*

The observable chemical potential is similar. Chemical potential arises through a transition from an ensemble of molecules to a thermodynamics context for such ensembles (chemical thermodynamics), where Gibbs states bearing temperature exist. Araki et al. (1977) and Műller-Herold (1980) have shown that this transition is nontrivial. In brief, the stability condition of thermodynamic equilibrium is implemented by the KMS conditions. So, chemical potential already presupposes the existence of Gibbs states and temperature. Furthermore, these states are disjoint for different chemical potentials. This means that states with different chemical potentials fall into different superselection sectors, hence chemical potential is a classical observable. Moreover, Gibbs states with chemical potential fall into equivalence classes corresponding to distinguished chemical species. Many arrangements of the molecules are possible that have the same temperature and chemical potential.

So, we have the same song, second verse. The textbook reductionist story of chemical potential being just the species of molecules present is misleading.[10] There must first be KMS states with temperature and chemical potential as relevant observables which is guaranteed by the KMS stability condition.

4.3.3 *Pressure*

Pressure, like temperature and chemical potential, is usually introduced phenomenologically and discussed as if pressure simply reduces to the collisions of particles with walls and each other. However, the story is more complex, as pressure is intimately related to temperature, meaning that the KMS stability condition is also at work.

The Gibbs state for a finite system at inverse temperature $\beta \in (0, \infty)$ is the unique state of a system defined by the density matrix $D_{H,\beta} = Z_{H,\beta}^{-1} e^{-\beta H}$, where $Z_{H,\beta} = \mathrm{Tr}(e^{-\beta H})$ and H is the Hamiltonian characterizing the system's total energy.[11] The pressure for a Gibbs state, then, is $P_{H,\beta} = \beta^{-1} \log Z_{H,\beta}$. As with temperature and chemical potential, the pressure is defined independent of any mechanical force interactions. Once there are any objects in the universe fulfilling the KMS conditions, such as a quark–gluon plasma, the stability condition guaranteeing the existence of Gibbs states induces pressure as an abstract property.

When the details are considered, it turns out that pressure is not reducible to the motions or collisions of particles in the straightforward sense presented in phenomenological derivations in textbooks.[12] We have the

[10] For the microcanonical ensemble, chemical potential can be defined as the change in entropy with particle number, $-T\left(\frac{\partial S}{\partial N}\right)_{f.e.p.}$, where the internal energy and volume are constant.

[11] In cases where the Hamiltonian has a continuous spectrum, one either transforms it or can switch to a C*-algebraic formulation of the state.

[12] For the microcanonical ensemble, pressure can be defined as the change in entropy with respect to volume, $T\left(\frac{\partial S}{\partial V}\right)_{f.e.p.}$, where the internal energy and particle number (or spins) are constant.

same song, third verse. There must first be KMS states with temperature and pressure as relevant observables which is guaranteed by the KMS stability condition.

4.4 Emergent causal relations: white dwarf stars

White dwarf stars are stellar remnants left over when a star has exhausted its nuclear fuel. The star forms a so-called planetary nebula (which has nothing to do with planets!) and sheds its outer layers, while the core, usually made of carbon and oxygen, contracts to form a white dwarf. Since white dwarf stars are not hot enough to continue fusing elements together, they have no source of energy generation. Normal stars are in a state of balance known as hydrostatic equilibrium in which a thermal pressure gradient counterbalances the inward pull of the star's self-gravity. In white dwarf stars, this is not possible. As the star contracts, it becomes extremely dense.[13] The remaining core is highly ionized and is made of heavy nuclei and free electrons. The star reaches such densities that it becomes electron-degenerate, meaning that all available lower-energy states (below the top one) are filled by electrons. This degeneracy creates a degeneracy pressure which stabilizes the white dwarf and prevents further collapse.

The Indian astrophysicist Subrahmanyan Chandrasekhar (1957) was the first to theoretically predict the relativistic upper limit on the mass of a white dwarf star. He calculated that for masses above about 1.4 times the mass of the Sun ($1.4M_\odot$), degeneracy pressure is not strong enough to halt stellar collapse. Subsequently, it was discovered that for stars above this Chandrasekhar limit, but below about $3M_\odot$, collapse continues until neutron degeneracy pressure counterbalances gravity. The difference in the two cases is that above the Chandrasekhar limit the collapsing star has enough gravity to produce a process called electron capture, whereby electrons are "absorbed" by protons, producing neutrons. If the collapsing star has a mass above $3M_\odot$ then even the neutron degeneracy pressure cannot balance gravity, and a black hole forms.

Summarizing the three different regimes: white dwarf stars form when the original collapsing star has a mass below $1.4M_\odot$; neutron stars form when the star has a mass between $1.4M_\odot$ and $3M_\odot$, and a black hole forms if the star's mass is above $3M_\odot$. If a white dwarf accretes enough matter after it has formed to breach the Chandrasekhar limit, say from a binary partner, then it tends not to form a neutron star. Instead, it explodes as a type Ia supernova, the violence of the explosion blowing the star apart. By contrast,

[13] Approximately 1 ton of white dwarf material would barely fill a matchbox.

if a neutron star accretes enough matter to tip it over the limit of neutron degeneracy pressure, then it will form a black hole.

It is important to stress that this degeneracy pressure results from the restriction on energy level filling due to the Pauli exclusion principle (PEP); it is not due to collisions between electrons. The PEP states that no two fermions can be in the same state, so electrons as fermions are forbidden to occupy the same energy level in the same state (Appendix 2). Degeneracy pressure (Appendix 3) is present in all matter and is distinct from the thermal pressure discussed previously. In ordinary situations, degeneracy pressure is minuscule in comparison with thermal pressure, but in white dwarf stars the lack of thermal pressure and extreme densities mean that it is large enough to halt the collapse and make the star stable.

4.4.1 *White dwarfs as contextually emergent*

Now we are able to see white dwarfs as an example of contextual emergence. More specifically, they illustrate how stability conditions at a larger scale are relevant for the emergence of the essential causal interactions at smaller scales producing white dwarfs.

Note that degeneracy pressure does not result from forces acting upon the electrons (Appendix 3). They do not push each other away electrostatically, for instance, and as such it is a very different sort of pressure from thermodynamic pressure. In this sense, the term "degeneracy pressure" perhaps seems somewhat misleading. It is the fact that the energy distribution of electrons is *not degenerate*—electrons are restricted by the PEP from occupying the same state[14]—that creates the relevant pressure.[15]

In white dwarf stars, the PEP, combined with the uncertainty principle, means that electrons must occupy higher and higher energy states as the density of the star increases. These electrons contribute to the internal energy of the star and hence the degeneracy pressure (Appendix 3). This pressure in the unusual state of matter found in white dwarf stars dominates anything, due to mechanical interaction. The degeneracy pressure reflects the capacity of the internal energy of a system to respond to changes in volume. In a white dwarf, the volume cannot be further minimized because all the energy levels available to the electrons are filled. Furthermore, there is a shift in the degeneracy pressure from the classical to the relativistic regimes (Appendix 3).

[14] Occupying the same state with the same quantum numbers is what physicists mean by a degeneracy.

[15] Bosons—which have no restrictions on how many can occupy the same state—can gather in the lowest energy state of a system, and so create no such pressure because their energy stays low. Physicists say that their energy state is degenerate—all the bosons are in the same energy state. Hence, no pressure due to PEP is generated.

At relativistic speeds the electrons use energy differently, so the ensemble property of pressure behaves differently, restricting any further decreases in volume due to gravitational attraction.

This is unlike how you normally think of pressure, such as the air pressure in an aircraft. Yet as we have seen, thermodynamic properties such as pressure and temperature are not straightforwardly the summation and averaging of the forces acting on atoms bouncing off container walls or the average kinetic energy of molecules. Indeed, one might be tempted to say that degeneracy pressure is a fictitious pseudo-pressure, a folk tale for physics students, a pressure in name only, since it is so different from the collisions described by kinetic theory. In contrast, Bradford Skow (2014) argues that such fictionalism about degeneracy pressure is not consistent with how pressure actually is used in other areas of physics. There are many different degrees of freedom that can contribute to the abstract statistical definitions of pressure and temperature given the KMS stability condition. Electron degeneracy pressure is one of these, accounting for why the volume of a solid or liquid does not collapse under normally applied mechanical pressures.[16]

In white dwarf stars, the PEP means that electrons must occupy higher and higher energy states as the density of the dwarf increases. These electrons contribute to the internal energy of the star and hence the degeneracy pressure. To claim the fictionality of degeneracy pressure because of a lack of causal-mechanical mechanisms is erroneous. Just as temperature in a paramagnet is not determined by kinetic energy, so pressure in the unusual degenerate state of matter found in white dwarfs is not dominated by mechanical processes. Once pressure is understood as defined through the KMS stability condition, degeneracy pressure is no more illusionary than any other type of pressure. It merely reflects the capacity of the internal energy of a system to respond to changes in volume. The shift in the dependency of degeneracy pressure on density, from the classical to the relativistic regimes, shows that how internal energy is distributed matters for degeneracy pressure.[17]

David Lewis (1986) famously thought that nothing halts the collapse of a white dwarf; rather, the star simply has nowhere else to go physically: "It's not that anything caused [the star] to stop [collapsing]—there was no countervailing pressure ... There was nothing to keep it out of a more collapsed state. Rather, there just was no such state for it to go into. The state-space

[16] As a point of comparison, standard atmospheric pressure is 10^5 Pascals, while the degeneracy pressure of the electrons in a block of copper is on the order of 10^9 Pascals, since they have no choice but to occupy all available energy states. Hence, electron degeneracy pressure can be a significant contributor to the behavior of a system!

[17] For neutron stars, collapse is halted by the symmetry pressure due to neutrons in the relevant mass range. Another nonmechanical pressure.

of physical possibilities gave out" (1986, p. 222).[18] While this connects with what we have said about possibility space and accessibility in Chapter 2, Lewis's characterization of the state space running out is incongruent with the astrophysical picture. There are other physical possibilities for the system to explore: A star could have become a white dwarf, or a neutron star, or a black hole, or a supernova, depending upon the time at which it accreted more mass. It is simply not true that the state space of the system "runs out." Instead, the collapse is halted by electron degeneracy pressure.

Still, one might say, "I don't see anything other than bottom-up microscopic effects, here. Looks like reductionism to me." To see that the emergence of a white dwarf star exemplifies contextual emergence rather than ontological reduction, let us start with another global stability condition: three-dimensional space. The reductionism reaction ignores the role that the dimensionality of space plays in the existence of stable states of matter.[19] Note that spatial dimensions are not part of the microphysics of QM. Rather, whatever microphysics there is must come to expression in a space of three dimensions.[20]

It is true that the explanation of the white dwarf's stability makes irreducible reference to electron degeneracy pressure as the difference maker, so it is a stability condition crucial to the existence of such stars. But degeneracy pressure is not a mechanical property—it is a statistical property of the ensemble of fermions. Hence, it is not possible to reduce white dwarf stability to microscale mechanical properties. Moreover, a change in the degeneracy pressure can impact the white dwarf's stability. An example would be type Ia supernovae, where the accretion of matter in binary systems tips white dwarves over the $1.4M_{\odot}$ limit. This leads to a huge increase in degeneracy pressure such that the star explodes violently. The statistics of fermions is not an insignificant difference-maker!

Is the electron degeneracy pressure macroscale or microscale? As we have discussed, thermodynamic properties such as pressure and temperature are not necessarily microscale and do not have to be causal–mechanical. The electron degeneracy pressure is related to the energy of the electrons, which, in turn, is related to the macroscale properties of the system through the density which involves system volume (Appendix 3). Indeed, the total

[18] For Lewis this is still a causal explanation, since the information that the halting of collapse has no cause is itself "causally relevant" information. He allows that providing causally relevant information is all an explanation must do to count as causal. Notice that Lewis does not argue that the star is stable because PEP provides a cause of the collapsing. Rather, it is that being able to say something is *uncaused* is itself a form of causal explanation, since he has ruled out the possibility of any causal history making a difference.

[19] Recall our discussion of spatial dimension as a stability condition in Chapter 2, Section 2.3.2.

[20] From this point forward we will rarely mention this global ancausal stability condition except where it is important to be pedantic in the face of reductionistic tendencies.

energy of the electrons in the white dwarf is proportional to the degeneracy pressure multiplied by the volume that the star exercises—all macroscale properties.

The electron degeneracy pressure is induced by a macroscale property of white dwarfs, but it has a dependency on PEP and Heisenberg uncertainty. Furthermore, PEP and Heisenberg uncertainty are not causal in any of the usual senses that philosophers discuss; rather, they are structural constraints on allowable states in QM—stability conditions. For example, the asymmetry of the fermionic wavefunction lies at the heart of PEP, but this is a feature that is non-manipulable, exerts no force, nor fits into any of the other typical philosophical accounts. It is a stability condition restricting fermions and their behavior to specific portions of possibility space. Yet electron degeneracy pressure is a macroscopic difference-maker that can be manipulated by changing one parameter: the mass of the star.

This is an unusual situation, to be sure. There is a necessary condition for microscale determination in as much as the degeneracy pressure is dependent on the antisymmetric properties of fermion wavefunctions. But, given several widely accepted accounts of causation, the macroscale explanation is causal while the microscale explanation is acausal. Therefore, we have *the contextual emergence of a causal relation*. QM at the microscale contributes some necessary conditions for the existence of a white dwarf star, but the relevant causal determinant of its existence is the macroscale degeneracy pressure counterbalancing the gravitational pull of the star's mass. Here, we see the degeneracy pressure and mass of the star acting as a stability condition for making the existence of white dwarfs possible.

This is to say, with particular contextual features such as stability conditions and system/environment distinctions, the property of degeneracy pressure can manifest as a property of the white dwarf star. Such a property can be linked back to the microscale property of fermion wavefunction asymmetry, but it only results in something that is a well-defined manipulable property that can respond as a pressure given a particular context that provides additional modal constraints beyond those found purely at the microscale. Nonetheless, when those constraints exist—when there is such a context—the resulting property can be identified as a causal property even though the candidate property necessary for it is an acausal or purely structural property as standardly conceived.

As a final observation, note that the contextual emergence of the causal relation associated with the degeneracy pressure, here, is not a product of realization. Consequently, this emergent causal relation does not receive its causal effectiveness from any microscale realizers. Stability conditions from a larger scale than QM play an ineliminable role in the contextual emergence and existence of white dwarf stars.

4.5 The contextual emergence of the macroscopic world

Simplified textbook discussions present QM as reducing our best description of the macroscopic world of our experience, classical mechanics (CM), in the limit as $h \to 0$, where h is Planck's constant. At best, this is a heuristic device, since h is a fundamental constant and so never changes magnitude. Going in the other direction, the Dirac quantization procedure starts with the equations of the Hamiltonian formulation of classical electrodynamics for two charged particles, say, and quantizes the observables to obtain Schrödinger's equation for hydrogen. Also, a heuristic approach, but notice that it makes use of information from the large-scale macroscopic world to reach a quantum formulation at the microscopic scale. Similarly, machine learning approaches to developing quantum models and solutions to Schrödinger's equation make unavoidable reference to macroscale phenomena (e.g., Zhang, Whang, and Weinan 2018; Dral 2020).

The actual relationship between quantum and classical descriptions is much more complex. It involves a series of limits involving ratios of h and other physical observables going to zero (e.g., the classical and quantum actions), or limits involving the separation of nuclear and electronic frames of motion (discussed subsequently), or the emergence of temperature and chemical potential (Section 4.3), among others. All these limits involve singular asymptotic series with distinct algebras of observables on either side of the singularity—a pattern discussed previously in Chapter 3. The algebra of observables for molecules contains both quantum and classical observables.[21] Observables point us to significant ontological differences between the quantum scale of particles and atoms, on the one hand, and the large-scale world of rocks, trees, and planets, on the other. Moreover, each of these limits points to physical transitions (e.g., the emergence of temperature, the disentanglement of nuclear and electronic frames of motion, etc.).

This implies that the relationship between quantum phenomena and classical phenomena is not one involving something like bridge laws relating the two domains as Nagelian and other forms of reduction would require (Chapter 3). For one thing, no such bridge laws survive across the singular limits separating classical and quantum domains. More importantly, there is no bridge law relationship between the states and observables at the microscale of QM, and the new states and observables at the macroscale relevant for molecules and larger structures. This implies that the new states and observables are neither a function of, nor "governed" by, QM's intrinsic states and observables—the same contextual emergence pattern we have met repeatedly.

[21] For details, see Bishop (2019) and Primas (1983).

Moreover, given the quantum domain, there are different classical worlds that result from taking these various limits in different orders. Taking these limits in different orders corresponds to different physical transitions happening in different orders (e.g., the different transitions leading to temperature, chemical potential, molecular shape, etc.). The resulting different classical domains are incompatible with each other, meaning that the limits—marking out actual physical transitions—must occur in a precise order to recover the macroscopic world of our experience from the "first-principles" microscopic scale of the quantum world. There are no resources in QM determining the sequence of transitions that must take place to get to the macroscopic world of our experience. That sequence of transitions was a matter of historical contingence.

The actual relationship between QM and CM, then, is much messier—and more interesting—than Michael Berry (1994, 2002) and Alisa Bokulich (2008) indicate. Like Batterman (2002), they focus on singular limits and the "no-man's land" of semi-classical mechanics. But all these approaches fail to deal with the sequence of transitions, referred to previously, that must take place to produce the macroscopic world.[22]

For those who have read the quantum foundations literature, note that this has nothing to do with the so-called basis problem—the problem of choosing the "correct" or "best" observables to solve the quantum measurement problem. The emergence of the classical world is not merely a matter of environmental decoherence, as Wojciech Zurek (1991), Roland Omnès (1999) and David Wallace (2012), among others, would have it. First, there is no context-free limit of infinitely many degrees of freedom. As in the previous case of temperature, this limit always has uncountably and infinitely many physically inequivalent representations making inescapable reference to classical macroscopic observables.[23]

Second, it is false that an improper mixture of quantum states "allows one to interpret the state of the [quantum] system in terms of a classical probability distribution," so that "it is useful to regard 'mixed states' as effectively classical," such that "one can interpret the system described by [a nonpure density operator] in terms of a classical 'mixture' with the exact state of the system unknown to the observer" (Zurek 1991, 46–47). Impure quantum states can be interpreted as classical mixtures if and only if their components are described by *disjoint states*. For a classical mixture of two pure states (e.g., water and oil), the pure states are disjoint if and only if

[22] It is unfortunate that Bokulich's book pits reductionism against pluralism in trying to seek a middle ground, as this leads her to roughly the same unification ideas as Batterman (not surprisingly, since she draws heavily on Batterman).

[23] For more discussion of distinctions between quantum and classical observables, see Bishop (2019).

there exists a classical observable such that the expectation values with respect to these states are different. It is this disjointness that makes it possible to distinguish states in a classical manner.

Moreover, the "approximately classical" or "quasi-classical" trajectories for quantum systems motivating such accounts of emergence of the classical macroscale world from QM only exhibit such quasi-classical behavior for limited times (except for overly idealized models) and under very special initial conditions (Pauli 1933, p. 166). In particular, this applies to ground states only (energy eigenstates never show classical behavior). Appeals to Ehrenfest's theorem would not help here, because all this theorem guarantees for such very special, short-lived dynamics is that the usual physics practice of averaging the values of the QM observables tends to wash out the errors or differences between the classical and quantum calculations for contextually relevant situations and times. It is neither necessary nor sufficient for classical behavior. For instance, applying Ehrenfest's theorem to the textbook problem of a quantum harmonic oscillator yields average quantities for the position and momentum that track with the classical quantities. Yet the quantum oscillator's discrete states yield thermodynamic properties very different from a classical oscillator. Thus, satisfying the theorem is insufficient to guarantee classical behavior.

Most discussions of decoherence try to treat the phenomenon as if it is wholly quantum (e.g., treating the environment as quantum but "for all practical purposes" as classical using some of the tricks just referred to). They presuppose that the kind of "entanglement" between QM systems and their environment is the quantum entanglement discussed in Section 4.7 (e.g., Wallace 2012; Zeh 2000). Yet as we shall see, classical observables associated with the macroscale play important roles in decoherence. Hence, there is a category mistake in decoherence approaches regarding the kinds of interactions involved (similarly for Everettian/many worlds interpretations). Failure to notice this difference probably comes from focusing abstractly on the QM formalism instead of investigating the detailed nature of the states and observables involved in concrete contexts as well as from assuming that 'fundamental physics' is causally closed.[24]

4.6 Contextual property emergence: molecular shape

One important example in this story of the relationship between QM and the macroscopic world is the emergence of molecular shape.[25] The received view, articulated by Paul Dirac (1929), is that chemistry has largely been reduced by QM:

[24] We will have more to say on causal closure in Chapter 7.
[25] For more details, see Bishop (2019, section 4.3).

> The underlying physical laws necessary for the mathematical theory of a large part of physics and the whole of chemistry are thus completely known, and the difficulty is only that the exact application of these laws leads to equations much too complicated to be soluble.
>
> (p. 714)[26]

It is because of these mathematical difficulties that we must adopt "approximate practical methods of applying quantum mechanics" (p. 714). By now, the reader should be alert that not everything that is called an "approximation" is what the literature claims it to be. Chemistry is no different. For example, Guy Woolley has observed that for chemistry "new concepts which cannot be inferred from the underlying 'fundamental' laws [of QM] are required" (Woolley 1976, 32).[27]

The plausibility of the reductionist story for molecular chemistry stems from what appears to be a straightforward recipe for how to use QM to "build" molecules and model their behaviors: Enumerate the relevant particles and their properties (nucleon, electron, charge, mass, etc.). Invoke the relevant "fundamental" physical interactions (electromagnetic, strong and weak-nuclear, etc.), and list the pairwise interactions among the relevant particles. Identify the kinetic and potential energy operators and add them to derive the Hamiltonian describing the physical system. Proceed to derive the properties and behaviors of the chemical system in question. This appears straightforwardly ontologically reductionistic, but this recipe almost immediately runs into the mathematical difficulties to which Dirac refers. But is this all just a problem for epistemic reduction?

Although many discussions of molecular properties in the philosophy literature on reductionism treat molecules as if they are isolated systems simply composed of protons, neutrons, and electrons, there is no such thing as an isolated molecule. All molecules are always coupled to the radiation field of the environment as a molecule + field system (what are known as "dressed states"). Interactions with electromagnetic and gravitational fields can never be fully screened from molecules. Chemists are always dealing with environmentally determined aspects of molecular structure, such as chirality in amino acids or drugs such as thalidomide.

Indeed, implicit in Erwin Schrödinger's work but made explicit in the famous Einstein, Podolsky, and Rosen (1935) paper, QM is a "holistic" theory, meaning that, traditionally understood, absent measurement and environmental effects, an isolated molecule is a whole lacking any identifiable parts. This is to say that, undisturbed, a molecule is not a separable

[26] A rather bold claim given that all Dirac had available to him as supporting evidence was the famous Heitler and London (1927) theory and the work of Born and Oppenheimer (1927). Not that such a thin basis seems to blunt the reductionistic attitude.

[27] See also Primas (1983).

system of nucleons and electrons. Quantum entanglement is present, and until it is broken or otherwise suppressed, there are no subsystems as individual entities.[28] Therefore, at least on most interpretations, isolated quantum wholes are not reducible to or explainable by their "parts," because there are no proper parts. As a quantum system, a molecule only has proper parts if its subsystems are in pure states.

As an example, consider a multielectronic atom in a pure state. While the whole is in a pure state, the electrons are in mixed states (describable by density operators), meaning that the electrons do not admit of a description as individual objects. Only the atom can be described as an individual. There is no meaning even to a proportion of pure electron states, since mixed states have no unique decomposition into pure states. If the electrons were in pure states, then the atom would be describable as a product state, indicating that the EPR correlations between the electrons and the nucleus have been suppressed.

Chemists never work with such "wholes" in their work with molecules. Given such basic facts, we should be prepared to see that the contextual emergence pattern recurs for molecular structure.

4.6.1 *The problem of molecular structure*

The history of molecular structure begins with the publication of Jacobus Henricus van't Hoff's booklet launching stereochemistry (1875). His proposal was that molecules' physical properties were due to their three-dimensional shapes (e.g., optical activity of molecules could be explained based on their three-dimensional structure). One remarkable thing about this suggestion is that molecular structure was the explanatorily relevant factor with no understanding of what the underlying physics might be. In other words, any underlying physical factors could be at play so long as the specific molecular shape was manifested at the scale of molecules. The "structural hypothesis" proved immensely fruitful, as it brought order and coherence to a large set of heretofore heterogeneous observations (Lewis 1923). Molecular structure is the explanation for chemical phenomena such as optical activity, single biomolecule spectroscopy, chemical and biological self-assembly, functioning of DNA, and disease mechanisms and medications. Chiral, or handed, molecules have structures that are not superposable as mirror images. For instance, (R)-bromochlorofluoromethane is not superposable on (S)-bromochlorofluoromethane. The list of actual-world effects of molecular shape could go on for the rest of this book.

Nevertheless, rather than being a good example of reductionism, molecular structure is a problem for QM. The structural formula that chemists write

[28] We have more to say about quantum entanglement in Section 4.6.3.

down for a molecule summarizes the relationship between the spatial organization of the constituent atoms and the kinds of chemical reactions in which the corresponding molecules may participate (e.g., reactions, synthesis, potential for new unexplored reactions and syntheses, etc.). In other words, the chemical properties of molecules exist in the context of the network of relations in a context. At this point, QM runs into a problem.

The quantum Hamiltonian, constructed by the "first principles" approach outlined previously, is as context-free as possible (ACFAP). However, the eigenstates and observables of the ACFAP quantum Hamiltonian evidently do not correspond to molecules with structure:

1. The structure of molecules is associated with their electric dipoles which are crucial to distinguishing polar vs. non-polar molecules and their behaviors. Yet the ACFAP quantum Hamiltonian commutes with the inversion operator, meaning that it has parity eigenstates, so the expectation values for the static electric dipole operator—the quantum observable corresponding to electric dipoles—are zero.
2. Molecular structure is the basis for stereoisomers which exhibit optical activity such as the rotation of light. Yet, the ACFAP quantum Hamiltonian cannot account for the optical rotation angle of stereoisomers, because this is a pseudoscalar.
3. Chemists work with interesting molecules such as hydroxyl hydrogens, methyl hydrogens, and aromatic hydrogens. Yet the ACFAP quantum Hamiltonian exhibits a permutation symmetry for the constituents of the system it describes. Identical atoms can play distinguishable roles, but the permutation symmetry of atoms often is not preserved in molecular structures. Hence, the ACFAP quantum Hamiltonian cannot distinguish workhorse molecules of the chemist.
4. The handedness of chiral molecules is part of the everyday work world of chemists, as well as making life possible through DNA. Yet the ACFAP quantum Hamiltonian has no eigenstates corresponding to chirality.
5. Isolated improperly "chiral" molecules, such as ammonia, have two distinct pyramidal states and superpositions of those distinct states form the ground state and first excited state.[29] Both pyramidal states are eigenstates of the ACFAP quantum Hamiltonian. Nevertheless, neither eigenstate has a nuclear frame, meaning neither represents a molecule with structure. No nuclear frame means no molecular structure, since there is no fixed position for nuclei relative to electrons.
6. Many of these problems are related to the fact that the ACFAP quantum Hamiltonian has too many symmetries missing from actual-world molecules (e.g., nuclear permutation and rotational symmetries, spherical and inversion symmetries).

[29] We will have more to say about superpositions in Section 4.7.

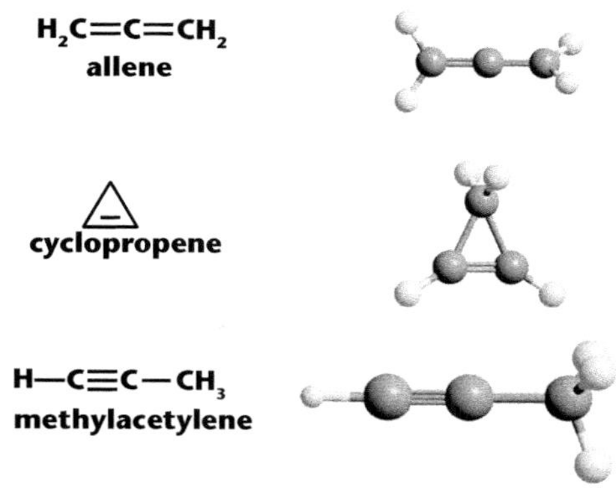

Figure 4.1 Chemical representations of allene, cyclopropene, and methylacetylene (courtesy of Timothy Wilkinson).

Chemical isomers provide the most striking examples of the failure of the ACFAP quantum Hamiltonian. Isomers are molecules having identical chemical formulas, but different chemical properties due to their differing structural arrangements. An oft-discussed example is C_3H_4, which is the formula for three distinct chemical compounds: allene, cyclopropane, and methylacetylene (Fig. 4.1).

The ACFAP quantum Hamiltonian cannot distinguish these three isomers—indeed, any chemical isomers—for a fundamental reason: it only accounts for the *total number* of nucleons and electrons and their Coulomb forces, *not their specific spatial arrangements*, because the electrons and nuclei are in entangled states and the effects of contexts are missing.[30]

Some, such as Hinne Hettema (2017), have argued for a deflation of the problem of molecular shape by denying its "significant relevance for chemistry" (2017, p. 67).[31] Yet many chemical properties depend crucially on molecular shape, and such structure plays crucial roles in chemistry. For instance, 2-butene is a stereoisomer, meaning that its two forms have the same chemical formula, but different spatial orientations that cannot be related by translation or rotational transformations. The two forms have different properties, such as different boiling points and densities.

One might try to construct quantum counterparts to the classical shape observables by taking the interparticle distances and using the symmetric sum over them to produce an observable that is represented by a self-adjoint

[30] The shortcomings become worse if two hydrogen atoms are replaced by two deuterium atoms (which chemists have done). This generates seven distinct species of isomer, two of which are optically active (which chemists have synthesized in the laboratory). Naturally, the ACFAP quantum realm contains no resources specifying how such replacements can be accomplished, nor how to synthesize the compounds.

[31] Hettema claims, regarding molecular structure: "All that is required for reduction, indeed, all that scientists happily accept as an explanation in most cases, is a Born–Oppenheimer Hamiltonian and some form of bond topology as chemical structure" (2017, p. 67), without realizing that this is to adopt stability conditions not given at the microscale of QM.

operator on the Hilbert space of functions.[32] The idea would be to try to create quantum observables corresponding to the geometry of each distinctive isomer. Yet as is well known, this procedure leads to problems with defining molecular structure. Consider the case of C_8H_8. The expectation values of these quantum counterpart observables imply that all the carbon–hydrogen interparticle distances are the same on average, meaning that no structural interpretation can be produced. The expectation values of what appear to be appropriate spatial observables do not lead to molecular structure! Therefore, the only quantum observable that can be formed in an ACFAP approach corresponding to classical observables associated with molecular structure yields unphysical results.[33]

4.6.2 *Ontological emergence of molecular shape*

If molecular shape is not accounted for by the microscale resources of ACFAP QM, then where does it come from? The first precise definition of molecular structure was given by Max Born and Robert Oppenheimer (1927) through their famous procedure. Although often mischaracterized as an "approximation," this procedure was one of the first clues that there was a distinct difference between the states and observables of the microscale quantum realm and that of "chemists" molecules.[34] The rough idea is to treat the nucleus as very much more massive than the electron mass, and then consider the nucleus as if it is (almost) stationary with respect to electronic motions. Mathematically, this corresponds to an asymptotic series expansion in powers of a parameter $\varepsilon = (\text{electron mass/nuclear mass})^{1/4}$, where $\varepsilon \rightarrow 0$ as the nuclear mass becomes indefinitely large with respect to the electron mass.

Slightly more accurate results can be achieved with other adiabatic procedures, where the electrons are considered to move much faster than the heavier nuclear frame of a molecule. All these adiabatic procedures are ways of implementing something in the direction of the clamped-nucleus assumption; these are the only approaches under which the singular limits dividing the quantum states and observables from chemistry states and observables can be regularized. Moreover, assuming the nucleus is clamped is the only assumption that leads to self-adjoint Hamiltonians for quantum chemistry (Sutcliffe and Woolley 2012). The clamped nucleus represents a context where:

[32] Observables in quantum mechanics are required to be self-adjoint, and this is the only such self-adjoint operator that can be constructed from interparticle distances.

[33] Attempts involving the charge density functions fair no better for capturing molecular structure (Sutcliffe and Woolley 2012, pp. 388–426).

[34] Recall the discussion of idealization vs. abstraction in Chapter 3, Section 3.2.4.

1. The nuclear and electronic motions are distinguished, so entanglements between the nucleus and electrons are broken.
2. The nuclear permutation, rotational, and translational symmetries of the microscale QM domain simpliciter are broken.
3. The electrons composing molecules are distinguishable.

These are all properties of actual-world molecules and are absent from the microscale ACFAP quantum domain. Chirality and other features of molecular structure as classical observables exist in the larger-scale domain of chemistry (Amann 1993; Bishop 2019; Primas 1983, pp. 335–341). Indeed, all the ACFAP Hamiltonians and their quantum chemistry counterparts typically yield is energy-level structures and spectra for translational, vibrational, and rotational phenomena, not molecular structure.

The regularization of the singular limits tracks with ontological changes that occur, such as the broken entanglements among nuclei and electrons and broken symmetries, leading to distinguishable particles as well as new states and observables. The clamped nuclei condition (or the $\varepsilon \to 0$ condition of Born and Oppenheimer) is a stability condition representing the physical effects of the nucleus being significantly more massive than the electrons in an environment where molecules are never isolated. This condition leads to distinguishing nuclear and electronic frames in the quantum chemistry context containing the relevant states and observables missing from the microscale quantum domain. For instance, there are new observables corresponding to molecular shape, the electric dipole moment, and optical rotation angle while states corresponding to individual atoms vanish and are replaced by states corresponding to molecules. We have a substantial change in ontology, including fundamental physical properties absent from the microscale QM domain alone.[35]

The stability condition of massive nuclei (leading to electronic motion much faster than the nucleus) defines a context in a radiation field (more on the field in Section 4.6.3) where:

1. Quantum entanglement between nucleons and electrons is broken, distinguishing nuclear and electronic motions.
2. The breaking of nuclear permutation, rotational, translational, among other symmetries of the microscale quantum domain.
3. Electronic states in molecules are distinguishable.
4. Nuclear and atomic orbital equivalence classes (Kryachko and Ludña 1990) are induced which play important roles in nuclear magnetic resonance and electron spin resonance spectroscopy (e.g., Balasubramanian 1991, 1992). These equivalence classes are distinguished in

[35] For technical details on the observables, see Bishop (2019, section 4.4.3).

terms of the new states and observables of the larger-scale chemical context.

In the macroscale chemical context, the new states and observables yield behavior where the nuclear frame motion is roughly Newtonian. There are two important correction factors keeping nuclear frame behavior from being fully classical. First, the electrons contribute a small stochastic feedback to nuclear motion. Second, the algebra of observables for the nuclear frame is a mixed quantum/classical algebra, meaning it contains both classical and quantum observables, capturing both the microscale and macroscale. Hence, the position and momentum of the nucleus fulfills a stochastic version of Heisenberg's inequality. These fluctuations are much more rapid than the collective translational, vibrational, and rotational motions of the nuclei.

4.6.3 *The molecular environment*

As we have stressed in the discussion, molecules are not isolated as the AC-FAP quantum Hamiltonian represents them. The environment does play an important background role in the emergence of molecular structure—a role that is underappreciated in the reductionism literature. One reason why this role is neglected is that it is typically chalked up to decoherence (e.g., Zeh 2000, pp. 27–29). But things are much more subtle than this easy reductionist dismissal. The environment involves heat baths with contextually emergent properties such as temperature, chemical potential, and pressure, as we saw previously. The KMS stability condition leads to chemical species in different superselection sectors, meaning that molecules have classical properties rather than being wholly quantum-mechanical.

More importantly, the electromagnetic field in quantum electrodynamics (QED) has uncountably many superselection sectors that arise because of the degrees of freedom induced by the far-field consistency condition for the electromagnetic field—a stability condition inducing classical states and observables in the electromagnetic field (e.g., Buchholz 1982; Gervais and Zwanziger 1980). One reason why this gets ignored is that in typical field-theoretic treatments of QED the most convenient subset of states is selected while the abundance of superselection sectors is ignored. But physically, the far-field stability condition structures the electromagnetic field and forces yielding equivalence classes of classical states in QED. Note that the far-field consistency stability condition exists at the largest length scales of the electromagnetic field, so guaranteeing relevant states and observables for microscopic and larger scales. Just as with the KMS condition, the far-field consistency global acausal stability condition.

Consider electrons with speeds small with respect to the speed of light. They will eventually enter a forward light cone C_+ if they are not annihilated. Yet the asymptotic electric flux-distribution in C_+ is not determinant, since photons coming from the remote past cannot be observed in that cone even though they affect the flux distribution in general and in C_+ in particular. This implies that the electric charge of a state is determinable in a light cone even though its sector is not, yielding equivalence classes of the states and observables in C_+ having the same charge state but arbitrary asymptotic flux distributions. These equivalence classes are disjoint, so charge is a dispersion-free, classical observable. States can have the same charge while belonging to different charge equivalence classes differing by radiation fields. Hence, the radiation field molecules are always interacting, which provides a classical background for their classical macroscopic behavior.

The implication is that molecules are always coupled to some classical states in the environment, and these states are not a result of the process of environmental decoherence. Decoherence presupposes a richer context than is usually assumed (e.g., Joos and Zeh 1985; Omnès 1999; Wallace 2012; Zeh 2000; Zurek 1991). Robust molecular structure cannot be dismissed as an effect of environmental decoherence.[36]

4.6.4 *Molecular structure and the contextual emergence pattern*

When the details are made explicit for molecular shape, we see the contextual emergence pattern again. While the microscale quantum laws and properties contribute some of the necessary conditions for molecular structure, it is stability conditions at the macroscale molecular context that provide the rest of the necessary and sufficient conditions for the molecules of the chemist's laboratory and living systems. The ACFAP quantum Hamiltonian captures a very wide possibility space but encompasses far more possibilities than the actual molecules of the macroscopic world. It is the stability conditions of the macroscopic world that provide the accessibility conditions for actualizations for the chemical isomers of C_3H_4 or DNA.

The context into which the laws and properties of QM come to expression is crucial to the kind of macroscopic world we experience. Yet we also see that, as with thermodynamics and Rayleigh–Bénard convection, the macroscopic context of molecules does not violate any of the basic laws and properties of QM; instead, what we observe is multiscale stability conditions making classical molecules possible where the constraints provided by that molecular context channel or shape possibilities provided by the microscopic quantum

[36] Indeed, environmental decoherence itself should be thought of as an example of contextual emergence (Bishop 2019, section 6.2).

context. Molecular structure is a compelling example of the order and unity of the contextual emergence pattern uncovered by the sciences.

4.7 Quantum superpositions, entanglement, and contextual emergence

The history of QM is the history of multiscale context. From Planck to Bohr to Einstein, to Schrödinger to Heisenberg to von Neumann to Wheeler and on, the large-scale context of the macroscopic world has been inseparable from QM. This historical fact has long been illustrated by the twin-slit experiment.

When a "momentum measurement" is made with both slits open, the quantum system in question, electrons say, behaves in a wave-like fashion (though the interference pattern is made up of discrete points and can be built up by "firing" one electron at a time from the "source"). When a "position measurement" is made with one or the other slit closed, the quantum system behaves in a particle-like fashion. Although we call this wave–particle duality, such a phrase is merely descriptive not explanatory. Quantum textbooks state that an electron with a definite momentum is in a superposition of various position states (and vice versa). Whether the electron has definite momentum or definite position is determined by the experimental context, meaning that the slit configuration is a macroscopic stability condition determining whether wave-like or particle-like behavior manifests at the microscopic scale.

If one places detectors at the slits to observe which slit individual electrons supposedly pass through or attempts to obtain which-way information, the experiment yields particle-like behavior even though the twin-slit condition normally yields wave-like behavior. Hence, the presence or absence of detectors at the two slits also functions as a macroscopic stability condition determining whether wave-like or particle-like behavior manifests at the microscopic scale. Indeed, merely threatening to obtain which-way information is often enough to destroy the interference pattern.

Although these stability conditions are clearly present in the laboratory, most discussions of twin-slit experiments focus on how wave–particle duality is a manifestation of *quantum superposition*: Quantum systems are allegedly "existing in multiple states simultaneously," whatever that means. Admittedly, this counter-intuitive claim (itself an interpretation) typically captures everyone's interest. Nonetheless, we should not let that obscure the fact that there are macroscale stability conditions for when wave-like or particle-like behavior manifests at the microscale. As Bohr, Wheeler, and so many others have emphasized, large-scale macroscopic conditions (e.g., "classical objects" such as slits and detectors) are crucial to creating and determining the behavior of quantum objects.[37] This is an experimental fact that exists independently of any particular interpretation or physical model of what is happening, such as Bohmian mechanics. Interpretations aside, the type of measurement being made (e.g., no slits open, versus one slit open, versus two slits open) makes all the difference to the outcomes and the behavior of the microscale quantum behavior.

Take the following descriptions of the relationship between quantum and classical mechanics:

> Thus quantum mechanics occupies a very unusual place among physical theories: it contains classical mechanics as a limiting case, yet at the same time it requires this limiting case for its own formulation.
>
> (Landau and Lifshitz 1977, p. 3)

> Another unappealing aspect of quantum mechanics is that by referring to measurements, the axioms assume the existence of macroscopic objects—detectors, computers, brains, et cetera—and this is a blow to reductionism. A fundamental theory should explain the emergence of the macroscopic world and not assume its presence in the axioms ... Quantum field theories inherit the problems from quantum mechanics. The standard model, therefore, has the same difficulty with explaining the macroscopic world.
>
> (Hossenfelder 2018, p. 122)

As these quotations suggest, we have laws for classical mechanics and postulates for quantum mechanics, leaving us with two theories that are codependent. The most common reaction to this state of affairs is that we are missing an interpretation of quantum mechanics or some theory "fundamental" to classical and quantum mechanics, that would explain this codependency; i.e., the common reaction is to seek a reductionist explanation. As Weinberg put it:

> You can very well understand quantum mechanics in terms of an interaction of the system you're studying with an external environment which includes an observer, but this involves a quantum mechanical system interacting with a macroscopic system that produces the decoherence between different branches of the initial wavefunction. And where does that come from? That should be described also quantum mechanically. And, strictly speaking, within quantum mechanics itself there is no decoherence.
>
> (Weinberg 2013, pp. 126–127)

Unsurprisingly for us, this state of affairs is evidence that contextual emergence is the nature of reality itself. (See also Aharonov, Cohen, and Tollaksen 2018, pp. 11,730–11,735, for a similar suggestion). As Guido Bacciagaluppi puts it: "On the other hand, Bohr's intuition that quantum mechanics as

[37] We will discuss the ontological implications of this more in Chapter 7.

practiced requires a classical domain would in fact be confirmed by decoherence, if it turns out that decoherence is indeed the basis for the phenomenology of quantum mechanics" (2020). We agree completely, so long as environmental decoherence is understood in terms of contextual emergence (more on this subsequently).

4.7.1 *Quantum entanglement and nonlocality*

To this point we have been discussing the behavior of a single quantum system. When one quantum system interacts with another one in a particular way, what obtains is "quantum entanglement." In entangled states, particle 1 has no measurable property alone, and particle 2 has no measurable property alone—there is no such thing as individual particles having any definite values, as illustrated previously by the ACFAP Hamiltonians for molecules. Only the two-particle system or entangled state can have a definite value of any measurable property. To get slightly more technical, a set of entangled quantum systems compose a system whose quantum state is represented quantum-mechanically by a tensor-product state-vector which does not factorize into a vector in the Hilbert space of each individual system. Entangled states, then, are not Cartesian products of their components, and thereby appear to violate, among other things, classical probability theory, as with the so-called Einstein–Podesky–Rosen (EPR) correlations.[38]

Here is where quantum nonlocality as such comes into the picture. A measurement on one particle in a superposition may "affect" the state of another particle in the superposition instantaneously (measurement outcomes), even if the measurements are spacelike separated, meaning that the two events—the two correlated measurement outcomes—would require a signal traveling faster than the speed of light between them if there is a causal connection. Hence, the intrigue raised by spacelike-separated measurements of entangled particles wherein upon the measurement of one particle property, the other entangled particle instantaneously registers an almost perfectly correlated measurement outcome (or anticorrelated depending on the property in question). This, at least, violates the spirit of special relativity (SR). For example, if the explanation for such EPR correlations were to be causal, in a relativistic setting there would be no frame-independent fact of the matter about which particle outcome were measured first and which second. That is, there would be no frame-independent fact about what was cause and what was effect.

[38] Though some interpretations of QM might dispute this supposed violation.

Quantum entanglement contrasts with the behavior of linear compound systems in classical physics. The latter consist of separable, distinct constituents that interact by means of forces encoded in the Hamiltonian function of the overall system.[39] In principle, if the full Hamiltonian is known, maximal knowledge of the values of the physical observables pertaining to each one of these constituents yields an exhaustive knowledge of the whole compound system. Linear classical systems appear to obey the following compositionality principle:

> **Compositionality Principle** The states of any spatiotemporally separated subsystems S_1, S_2, ..., S_N of a compound system S are individually well defined, and the states of the compound system are wholly and completely determined by them and their physical interactions including their spatiotemporal relations.

Many people have argued that quantum entanglement entails, in one way or another, a failure of this compositionality principle *even though QM deals with linear systems.* To examine this failure we shall be concerned with the following questions. Does quantum entanglement imply that wholes such as entangled states have properties that are not determined by the properties (whether intrinsic or relational) of the particles that enter into such states? That is, does quantum entanglement violate the compositional and realization picture of the physical world, the reductive atomism (Chapter 2, Section 2.4.3) so widely assumed in philosophical and scientific discussions of reduction? More specifically, does quantum entanglement tell against either or both of the following principles?:

- **Separability principle** Any two systems A and B, regardless of the history of their interactions, separated by a non-null spatiotemporal interval have their own independent real states such that the joint state is completely determined by the independent states.
- **Locality principle** Any two spacelike separated systems A and B are such that the separate real state of A, let us say, cannot be influenced by events in the neighborhood of B.

Western metaphysicians have often opted for atomistic or Lego-like conceptions of reality. Consider two such philosophical statements of classic modern atomism, the first from David Lewis:

> Humean supervenience is ... the doctrine that all there is to the world is a vast mosaic of local matters of fact, just one little thing and then another ... We have geometry: a system of external relations of spatiotemporal distance between points. Maybe points of spacetime itself, maybe point-sized bits of matter

[39] This notion of separability breaks down for systems where nonlinearities are relevant, as in Rayleigh-Bénard convection.

> or aether fields, maybe both. And at those points we have local qualities: perfectly natural intrinsic properties which need nothing bigger than a point at which to be instantiated. For short: we have an arrangement of qualities. And that is all. All else supervenes on that.
>
> (Lewis 1986, p. ix)

Humean supervenience says that the complete physical state of the world is determined by (i.e., supervenes on) the intrinsic physical state of each spacetime point or each point-like entity such as classical particles and the spatiotemporal relations between those points. Thus, all "fundamental" properties are local properties and spatiotemporal relations are the only "fundamental" external physical relations. Humean supervenience is an affirmation of both the locality and separability principles given previously. Yet note that nonlinear systems, such as Rayleigh–Bénard convection, violate such supervenience by violating the separability principle.

The second statement is from Jaegwon Kim:

> Mereological supervenience: systems with an identical total microstructural property have all other properties in common. Equivalently, all properties of a physical system supervene on, or are determined by, its total microstructural property.
>
> (Kim 1999, p. 7).

The idea behind mereological supervenience is that the intrinsic properties of the basic parts and/or their relations determine (e.g., realize) the properties of wholes. Kim's principle also adds a modal claim involving microphysically identical systems. As with Humean supervenience, note that Rayleigh–Bénard convection violates mereological supervenience by violating the separability principle.

There is little doubt that, on almost every interpretation (see later), quantum entanglement tells against Humean supervenience and mereological supervenience in some essential way, as many of us have argued in recent years.[40] Quantum-entangled states are neither formally (i.e., the non-factorizability of such states) nor empirically (i.e., experimentally confirmed distinct quantum probabilities for outcomes in EPR-correlations) a function of the properties of the particles that "make them up" (Hawthorne and Silberstein 1995; Maudlin 2007; Silberstein 1999 and 2002). It is agreed that quantum entanglement tells against separability and/or locality depending on one's interpretation of quantum mechanics, or some other highly intuitive principles, as we discuss subsequently. Einstein was most concerned about this feature of QM, as the following letter to Max Born makes clear:

[40] For details, see Silberstein (2017) and Silberstein, Stuckey, and McDevitt (2018).

> If one asks what, irrespective of quantum mechanics, is characteristic of the world of ideas of physics, one is first of all struck by the following: the concepts of physics relate to a real outside world, that is, ideas are established relating to things such as bodies, fields, etc., which claim "real existence" that is independent of the perceiving subject-ideas which, on the other hand, have been brought into as secure a relationship as possible with the sense-data. It is further characteristic of these physical objects that they are thought of as arranged in a space-time continuum. An essential aspect of this arrangement of things in physics is that they lay claim, at a certain time, to an existence independent of one another, provided these objects "are situated in different parts of space." Unless one makes this kind of assumption about the independence of the existence (the "being-thus") of objects which are far apart from one another in space—which stems in the first place from everyday thinking—physical thinking in the familiar sense would not be possible. It is also hard to see any way of formulating and testing the laws of physics unless one makes a clear distinction of this kind. This principle has been carried to extremes in the field theory by localizing the elementary objects on which it is based and which exist independently of each other, as well as the elementary laws which have been postulated for it, in the infinitely small (four-dimensional) elements of space. The following idea characterizes the relative independence of objects far apart in space (A and B): external influence on A has no direct influence on B; this is known as the "principle of contiguity," which is used consistently in the field theory. If this axiom were to be completely abolished, the idea of the existence of (quasi-) enclosed systems, and thereby the postulation of laws which can be checked empirically in the accepted sense, would become impossible
>
> (Born 1971, pp. 170–171).

Let us take a moment to appreciate in more detail why QM is believed to have these anti-atomistic implications. Recall that the British emergentists maintained that resultant properties share an additive parts/whole relationship, where the property of the aggregate is simply the sum of properties of its constituents. In contrast, emergent properties are non-additive. For instance, four hydrogen atoms combine to form a helium atom in the sun. The mass of the helium atom is non-additive or emergent, with respect to the masses of the hydrogen atoms, according to Newtonian mechanics. However, according to SR, additivity is restored, insofar as mass and energy are interchangeable (the mass deficit of the helium atom is converted to energy in the fusion of the hydrogen atoms).

Most consider the traditional distinction along the lines of additivity and non-additivity to be at best oversimplifying, if not outright misleading. For example, Fred Kronz and Justin Tiehen (2002) state:

> [It] appears that a central claim of the British empiricists, that additivity is the mark of resultant (i.e., non-emergent) properties, is wrong ... the mark

> of a non-emergent property of composite systems in quantum mechanics crucially involves a *multiplicative* operation ... [However] [t]he situation is different for evolution. A non-separable evolution is a *product* rather than a superposition [i.e., addition] ...This may provide a way to partially vindicate the British emergentists.
>
> (2002, p. 333)

To illustrate this point in greater detail, Kronz and Tiehen examine quantum entanglement in the case of a triplet state or three spin-1/2 systems. "Entangled" states (a term first coined by Schrödinger in the 1920s) are quantum phenomena arising out of combinations of addition and multiplication of state vectors in the formalism of QM. Contrast this with CM, where states are represented by points and not vectors; talk of adding or multiplying points is nonsensical. QM has cases in which a state description or an observable quantity cannot be "factored" into the product of simpler constituents thus violating the compositionality principle. Such states are referred to as "entangled": If no such factoring can take place, then no information can be extracted concerning the properties of the system's basic constituents. In such cases one can only have information concerning the properties of a whole system, not of its basic parts. In contrast to CM, QM has a non-Boolean logical structure, and quantum properties have a noncommutative algebraic structure (e.g., Primas 1983). Such entangled states exhibit stronger degrees of correlation than the rules of classical probability theory allow, as delimited by the Bell Inequalities. Experimentally confirmed EPR correlations tell us that such correlations apparently do not drop off with distance, obtaining even at spacelike separation. This is the phenomena of *quantum nonlocality*.

More formally, in the case of QM, the three state spaces of particles 1, 2, 3 (described respectively by the two-dimensional spinor spaces (H_1,H_2,H_3)) combine to form a tensor product eight-dimensional composite space: $H_1 \otimes H_2 \otimes H_3$.[41] Similarly, the system Hamiltonian H_S combines via the rules of tensor product and superposition. Now, in principle, H_S can evolve in time to become fully entangled; that is to say, the (eight-dimensional) matrix representing H_S cannot be factored into the (eight-dimensional) representations of the Hamiltonian matrices representing particles 1, 2, 3 respectively (represented accordingly by $\widehat{H}_1 = \widehat{h}_1 \otimes I_{2+3}$, $\widehat{H}_2 = I_1 \otimes \widehat{h}_2 \otimes I_3$, $\widehat{H}_3 = I_{1+2} \otimes \widehat{h}_3$). In other words, in such a case there is no factorization which would allow a state such that $\widehat{H}_s = \widehat{H}_1 \otimes \widehat{H}_2\, \widehat{H}_3$ Other possibilities include H_S evolving into a superposition of partially entangled mixed doublet states, with respect to, say, systems 1 and 2: $\widehat{H}_s = \widehat{h}_{1+2} \otimes I_3 + \otimes I_{1+2} \otimes \widehat{h}_3$. Finally, the system Hamiltonian can evolve into a superposition of (fully non-entangled) Hamiltonians: $H_S = H_1 + H_2 + H_3$.

[41] This contrasts with the direct sum $\oplus$ in the case of CM.

Again, these facts about entangled particles obtain even at spacelike separation! Although each particle in our entangled system presumably occupies a region disjoint from the other, it is not the case that each has its own intrinsic spin state. Nor is it the case that the spin state of the composite system is determined by the states of the particles taken individually, together with the spatiotemporal relations between them. No "pure" state for a single particle yields the same predictions as the "singlet" state. Were one to ascribe a pure state to each of the electrons, their joint state would be a product state rather than an entangled state. The joint state of the pair cannot be analyzed into pure states for each of the components. There is very little disagreement about anything said so far, but there are several differences of opinion about exactly how to interpret quantum entanglement, how to model it physically, and what it implies (Lewis 2016). We will return to this discussion shortly, but first we detour through the so-called no-go theorems.

4.7.2 *The no-go theorems of Bell and Kochen–Specker*

Bell's theorem and the Kochen–Specker theorem are the two most famous no-go theorems in QM. First we will deal with Bell's theorem. As is well known, Einstein and many of his colleagues were concerned about the "completeness" of QM. The theory seemed to entail that to make the right predictions about the behavior of entangled states at spacelike separation, not every property of entangled particles, such as spin, can have definite values prior to measurement. In 1935 the famous EPR paper was published arguing for the incompleteness of QM and the need to supplement it with "hidden variables." "Hidden" means adding definite properties not given by textbook theory. EPR argued that the quantum system must have pre-existing definite properties corresponding to any measurement whatsoever that someone might perform on the system in question. This classical assumption is sometimes known as "realism."

Given two explicit assumptions, what Bell's (1964) theorem shows is that no assignment of spin properties, say, to the two particles can reproduce the measurement results we observe that are correctly predicted by QM. Hence this result is known as a no-go argument against any hidden-variables interpretation of QM.

Below we follow Peter Lewis's (2016) presentation as inspired by (Mermin 1981). Nevertheless, both the explicit (classical/macroscale) assumptions that Bell made can be questioned, and they are as follows:

- **Independence assumption** Measurements performed on a quantum system in the future cannot affect the current state of its physical properties. For example, the spin properties of electrons now are completely independent of the measurements to be performed on them in the future.

- **Locality assumption** Measurements performed on one particle at spacelike separation from another cannot influence the state of the other particle, regardless of what QM seems to imply. For instance, the spin properties of entangled electrons at spacelike separation cannot affect one another.

The idea here is straightforward. The assumptions are that the properties of particles "now" cannot be affected by future measurements or by events at spacelike separation "now."[42] This is exactly what the light-cone structure of SR would seem to suggest. The point, however, is that one can devise a hidden-variable account of QM if one is willing to reject one or either of the independence and locality assumptions. Bohmian mechanics and spontaneous collapse theories both reject the locality assumption, which raises issues with SR.[43]

The Kochen–Specker theorem (Held 2013), another no-go theorem against hidden variables, is formally non-trivial. In a nutshell, it shows that it is impossible that every property for which QM makes predictions can simultaneously be ascribed a definite value. Any attempt to do so results in ascribing contradictory physical properties to the system and does not recover the predictions of QM. The Kochen–Specker theorem also relies on the independence assumption. While there are ways around this no-go result as well (see later), the consensus regarding the implication of the Kochen–Specker theorem is as follows:

> Independence is essentially an assumption that the physical properties in question are non-contextual ... choosing what measurement to perform on a system is tantamount to choosing the physical environment of the system. So if physical properties depend on the physical context of a system as well as its intrinsic state, then they can depend on the measurement performed on the system. Taken at face value, the lesson of the Kochen–Specker theorem seems to be that if there are any physical properties responsible for the measurement results predicted by quantum mechanics, they must be contextual properties.
>
> (Lewis 2016, p. 44)

Just as we saw with the twin-slit experiment, the no-go theorems imply that the measured properties of quantum-entangled systems are inescapably macroscale-context-dependent. This context dependence threatens either locality and/or independence: The behavior of quantum systems, even given

[42] This claim must be nuanced when dealing with spacetime, given the relativity of simultaneity.

[43] For discussion and references, see Silberstein 2009a.

definite properties prior to measurement, is going to be determined partly by events at spacelike separation or measurement settings in the future! For instance, while it is true that Bohmian particles have definite position values prior to measurement, those values are given partly by other events at spacelike separation. In Bohmian mechanics, the behavior of the particle depends on the positions of all the particles in the system at the time in question (i.e., the other particles with which any given particle is entangled), even at spacelike separation. Furthermore, at least as David Bohm and Basil Hiley tell the story, Bohm's interpretation also entails that the macroscale measurement setup determines the context:

> What all this means is that quantum properties cannot be said to belong to the observed system alone and, more generally, that such properties have no meaning apart from the total context which is relevant in any particular situation. (In this case, this includes the overall experimental arrangement so that we can say that measurement is context dependent.) The above is indeed a consequence of the fact that quantum processes are irreducibly participatory.
>
> (Bohm and Hiley 1993, p. 108)

As we have said, another way to attempt a local hidden-variable theory is to reject the independence assumption. One class of such interpretations is the retrocausal interpretations of QM. The Kochen–Specker theorem can be avoided by giving up the assumption that the properties of a quantum system now are statistically independent of any measurements that will be made in the future. The claim, then, is that future events can causally/dynamically influence earlier events, hence the name "retrocausal" (Cramer 1986; Price 1996). Presumably, then, if we are willing to accept retrocausation—"backwards causation"—then there is no formal reason why we cannot have a hidden-variable theory in which every determinable property of a quantum system always has a determinate value. But again, the price of the said properties with definite values before, during, and after measurement is that the values of those properties are not in any way intrinsic; rather, the macroscopic context for the quantum system now includes future measurement settings. Such properties might always be definite but clearly are relational.

Therefore, whether it is the macroscopic measurement setup in the present or future, or physical events at spacelike separation, explaining quantum weirdness inescapably involves stability conditions at the macroscale. The only other non-instrumentalist interpretation to preserve both locality and independence, the Many-Worlds interpretation, forces us to accept that "measurements" have no unique outcomes! This is quite a blow to dynamical explanation as ordinarily conceived: The past (indeterministic) state of a system plus a dynamical law gives rise to a unique outcome rather than multiple branches of the supposed universal wavefunction. And of course, such interpretations rely heavily on environmental decoherence as the mechanism for why branching of the universal wavefunction happens at all.

As Peter Lewis notes, many interpretations of QM, such as by Bohm and that of Girardi, Rimini and Weber (GRW) are both nonlocal and irreducibly "holistic"—dependent on macroscale context:

> So the fact that entanglement holds between a specific pair of particles means that non-local theories still need emergent properties. Bohm's theory and the GRW theory are non-local, but non-locality cannot recover Humean supervenience. In Bohm's theory, the entangled state cannot be fully captured by an ascription of non-relational properties to the two particles involved, or to the non-relational properties of the two regions of space occupied by their respective wave packets. In the GRW theory, the entangled state cannot be fully captured by an ascription of non-relational properties to the two regions of space occupied by the particle 1 and particle 2 wave packets. In each case, specifying the properties of the parts doesn't suffice to specify all the properties of the system: we also need to specify irreducible properties of the whole.
>
> (Lewis 2016, p. 197)

Aside from all the interpretations of QM regarding quantum superposition and entanglement, there is also the ontological contextual emergence in environmental decoherence referred to previously. Decoherence is a phenomenon for which every interpretation must account. Roughly, when quantum systems interact even slightly with the environment they very quickly start behaving classically for all practical purposes:

> The lesson is that if you want to observe interference effects, you need to keep the system in question isolated from outside influences, since even an interaction with a single particle can destroy the effect. This is difficult for microscopic systems, but effectively impossible for macroscopic ones: such systems will inevitably interact with photons, air molecules, and other aspects of their environment in complicated and untraceable ways ... if the state of a macroscopic system comes to have two components, those components will not exhibit any appreciable interference effects. And since interference is just a name for interaction between two wave components, this means that for all practical purposes the two components do not interact. They can be regarded as separate branches of physical reality.
>
> (Lewis 2016, p. 109)

The ubiquity of environmental decoherence phenomena for complex systems on relevantly long time scales means that one may, for all practical purposes (FAPP), treat the quantum system as classical. That is, the interaction of a quantum system with its environment entangles the two and distributes the quantum coherence over so many degrees of freedom as to render it unobservable, on the usual account. An intuitive picture of the interaction between system and environment can be provided by the analogy with a measurement interaction. The environment is monitoring the system; it is spontaneously "performing a measurement" (rather, it is letting the system undergo an interaction as in a measurement) of the preferred states, but it is not actually an irreversible measurement of course. The point is environmental decoherence alone, and however you explain or characterize it, it is an example of contextual emergence.

4.7.3 *Quantum mechanics and contextual emergence*

Stepping back, we return to the interpretative question regarding quantum superposition, entanglement, and non-locality. It seems very clear that on almost any account, quantum entanglement violates the compositionality principle and realization, where the properties of parts determine the properties of wholes (whether that be functional, token–token, mechanistic, or some other account). But beyond that, what one says about such cases will depend on various metaphysical conceptions of properties, states, systems, laws, and entities, as well as various explanatory assumptions or preferences.[44]

Consider a system of two particles. The fixing of the state of both particles upon measurement is not a reflection of epistemic ignorance. As we have seen, EPR correlations suggest a failure of reductive atomism of one sort or another. Furthermore, again, SR tells us that there is no absolute temporal ordering of such superluminal non-local influences unless there is a preferred frame of reference not found in SR. Otherwise, EPR correlations are best explained in terms of non separability of some sort, or more specifically, adynamical global/spatiotemporal constraints (Silberstein and Stuckey 2020; Silberstein, Stuckey, and McDevitt, 2021). Silberstein and McGeever (1999) follow Paul Humphreys (1997), Richard Healy (1994), and Paul Teller (1986) in suggesting that the appropriate ontology of entangled states is to regard the particle pairs as forming, in some sense, a spatially extended "whole," meaning that there is system non-separability. Instead of one particle influencing the other, any measurement records the properties of this new "whole-entity."

One must also fully understand the classical–quantum contextual implications of this view. The "quantum system" is in fact the totality of the entire experimental setup such that different setups or configurations are not probing some autonomous quantum realm, but actually constitute different "systems." QM, then, is a theory that has quantum–classical contextuality at its heart—that is, its deepest lesson about the nature of physical reality. As

[44] We will follow up on the implications of contextual emergence for metaphysics in Chapter 7.

Philip Ball puts it, "the quantum experiment is not probing the phenomenon but is the phenomenon" (2018, p. 90).

This is a clear failure of mereological supervenience and the like. As we have seen throughout this section, the macroscopic stability conditions for system behavior come from the experimental setup and the environment. For example, the experimental setup provides the stability condition, making the wholistic state possible, not the individual states of the entangled particles.[45]

Are there length scales over which quantum superpositions and nonlocality cease? According to textbook quantum theory, which assumes the completeness of QM, the applicability of QM extends to all objects such as macroscopic measuring devices, planets, and the universe. Supposedly, if one could screen off an elephant, the said pachyderm would exhibit quantum interference effects. Does the quantum-elephant suddenly become something at a smaller length and time scale like a subatomic particle? No. Think also of environmental decoherence, which works across various length and time scales. It appears that quantum phenomena can exist on many scales that range from the microscopic to the macroscopic. The weirdness of QM is not about small versus large scale as such; it is about multiscale contexts or the removal or screening off of such multiscale interactions.

For instance, as we saw previously, actual-world molecules maintain very sharp positions and momenta as well as semirigid structures in chemical experiments, medicines, and so forth. On the other hand, interferometer experiments with specially prepared organic molecules have revealed quantum interference for molecules (e.g., of up to 2,000 atoms; Ball 2021). Soon such experiments will be done with micro-organisms. Such interference implies that molecules are "temporarily in a superposition state of being in both arms of the interferometer at the same time" (whatever that means). Recently, single-, double-, and triple-slit experiments have been performed on macromolecules up to 515 atomic mass units, revealing interference in the latter two conditions (Cotter et al. 2017). The key to these kinds of experiments is the macroscopic preparation and screening off as much environmental decoherence as possible for the molecular beams over the brief duration of the interferometer and twin-slit experiments. Producing wavelike behavior still depends on macroscopic diffraction gratings and the absence of macroscopic detectors at the gratings to determine which-way information.

Recall that the algebra of observables for molecules is a mixed quantum–classical algebra, meaning that the ontology of molecules spans both the micro and macro scales. So, it is not surprising that if the appropriate

[45] For discussion of a variety of positions on the failure of separability, see Humphreys (1997), Kronz and Tiehen (2002), Peter Lewis (2016), Tim Maudlin (2007), and Silberstein (2017).

screening is set up, that one might be able to engineer superpositions of molecules in a beam and achieve some interference effects. As we have seen, the superpositions characterizing EPR correlations depend crucially on macroscale conditions in the environment of quantum systems such as photons, electrons, and molecules.

Something similar is the case for recent demonstrations of macroscopic entanglement of mechanical oscillators via microwave excitations (Kotler et al. 2021). The basic idea is to build a circuit where a membrane forms the capacitor so that microwave photons can be input into the membrane, causing a force related to the number of photons to act on the latter. Two mechanical drumhead resonators, also acting as capacitors, are coupled to the membrane (see Kotler et al. 2021, figure 1). Applying two simultaneous appropriately adjusted microwave pulses to the membrane, the two mechanical oscillators can become correlated such that the vibrating motion of the drumheads becomes entangled. Basically, one pulse entangles drumhead 1 with the cavity within which the device is suspended via photon–phonon pair correlation (phonons in drumhead 1's membrane). The second pulse sets up an analog of a beam-splitter interaction between drumhead 2 and the cavity swapping the cavity photons at their frequency with the phonons of drumhead at their frequency. It is important that the two pulses be delivered simultaneously to ensure energy flow from the cavity to both drumheads, otherwise no superposition would be created. The superposition of the "quantum level" of the drumheads was observed by measuring the correlations of the dimensionless quadratures of the quantum position and momentum observables (Braunstein and van Loock 2005; Caves et al. 1980; Mercier de Lépinay et al. 2021), where correlations due to superposition will be distinct from any classical correlations.

As we saw previously, there are distinct states and observables at the microscale and macroscales (e.g., a whole range of new observables appear at the macroscale that are not functions of the microscale observables; observables due to superpositions disappear when those superpositions are broken). What these *tour de force* experiments demonstrate is that it is possible to produce enough isolation and macroscale control that some of the microscale observables in the mixed quantum–classical algebra can be excited and a superposition among them formed.[46]

Even the observation of such correlations among the quadratures is quite interesting. Such observations require macroscale preparation of Gaussian states with continuous variables. Hence, the entanglement observed is a form

[46] It is currently unknown whether there is a length scale at which the remaining quantum observables of molecules ever vanish. What is known is that there are length scales where any effects of such observables are utterly swamped by the macroscopic states and observables.

of continuous-variable entanglement. Detection of the quadrature mode of the electromagnetic field requires a beam splitter and detector—both macroscale objects. The quadrature of the mode is measured through the difference in the photocurrents output from the beam splitter due to a phase introduced by a local oscillator—another macroscale object (Braunstein, and van Loock 2005). Beam splitters, oscillators, and phase shifters preserve the photon number and so only introduce linear transformations to the Gaussian states. Again, we see the contextual emergence of interesting quantum phenomena—superposition states are either created or destroyed due to macroscale stability conditions.

One way to think about these cases of superposition/interference involving molecules and other macroscopic objects is in terms of possibility space. Whether thinking about twin-slit experiments, interferometers, or 70-picogram drumheads, various kinds of superposed states represent quantum possibilities in possibility space as the laws and properties of QM clearly allow. Concrete realizations of these possibilities, however, require specific macroscale stability conditions not given or present at the microscale of QM. We see the contextual emergence pattern again: Macroscale stability conditions function as gatekeepers in possibility space, determining when various cases of superposition are actualized and when not. As we have seen with the contextual emergence pattern, the quantum microscale provides some of the necessary conditions for quantum and macroscale phenomena, while macroscale stability conditions provide the rest of the necessary and sufficient conditions for macroscale and even microscale phenomena such as superposition.

Although we are accustomed to thinking of QM as our most fundamental atomistic theory (e.g., quantum field theory, quantum electrodynamics), we see that *contextual emergence is at the ontological heart of reality*. Contrary to how most metaphysicians conceive them, intrascale and multiscale contexts are at least as fundamental, if not more so, as (microphysical) properties, entities, and laws. That is, what metaphysicians typically call properties, entities, and laws, arise dynamically out of the multiscale stability conditions defining contexts (see Chapter 7 for more details). One implication is that there is no autonomous "quantum system" living in configuration space, with specific properties that pre-exist concrete contexts such as the experimental twin-slit setup.[47] If one insists on talking about "quantum systems," then contextual emergence says that the experimental setup is also part of the "system." But of course, this is just a rejection of the idea

[47] Hence contextual emergence offers a framework for understanding Bohr's contextuality through, for instance, identifying the stability conditions making electrons and other quantum objects exhibit wavelike or particle-like behavior in the twin-slit experiment.

that the classical world in spacetime somehow "emerges" from some *absolutely more "fundamental" purely quantum systems living in configuration space.*[48] Reality is fundamentally multiscale and intrascale in nature. There is no "Dedekind cut" between the quantum and the classical, the microscopic and the macroscopic, and there is no absolutely autonomous—context-free—"fundamental" entity such as the quantum wavefunction alleged to exist.

According to contextual emergence—and the sciences, we argue—the world does not come objectively and universally carved up into pregiven parts/wholes and systems/subsystems. Such carvings are always of explanatory interest—and theory-relative because they depend on which setup of all possible contexts and relations ones chooses to explore. For example, do you want to examine the twin-slit context that yields quantum interference and "wave–particle duality" (both slits open)? Or another such as changing the context of twin-slit by adding some element to the slits and allowing interference with which-way information? Note the fact that such explanatory judgments are perspectival does not make them unreal or even mind-dependent, just part of a concrete context. After all, physicists, out of just such judgments, actually produce wavelike or particle-like behavior in laboratories all the time![49]

Finally, there are very recent no-go theorems which allege that no account of QM can escape context-dependence, because it is necessary to reproduce the observed statistics of quantum theory (Spekkens 2005). More recent no-go theorems allege to show that not even accounts that give up measurement independence—such as retrocausal, superdeterministic, or even non local models of QM—can escape one or another strong form of contextuality, going so far as to claim that what is contextual is not just the QM state, but many other features of QM, such as what counts as a system, dynamical law, and boundary conditions (Shrapnel and Costa 2018). Going even further, Bong et al. (2020) allege to provide a new and more powerful no-go theorem that we *must* give up at least one of the following assumptions:

- **Assumption 1 (Absoluteness of observed events, AOE)** An observed event is a real single event, and not relative to anything or anyone (realism or non-contextuality).

[48] See Chapter 7.

[49] In an analysis complementary to ours, Sally Shrapnel and Fabio Costa (2018) have also argued that the context-dependent nature of quantum measurement is independent of any assumptions about possible causal structure and extends to states, dynamical laws, and boundary conditions. In other words, QM's contextual nature is pervasively relational. See also Aharonov, Cohen, and Tollaksen (2018), Primas (1983), and Silberstein, Stuckey, and McDevitt (2018).

- **Assumption 2 (No-superdeterminism, NSD)** Any set of events on a spacelike hypersurface is uncorrelated with any set of freely chosen actions subsequent to that spacelike hypersurface.
- **Assumption 3 (Locality, L)** The probability of an observable event *e* is unchanged by conditioning on a spacelike-separated free choice *z*, even if it is already conditioned on other events not in the future light-cone of *z*.
- **Assumption 4 (The completeness of QM, COMP)** QM unmodified applies to any and all macroscopic measuring devices including human observers.

Since, as we noted, there are many other formal and empirical reasons for giving up Assumption 1, Occam's Razor alone favors that choice. However, several new results and arguments also favor that move (Silberstein, Stuckey, and McDevitt 2021; Budroni, et. al 2021).

We are only arguing in favor of giving up non-contextuality and not realism. Recent experiments (*gedanken* or otherwise) and theorems in quantum mechanics (QM), such as new iterations on Wigner's friend and delayed choice, have led some people to claim that QM is not compatible with determinate and intersubjectively consistent experience (what some call absoluteness of observed events or "realism"), such as experiences of experimental outcomes being relative to branches or conscious observers in a way that allows two or more ideal observers to disagree and yet both be right. For arguments that these new no-go theorems against realism (i.e., universally definite and determinate experimental outcomes) fail and that non-contextuality is sufficient, see (Silberstein and Stuckey 2022).

4.8 Renormalization, universality, and contextual emergence

Discussions of the renormalization group (RG) method, universality, and the like have become prominent in discussions of emergence (e.g., Batterman 2002, 2018; Butterfield 2011b; Mainwood 2006; Morrison 2012). This is especially true, as many of these techniques and methods from physics are now being used in various special sciences such as biology, cognitive and neural sciences, and economics. We have already seen some examples of how universality enters into understanding the structure of the physical world in the equivalence classes that arise in thermodynamics, molecular structure, and convection. While there is much debate about the implications of universality for or against emergence of one sort or another, the role of multiscale stability conditions and contexts are rarely if ever mentioned.

4.8.1 *Renormalization group methods*

Renormalization techniques had their start in quantum field theory (QFT),[50] but have found practical utility in nonlinear dynamics, diffusion, turbulence, condensed-matter physics, and even economics—most any situation involving interactions among many degrees of freedom is amenable to RG analysis (McComb 2004).

Historically, renormalization had its start in analyzing many-degrees-of-freedom systems. Consider the dressed states referred to in Section 4.6. An isolated electron has a charge q and produces an electrostatic potential of q/r, where r is the distance from the electron. Electrons, however, are always interacting with other charges and the electromagnetic field in their environment. In the presence of enough electrons, the electrostatic potential of our electron of interest becomes

$$\frac{q}{r}e^{(-r/\lambda_d)}$$

where λ_d is the mean free path length before a collision with another electron. The dressed-state electrostatic potential of an electron interacting with other electrons is smaller than that of an isolated electron. Physically, the cloud of electrons surrounding the electron of interest shields the latter's normal electrostatic potential.[51] This is known as *charge renormalization*. So, renormalization is related to physical effects of interactions with multiple degrees of freedom. Such renormalization effects occur in systems with many degrees of freedom (e.g., each fluid parcel in Rayleigh–Bénard convection experiences a renormalization). In essence, renormalization is an example of collective effects in systems with many particles/many degrees of freedom.[52]

The RG "is a method of establishing scale invariance under a set of transformations (which form a simple group, hence the terminology) and the renormalization of, for example, coupling constants occurs as part of this process" (McComb 2004, p. viii). Whether looking at critical phenomena in statistical mechanics, thermodynamics, or fluid dynamics, the physical idea behind the RG is the same: the large-scale behavior of gases and fluids near criticality (e.g., the critical temperature leading to a shift from the quiescent state to the formation of Bénard cells) is dominated by mesoscale fluctuations

[50] Though renormalization perturbation theory thrived on a partially independent track in statistical mechanics before the emergence of the RG.

[51] Such shielding is important in contexts such as plasma physics (e.g., Krall and Trivelpiece 1986).

[52] Debye screening is one of the earliest examples of renormalization due to collective effects, and λ_d is known as the Debye length.

in the properties of the system (e.g., temperature and density). These fluctuations are much larger than the small-scale intermolecular forces, hence the physics of the fluctuations are insensitive to the small-scale constituents and forces. This is to say, the fluctuations in system properties are stable with respect to changes in the small-scale constituents and forces (e.g., one can generate Bénard cells in a variety of different kinds of fluids with exactly the same behaviors discussed in Chapter 2). Here we see a link with equivalence classes.

RG methods take advantage of this physical, multiscale situation. Begin by constructing a state space of points where each point represents the properties of a system (e.g., a gas or fluid) at each point. Then partition this space (e.g., by coarse graining), eliminating some of the degrees of freedom associated with each point replacing them with some average behavior for a set of points. In essence, create partitions representing equivalence classes of similar behavior for the molecules associated with each point in the state space. These equivalence classes then are treated as the new "molecules" or clusters. This procedure can be repeated by partitioning the space of these clusters where they fall into equivalence classes at a larger scale. What Kenneth Wilson (1971a, 1971b, 1972, 1983) and others discovered is that near criticality there is self-similar behavior on multiple scales that corresponds to the independence of the large-scale states and observables from the details of the small-scale constituents and properties—just the kind of pattern of contextual emergence that we saw in Chapter 3. Mathematically, the RG represents a group of similarity transformations, though these transformations would be powerless to represent physical systems if there were no underlying physical similarities at appropriate length and time scales for the RG to map onto.[53]

One can think of the application of the RG as throwing away superfluous information from the smallest scales, while preserving the behavior that matters independent of the small-scale details (universality). Indeed, RG methods allow the construction of a hierarchy of effective theories at larger length and longer time scales without knowing the details of the physics at the smallest length and time scale (if there is a "smallest"). This fits with the theme of this chapter—namely, that there is no "fundamental" scale or "fundamental" entities determining everything else at larger length and longer time scales.

[53] For details and examples, see McComb (2004). We should warn the reader that as far as the mathematics are concerned, these similarity transformations lead to the self-similarity of fractals (repeated exact similarity on all length scales). In contrast, the actual world only has prefractals, so that only two or three repetitions of self-similar structure are ever found in nature (e.g., Avnir et al. 1998). This does not affect the RG, as it is typically only applied across two or three transformations of scales.

4.8.2 *Universality*

Renormalization and the RG lead us directly to the phenomenon of universality—that large-scale critical phenomena, for example, behave in remarkably similar ways independent of the microscale properties. A wide variety of systems exhibit what are called critical points—values of order parameters at which phase transitions and other phenomena occur. Critical phenomena depend on the value of these order parameters, the spatial dimensionality of systems, the operative symmetries, and other system-wide details. We met an example in Rayleigh–Bénard convection in Chapter 2: The critical temperature, T_c, was the crucial parameter value at which the fluid system transitions from a metastable version of the quiescent state to a convecting state. A variety of different fluid types can exhibit Rayleigh–Bénard convection, and T_c is different for each of these different fluid types, as one might expect, since they are chemically distinct kinds of fluids. Nonetheless, these systems exhibit a remarkable degree of similarity in how they approach the critical point marking the transition to convection: as $T \to T_c$, the correlation length diverges in the same way.[54] Physicists characterize this behavior associated with order parameters in terms of critical exponents. No matter what fluid is used in Rayleigh–Bénard systems, all these systems have the same critical exponent. This similarity in behavior is independent of the microscale details of the different fluids, hence the reference to universality.

Physicists observe the same behavior in magnetic, condensed-matter, thermodynamic, mechanical, chemical, and biological systems (e.g., McComb 2004; Wilson 1983). For instance, the paramagnetism–ferromagnetism transition is in the same universality class as the liquid–gas transition in water. That so many different kinds of physical systems exhibit the same behavior approaching critical points where the critical exponents are independent of the microscale details is a striking example of macroscale independence. It is one of the great triumphs of RG methods that they have elucidated such universality.

More generally, we also observe universality-like phenomena in the special sciences that do not involve critical exponents, such as phenotypic plasticity, neural plasticity, and robustness in complex biological/cognitive systems. That such universal behavior occurs so broadly is understandable as the recurrence of the contextual emergence pattern we are arguing for in this book. Not all cases in physics and the special sciences include singularities, spontaneous symmetry breaking, and the like, but they do all exhibit universality in generalized, though not (currently) always formally specifiable,

[54] As we saw in Chapter 2, the physical meaning is that all fluid parcels become correlated with each other, producing general rigidity.

senses. What is observed in all these cases is some form of screening off or decoupling of larger-scale phenomena from the microscale details.

Aside from difficulties in formalizing universality across the sciences, there are differences in how to understand and interpret the significance of universality. Restricting our discussion to mathematically formalizable systems, here is the basic outline of how the RG explains why specific physical systems divide into distinct universality classes. Essentially, the explanation turns on geometric features of the fixed points and their relations in state space associated with the systems under RG transformations. A fixed point is defined as an ensemble in the recursive applications of the RG leading to no further changes in the values of parameters characterizing the system and the correlation length. These fixed points are associated with the new macroscale states and observables. The Hamiltonians describing heterogeneous physical systems fall into the basin of attraction of the same RG fixed point under these transformations. The state space of Hamiltonians contains numerous fixed points, each of which is describing different universality classes with different critical exponents and scaling functions. The microphysically diverse systems in the same universality class will exhibit a continuous phase transition near which their analogous macroscopic quantities will obey power laws possessing the same numerical values of the critical exponents independent of the microscopic details.[55] In other words, the fixed points are properties of the RG, not of the microscale states, observables, and laws.

This leads to the question: is universality and the existence of universality classes simply mathematics—epistemological emergence—or ontological emergence in some form? Appealing to in-principle explanation of universal phenomena in terms of ACFAP microscale properties and laws is unlikely, given contextual emergence.

One line of interpretation for universality is Batterman's (2000, 2002). He has argued for a form of explanatory emergence emphasizing that in our theoretical analyses we use sophisticated mathematical tools, such as RG methods, to eradicate irrelevant microscopic details. In brief, multiple realizability can be viewed as a subset of universality, and the RG explanatory strategy can explain universality (the geometric features associated with fixed points in state space serving as basins of attraction for system Hamiltonians under RG transformations). It is possible to extended RG explanations to multiple realizability more broadly across physics and across the special sciences. Therefore, physicists and other scientists are not pursuing reductionistic explanations as they are often conceived, but a form of

[55] Here one can see the analogy with state space partitions leading to stable equivalence classes discussed previously.

emergentist-type explanation so described. All the while, Batterman eschews any matters of ontology other than to point out that the microscale details do not determine the universal behavior *in the sense described by RG methods.*

However, he does defend the criterion that the explanation for universal behavior must be "completely grounded in the lower-level theory" (Batterman 2000, p. 136). It seems as if reduction in this sense must be a criterion for scientific explanation. Paul Mainwood (2006) emphasizes this apparent tension in Batterman, and further characterizes his view as follows: "In the context of his discussion, Batterman is concerned with opposing a very specific view, expressed in (Papineau 1993, chapter 2), which demands a list of lower-level properties sufficient and necessary to realize a higher-level one. His defense is the failure of sufficiency" (2006, p. 133). Here is context for Batterman's discussion:

> The explanation provided by the RG method in the case of universal critical behavior differs significantly in kind from Papineau's explanation of the temperature/pressure relation in a gas. For Papineau, one has an explanation and hence a reduction, when one provides a common physical property—shared mean kinetic energy—that is both necessary and sufficient for the observed common upper-level behavior. But, my point has been that the RG account explains the universality of critical phenomena without finding any such property.
>
> (Batterman 2000, p. 136)

In these discussions, Batterman does not seem to have the kind of "reductive grounding" in mind that Papineau had.[56] Perhaps what Batterman wanted was to take multiple realizability seriously—there cannot be phase transitions in thermodynamic, magnetic, or fluid systems without atoms and molecules, but the microscale properties of atoms and molecules are not sufficient for the macroscale behaviors at critical points. So, there is still grounding in the sense that there are microphysical entities composing macroscopic systems, but not grounding in the sense of sufficiency as Papineau seeks.[57]

This is consistent with Batterman's (2018) more recent discussion that engages ontology a bit more, but still appears to hew in the direction of explanatory rather than ontological emergence. What becomes clear in his discussion of multiple realizability, and the famous round peg/square hole example of Hilary Putnam (1975), is that reductive explanation will not

[56] It is worth reminding ourselves at this point that, as we observed above, the kind of reductive explanation that David Papineau wants to give, fails in physics even for temperature and pressure.

[57] Though one could question exactly how the terms "necessary" and "sufficient" are being used here. Are we referring to logical necessity or physical determination? Are we referring to type or tokens? Furthermore, Batterman is often more concerned with explanatory patterns than with the question of bottom-up determination as such. In short, as we read Batterman, he is not merely just demonstrating how and why the reduction base can be disjunctive.

do in the physical sciences, hence we should not expect it to work in the philosophy of mind and cognitive science.

Another interpretation of the universality observed in critical phenomena, such as phase transitions, is proffered in Tarun Menon and Craig Callender (2013, p. 200): "Even though statistical mechanics does not group phase transitions together the way that thermodynamics does, it is still able to fully explain what goes on in individual instances of phase transitions." What is on offer in this interpretation is a kind of local reduction: "Perhaps the existence of individual explanations in every case constitutes an adequate explanation of the nomic pattern described by thermodynamics. If this is the case, the core sense of reduction is satisfied." While Menon and Callender acknowledge that temperature is not simply the mean kinetic energy of system constituents, they offer the possibility that properties, such as the temperature of phase transitions, are functional in the sense that they are multiply realized, though they do not spell out what they mean by reduction aside from the that functional property at the larger scale is realized at the microscale of SM. Yet as we observed previously, this kind of functional reduction does not capture how the multiscale nature of temperature, chemical potential, or even pressure works in physics. Nor does this kind of approach deal with the role of larger-scale stability conditions determining the expression of microscale constituents and properties.[58]

Batterman and others' response is that universality classes are not about individual systems but are *facts* (not merely projections or eliminable idealizations) about the *relationship between kinds of systems*, or universal features that those systems have in common. The local reduction strategy does not help us to understand these facts, and therefore is irrelevant.

So far, we have universality and RG explanations as being neither intertheoretic reduction nor causal-mechanical reduction. What else is left? Perhaps Batterman ultimately is arguing that such explanations are geometrical in terms of the structure of the basins of attraction in state space. In other words, this might only be descriptive mathematics and nothing more, hence only epistemic emergence. But there is more, in the sense that universality and RG methods highlight the scale-invariant nature of disparate physical systems. What we observe in physics is that the diverse systems' order parameters exhibit power-law behavior with the same exponent at criticality in their universality classes, and this is linked to their equivalent phenomenological behavior. Think back to Rayleigh–Bénard convection. Different kinds of fluid all produce Bénard cells under appropriate conditions

[58] Menon and Callender's appeal to crossover theory, wherein systems approach a critical point and there is a deviation from the behavior of one universality class and a crossing over to another—but this does not help their cause. This theory does not explain the actual transition from microscale states and observables to the new macroscale states and observables.

with the same critical exponent. This implies that the larger-scale stability conditions constraining the degrees of freedom of motion of fluid parcels operates the same, independent of the microscale details of fluid chemistry.

On the other hand, according to Batterman and others, this scale invariance or the absence of any characteristic length scale is essential for universal or critical behavior; that is, it is essential for the appearance of a fixed point in state space. This scale invariance, in turn, depends on idealizing the system as infinite. If the number of molecules, atoms, particles, and so forth, is finite, then there will always be a characteristic length scale determined by the size of the system—in which case, repeated course-graining or rescaling will no longer yield statistically/behaviorally identical systems.

This raises a question—broached in Chapter 3—as to whether the singularities and infinities involved in the RG analysis are merely artefacts of the formalism that in principle could be eliminated by some more "fundamental" theory. If not, what do the said singularities and infinities denote in the physical world? As we indicated in Chapter 3, the best way to think about the singularities in mathematical models is that they mark out physical transitions to new states and observables at larger length and time scales that do not exist at smaller length and time scales. System do not have to have any genuine infinities for these transitions to occur (though some do, such as infinite conductivity in superconductors); rather, these infinities are pointing to important macroscale features (e.g., infinite correlation lengths indicating that system-wide constraints act to control microconstituents).[59]

In contrast to Batterman, Margaret Morrison (2012) urges a more ontological interpretation of RG explanations. After arguing that the RG has a relationship to, but is distinguishable from, probability-based approaches such as the central limit theorem[60], she draws on the fact that "What the fixed points do is determine the kinds of cooperative behavior that are possible, with each type defining a universality class" (p. 1149). In other words, the RG transformation fixed points locate specific subspaces of possibility in possibility space. This draws our attention to the fact that RG methods not only involve discarding degrees of freedom irrelevant to larger-scale

[59] In this sense, if by some clever means someone found a way to eliminate the infinities from RG and other forms of analysis, the physical transitions would remain to be explained. What contextual emergence points us towards is that these transitions require stability conditions at larger scales that do not exist at the lower scales. Hence, the objections that emergence in physics depends on infinities in the mathematics (e.g., Butterfield 2011b; Menon and Callender 2013) are red herrings.

[60] Indeed, Morrison notes that statistical averaging techniques associated with mean-field approaches based on probabilities lead to infinities when trying to couple interactions at different length and time scales (2015, section 2). These are precisely the infinities that RG methods overcome. The central limit theorem can be derived from RG methods, so in some sense one can think of the former as a special case of the latter in SM contexts.

behavior, but also point to the existence of collective behavior possibilities at larger scales. This is where Morrison locates the RG explanation for universality.

She argues that the relationship between the rescaling taking place in RG transformations determining the universality classes is linked to the scaling laws that characterize the universality arising in the approach to critical points. It is the scaling laws and their exponents that typically are the observables in experimental behavior. These are the critical exponents discussed in the literature. The signature difference between RG methods and statistical averages in SM is that RG transformations are related to the scaling properties of the universality classes characterized by these scaling laws rather than to probability distributions. Such transformations are equivalent to transforming the state space into a space of system models or Hamiltonians and a study of the structural relationships among these models (p. 1150). As Morrison notes: "The mathematical foundation of these types of structural transformations comes from dynamical systems theory, which deals with changes under transformations and flows toward fixed points" (p. 1152)—hence the previous observation that RG methods are associated with geometrical or topological structures and relations of ensembles or Hamiltonians.

In dynamical systems theory it is often not possible to calculate the specifics of a system's detailed dynamics. Instead, the focus is often on understanding the behavior of system trajectories in state space over time (i.e., as time increases towards infinity). A key question is system stability to small perturbations, and this question is answered by showing that a perturbed system can be transformed in such a way as to recover the original unperturbed system. The way this is accomplished is by defining dynamical systems in terms of group transformations on a manifold (a topological space), with time as the single parameter.[61] The aim is to model the fixed points—basins of attraction in the manifold—the values of observables for steady states of the system over time. One thus finds universality in dynamical systems (e.g., Feigenbaum 1978) like that observed in RG methods.

As with dynamical systems, the RG shares in the focus on conditions for system stabilities. In the dynamical systems case, it is the structure of the evolution map, the equation describing the trajectories in the manifold. In the RG case in SM and other physical systems, it is the structure of the Hamiltonian.[62] Structural features in the state space of Hamiltonians are the key:

[61] For details, see Primas (1983).

[62] In the RG derivation of the Feigenbaum number in dynamical systems, the dynamical map (e.g., the logistic map) substitutes for the Hamiltonian.

> Shifting away from individual "physical" models of the system with a specific microstructure defined on a [state] space and replacing it with an emphasis on the way the structural features (the space of Hamiltonians or evolutionary maps) change under RG transformations illustrates how we can understand the relation between stable macro behavior and the micro level from which it arises.
>
> (Morrison 2012 p. 1155)

In other words, RG transformations focus on the relevant structural constraints for stable macroscale behavior, where these constraints are geometric and/or topological in the state space of Hamiltonians, but the mathematics reflects physical features of the macroscopic systems and collective behavior. Specific values of the microscale states and observables are not relevant to either the structures showing up in state space under RG transformations, nor to the macroscale Hamiltonians related to the physical nature of the collective behavior at macroscales. In other words, Morrison argues that RG explanations are connected to physical constraints that are determinative of system behavior, implying that microscale states, observables, and laws are insufficient for the determination of macroscopic stabilities represented in universality classes.

Although differing with Morrison at some points, Alexander Franklin (2018) has also argued for a qualified ontological realism for universality and the RG. The authors surveyed in this section are representative of the disagreement over how to understand universality and its broader implications. We think that contextual emergence offers a way to situate these different interpretations.

4.8.3 *Renormalization, universality, and contextual emergence*

The recurring theme in discussions of universality and RG explanations, if you will, is the important similarities in large-scale phenomena that are independent of the microscale properties and laws. What perhaps does not receive enough focus in these debates is that *everybody* is talking about the fact that modern physics reveals that:

1. Microscale states, observables, and laws are distinct from the relevant large-scale states, observables, and laws.
2. The former only determine the space of possibilities for the latter.
3. The microscale details only contribute some of the necessary conditions for the existence and behaviors of large-scale phenomena.
4. The larger scales contribute the stability conditions for the existence and persistence of relevant macroscopic states and observables.
5. The larger-scale stability conditions and their contexts constrain the behaviors of microscale entities.

In other words, the debates are all dancing around the contextual emergence pattern without explicitly acknowledging it! The important physical features are the transitions from microscale states and observables to macroscale states and observables with their new behaviors. As we have seen, the macroscale stability conditions not given or determined by any resources at the microscale are as important as any of the latter resources, as illustrated by Rayleigh–Bénard convection. As evidenced by convection, the work stability conditions do in the formation and maintenance of the states, observables, and dynamics of the larger-scale dynamics and structures account for the universality of convection behavior across multiple distinct types of fluids.

Physical connections to the universality classes have been demonstrated. For example, James J. Binney and collaborators (1992) "have shown that any three-dimensional physical system whose Hamiltonian can be written as an even functional of a one-component scalar field should have the same critical behavior as the Landau–Ginzburg model" (p. 366). This means that any system of interest with an order parameter that can be written as a one-component scalar field, such as temperature, will fall in the same universality class (e.g., including systems as different as those composed of fluid molecules or magnetic spins). Hence, order parameters play an important role in the universal behavior seen in physical systems. The number of components, *n*, of the order parameter represents a symmetry group. What we see showing up in common for each universality class are geometric or topological features of the state space of Hamiltonians, the symmetry group of the physical order parameter, the spatial dimension of the system, and scaling laws with the same critical indices along with other systemic or global features. Indeed, we saw exactly these kinds of larger-scale stability conditions—both causal and acausal examples—at work in the Rayleigh–Bénard convection, in the case of properties such as temperature and chemical potential, and in the case of white dwarfs.

As we have indicated throughout this section, the mathematics of RG methods signpost important macroscale physical similarities. These similarities are difficult to pick out for systems as diverse as fluids and magnetic spins when examining systems and their constituents. For example, the microscale states, observables, and laws give no indications as to the kinds of collective behaviors that are actualized at macroscales. Meanwhile, the macroscale behaviors have enough phenomenological distinctiveness to make universality hard to observe (e.g., the critical temperature for Bénard cell formation is different for different fluids). But as with dynamical systems theory, RG methods give us mathematical precision in understanding the state space behavior of macroscale systems as they approach critical points allowing us to hunt for why such interesting physics exhibits universality classes. For instance, in SM and fluid dynamics we have sharpened our understanding of why there are partitions of the microscale state space matched to or constrained by the macroscale states, observables, and their dynamics, or why the order parameters for systems exhibit the symmetries at the macroscale that they do. In other words, RG explanations help us see the indispensable roles of multiscale stability conditions and macroscale contexts more clearly and how to further understand the kind of autonomy they have from microscale properties.

Building on this, universality and RG explanations help us see more clearly that "fundamentality" is not about scale (e.g., high energy and small length scale). For example, microconstituents are constrained by the macroscale scaling laws with critical exponents defining universality classes which, in turn, are directly related to the fixed points of the RG transformations. This is the pattern we have seen with contextual emergence through the determinative role that large-scale stability conditions play in small-scale constituent behavior. Actual physical conditions and structures at larger scales cannot simply be dismissed with the handwave of the "in-principle" ACFAP microscale explanations. We have seen that such explanations do not exist.

Let us return to possibility space. One way to connect the kinds of universality RG methods reveal in the Hamiltonian state spaces for systems is that the fixed points in these state spaces represent possibilities for macroscale collective system behaviors. RG methods point to the ways by which macroscale stability conditions structure access to these regions of possibility space. It is these macroscale stability conditions that function as the gatekeepers to these regions of possibility space, not the microscale laws and constituents. The modal relations in possibility space are not simply bottom-up. Once again, we see supervenience as philosophers traditionally understand it (e.g., Kim 1999) violated by universality as we have seen it violated in all the examples discussed in this chapter.

RG explanations fit the contextual emergence pattern: the microscale states, observables, and laws contribute some of the necessary conditions for the existence of emergents (e.g., temperature, phase transition phenomena). These necessary conditions mesh with the specific large-scale stability conditions defining the system context (e.g., the existence of relevant states and observables at the larger scale), yielding the set of necessary and sufficient conditions for the transition to the emergent phenomena. The stability of the behavior of universality classes is established by the large-scale stability conditions which are not given by the microscale details. This is the contextual emergence pattern at work.

4.9 Further brief examples

We draw our discussion to a close with a few additional brief examples of the contextual emergence pattern in physics.

4.9.1 *Lasers*

Einstein first proposed the physical possibility of the stimulated coherent emission process that would eventually lead to lasers in QM in the context of thinking about probability for quantum events in the quantum theory of radiation. However, it is one thing to demonstrate the possibility that in a collection of atoms in identical states, one atom producing a single photon of the right energy can stimulate another atom to emit another photon which stimulates another atom which emits another photon which stimulates another atom, and so forth, leading all the atoms to release their excess energy in a cascade. The stability conditions for this process (e.g., preparation of a collection of atoms all in the relevant excited states, precise triggering of the population inversion returning the atoms to their ground states producing photons, a designed optical cavity trapping the photons and stimulating more photons, appropriate isolation from the wider environment) are another matter entirely. Such conditions are not concretely realized in nature. Nor are these conditions given by the atoms, photons, and their law-like interactions alone. Yet with the right macroscale stability conditions, the lasing process can be realized.

An interesting practical issue with lasers is that they often exhibit spatiotemporal instabilities given the nonlinear interactions of the light field with active media. Wave-chaotic microcavities have provided a means for suppressing spatiotemporal lasing instabilities (Bittner et al. 2018). The suppression of coherent nonlinear processes typically leading to the formation of self-organized structures (e.g., filaments) was achieved by using cavities with a complex spatial structure. These cavities create many propagating waves with random phases so that complex interference of waves prevent the formation of the self-organizing structures that lead to instabilities.

The geometry of the wave cavities is crucial to promoting chaotic cavity modes directly affect lasing dynamics The lasing emission profile is constrained by the cavity mode structure. Global suppression of lasing instabilities is achieved through either induced chaotic ray dynamics or random refractive index fluctuations across the whole cavity. The geometry of the wave cavity functions as a stability condition for producing propagating waves with random phases, while the latter, in turn, function as a stability condition for disrupting nonlinear instabilities. The region of possibility space where a multimode lasing state with stable spatiotemporal lasing dynamics emerges is accessible through the geometric stability condition. Normally, this state is only accessible by tedious means of reducing the number of lasing modes.

4.9.2 *Coherent electrons in scanning tunneling microscopy*

Recent innovations have led to significant improvement in the resolution of scanning tunneling microscopy (Garg and Kern 2020). Electron tunneling through a barrier typically is not a coherent process. For instance, applying a voltage onto a scanning tunneling microscope tip near a metal surface usually produces a stimulated tunneling current lacking coherence. This is because the randomly evolving phase of the applied voltage affects the electrons.

In contrast, by combining scanning tunneling microscopy with a designed phase-locking train of optical pulses, coherent microscopy with attosecond control leading to subfemtosecond resolution is achieved. The induced tunneling current is entrained to the phase of the electron current. The carrier-envelope phase of the electron current acts as a stability condition on the tunneling current, leading to the emergence of both coherence and unprecedented resolution. The process involves a feed-forward locking of the phase imprinting the optical pulse phase onto the tunneling electrons. Such a stability condition can be achieved and applied to a wide class of instruments.

4.9.3 *Topological invariants in ideal fluids*

In an ideal fluid, the vorticity of fluid flow is a measure of the local rotation of fluid parcels. Helicity is a topological measure of the degree of linkage of the vortex lines of a flow (roughly the knottedness of vortex lines; Kleckner and Irvine 2013). In an ideal fluid, the vortex lines are frozen in the fluid flow and helicity is a conserved quantity. Neither vorticity nor helicity depend on the properties and dynamics of chemical molecules or fluid parcels. Rather, they are constraints on the behavior of parcels, and hence constrain chemical molecule behavior. Furthermore, helicity is a topological quantity that is invariant under the velocity field of the fluid.[63] Vorticity and helicity are stability conditions for the behavior of fluid parcels in ideal fluids.

A more subtle topological invariant, the Godbillon–Vey or GV invariant, is defined as the twisting and squeezing of the vorticity (the helical compression of the vortex lines; think of wringing a towel). Although a dimensionless

[63] More generally, helicity is invariant under a divergence-free vector field.

topological invariant, $GV \neq 0$ is sufficient for preventing the rate of change of the vorticity from vanishing both locally and globally (Machon 2020). In other words, $GV \neq 0$ is a stability condition on the behavior of the vorticity, even though it has a set of distinguished closed vortex lines as a necessary condition for its existence. In essence, GV is a topological constraint on steady flow in ideal fluids.

4.9.4 *Coalescence in metastable liquids*

Metastable liquid states, such as supersaturated solutions, and supercooled melts, are interesting states of matter. While the dynamics of metastable liquids are fully nonlinear, it is the phase and structural transitions that constrain and shape everything about the dynamics and even the final state.

Consider a metastable liquid supercooled (or supersaturated) such that nucleation and growth of solid-phase crystals can occur. The crystal/particle-radius distribution becomes crucial for how the metastable state evolves. As crystals form, the "tail" of the particle-radius distribution characterizes the metastable system states during the final coalescence stage of the phase transition to the solid phase (Alexandrov and Alexandrova 2020). The structure of the particle-radius distribution turns out to be the power-dependent distribution function to which all mass distribution mechanisms conform (Slezov 1978; Lifshitz and Pitaevskii 1981). Essentially, the way this distribution relaxes in time, during what is known as the Ostwald ripening stage of the transition, constrains the dynamics of the transition from metastable liquid to solid phase. In the Ostwald ripening stage, larger particles grow at the expense of smaller particles, hence the role of the distribution of particle-radius in space and time at the beginning of the ripening stage acts as a stability condition constraining the dynamics of chemical molecules in the transition to the solid phase.

4.10 Conclusion

In this chapter we have observed how the contextual emergence pattern shows up in physics, from QM to classical physics, from foundations to applications. The claims that physics is a thoroughly reductionist science, as referenced in the introduction, have proven to ring hollow. Nature is more subtle and more interesting than expected from reductionist intuitions.

We will continue discussing the richness and pervasiveness of the contextual emergence pattern through case studies in condensed-matter physics in Chapter 5, and biology, neuroscience, and beyond in Chapter 6.

Appendix 1: Classifying algebras

Algebras can be classified by their factors or central decompositions. An atom is a minimal nonzero element of a set, and so cannot be decomposed into two proper subsets. This means for $\alpha \subseteq A$, α is an atom if and only if for every β, either $\beta \cap \alpha = \alpha$ or either $\beta \cap \alpha = \emptyset$. A factor of type I, then, contains an atom. A factor is of type II if it is atom-free and contains some nonzero finite projection. A factor is of type III if it does not contain any nonzero finite projection (Takesaki 2002, p. 296). Factors of type I have normal pure states. Factors of type II and III lack normal pure states.

Appendix 2: The Pauli exclusion principle

The Pauli exclusion principle (PEP) is named after Wolfgang Pauli, who proposed it in 1925. It states that no two identical particles with half integer spins, known as fermions, can occupy the same quantum state. Electrons are fermions, so PEP applies to them. The principle is a consequence of the indistinguishability of quantum particles.

Quantum systems are described by the Schrödinger equation:

$$H\psi(x,t) = i\hbar\frac{\partial\psi(x,t)}{\partial t},$$

where H is the Hamiltonian and $\Psi(x,t)$ is the wavefunction whose square modulus gives a probability that the system is in a given state. Since the wavefunction is a function of time, the probability of finding a particle, say, at a particular point changes with time. In classical mechanics, two identical particles can be distinguished because we can follow their separate trajectories, but when we have two identical particles in QM we cannot label them as individuals in this way. We cannot distinctly follow the path of each particle, as the particles have no distinct position between measurements. Rather, when a measurement occurs at some time we can only say that one of the particles has been found at one point and the other at another, but we cannot say *which* of them is which in each given measurement. The wavefunction must remain valid, regardless of whichever particle has been measured at a point. If one particle is described by Ψ_a and the other by Ψ_b, then there are two possible ways of combining the wavefunctions to describe a system with two particles.

$$\psi(r_1, r_2) = A[\psi_a(r_1)\psi_b(r_2) \pm \psi_b(r_1)\psi_a(r_2)].$$

Here, r_1 and r_2 are the possible positions of the particles. Fermions have antisymmetric wavefunctions; that is:

$$\psi(r_1, r_2) = A[\psi_a(r_1)\psi_b(r_2) - \psi_b(r_1)\psi_a(r_2)].$$

If both particles are in the same quantum state, $a = b$, then

$$\psi(r_1, r_2) = A[\psi_a(r_1)\psi_a(r_2) - \psi_a(r_1)\psi_a(r_2)] = 0$$

meaning that the wavefunction disappears, which is a physically uninterpretable solution. From the indistinguishability of quantum particles, and the requirement of an antisymmetric wavefunction, we arrive at the exclusion principle: *no two fermions can be in the same quantum state*. In practice, since we can have a spin-up or a spin-down state, this means that only two electrons can occupy a given energy state. Both electrons and neutrons are fermions, so the exclusion principle is very important for understanding both white dwarf stars and neutron stars.

Appendix 3: Degeneracy pressure in a white dwarf

One can imagine the electrons in a star as being confined to boxes of smaller and smaller size as the star collapses. QM says that particles such as electrons are also waves in some circumstances (wave–particle duality). The PEP requires that the wave of each electron must stay inside the confines of its own box. As the boxes become smaller and smaller, the wavelength must become smaller to stay inside. A smaller wavelength means a higher frequency, implying a higher energy.[64] As the star continues to collapse, the electrons are forced into higher and higher energy states, and this generates a resistance to the collapse known as *the electron degeneracy pressure*. A full derivation would take up too much space here, but we can provide a very rough estimate of the resulting degeneracy pressure.

Let us say we have n electrons, which do not interact with each other in any way, confined to a volume V. We can fit two electrons into each box (with opposite spins), but no more, due to the PEP, so each electron occupies a volume of approximately $V/2n$, and occupies a lateral space of approximately

$$\Delta x \sim \left(\frac{V}{2n}\right)^{1/3}.$$

The Heisenberg uncertainty relation states that uncertainty in position, Δx, and uncertainty in momentum, Δp, multiplied together cannot be less than a constant, $\hbar$, divided by 2:

$$\Delta x \Delta p \geq \frac{\hbar}{2}.$$

So:

$$\Delta p \geq \frac{\hbar \Delta x}{2} \sim \frac{\hbar n^{\frac{1}{3}}}{2^{2/3} V^{1/3}}.$$

Let m be the mass of the electron. In the low-temperature limit, assuming the electrons are not moving near the speed of light, the average kinetic energy is given by

[64] Blue light is of shorter wavelength, and hence more energetic, than red light, so when metal is heated to higher temperatures it will glow red first.

$$E = \frac{\Delta p^2}{2m} \sim \frac{\hbar^2 n^{2/3}}{2^{7/3} V^{2/3} m}.$$

The total internal energy U is then

$$U = nE \sim \frac{\hbar^2 n^{5/3}}{2^{7/3} V^{2/3} m}.$$

In SM, pressure can be defined as a partial derivative of internal energy with respect to volume under constant entropy. In other words, pressure is a measure of how internal energy changes in response to changes in volume. Hence, our pressure due to degeneracy alone is given by

$$p = -\frac{\partial U}{\partial V} \sim \frac{\hbar^2 n^{5/3}}{2^{4/3} V^{5/3} m}.$$

This derivation is extremely simplified. A rigorous calculation would involve doing quantum statistics and calculating the "density of states"—the number of electrons with momentum between two infinitesimally close values. The density of states provides a statistical weighing of which energy values are used, and allows a calculation of the average momentum flux across a surface and hence pressure. In our derivation we have shown that degeneracy pressure scales as the electron density (n/V) raised to the power of 5/3. As the electrons are squeezed into smaller volumes their velocity approaches the speed of light, and we must use a relativistic expression for the relation between energy and momentum. If we do this, we find that the pressure scales as the electron density raised to the power 4/3. This is important. As electrons gain more and more energy, a shift occurs in how they can partition that energy. At low speeds, as in everyday life, if we add more energy to a moving body we increase its speed. But when an object reaches the speed of light it cannot keep going faster. The extra energy must be added to the mass of the electrons—that is, their inertia. In other words, the way energy is distributed changes the pressure. At low speeds, pressure increases with density faster than at higher speeds.

5

Contextual Emergence in Condensed-Matter Physics

I am drawn to condensed-matter physics because it is simultaneously useful and fundamental, simultaneously mundane and fantastical, simultaneously large and small, and most of all, some of the phenomena that are observed in condensed-matter systems are just flipping cool.

Inna Vishik

5.1 Introduction

On the standard view, at least in some explanatory sense, quantum mechanics (QM) underlies all condensed-matter phenomena. In this chapter we argue that condensed-matter physics (CMP) offers demonstrably many-bodied and multiscale/intrascale examples of the contextual emergence pattern. Most notably, we have many-bodied systems with long-range entanglements reminiscent of what we saw in Chapter 4. We devote a separate chapter to CMP because it offers a powerful and, in many ways, unique multiscale glimpse into the inherently relational nature of reality and the role of contexts as fundamental. In this way, we agree with Margaret Morrison that the wonderful weirdness of CMP is not merely about quantum entanglement. However, her focus has largely been on spontaneous symmetry breaking (SSB) as the key to such phenomena; instead, we want to focus on the role of contexts.

In the literature on emergence and reduction, the topic of universality and the like has become one of the main fronts in the debate, as witnessed in Chapter 4. As we saw there, the concept of universality began its life in thermodynamics but has been discovered in both classical and quantum cases (and in biology and neuroscience, as we shall see in Chapter 6). The presence of contextual emergence in many-bodied CMP systems demonstrates that ontological emergence is not some rarified phenomena occurring only in

Emergence in Context. Robert Bishop, Michael Silberstein, and Mark Pexton, Oxford University Press.
 DOI: 10.1093/oso/9780192849786.003.0005

highly isolated experimental setups involving a few entangled particles and the like.[1]

5.2 Condensed-matter physics and contexts

CMP deals with large collections of particles which all obey the well understood laws of physics. Yet properties of these collections of simple particles are themselves not well understood. Methodologically at least, CMP is not a simple matter of scaling up from small numbers of particles to large numbers. In this respect, there is overwhelming evidence from the practice of CMP that it involves, at the very least, a kind of epistemic emergence (e.g., Anderson 1972; Laughlin 2005; Laughlin and Pines 2000). Furthermore, condensed-matter physicists, such as P. W. Anderson and Laughlin, argue that CMP is as fundamental as any other sort of physics. Indeed, for most CMP systems the "theory of everything" is already known: it is QM! CMP systems are entirely composed of either classical or quantum particles, and the base laws are well understood. Nonetheless, knowing these facts about CMP in no way allows one to predict or account, epistemically at least, for the range of properties and regularities observed in CMP systems.

For some in the philosophical community, the idea that physicists might use emergence talk made the subject worthy of respectable investigation again. Emergence was no longer a defunct philosophical idea allegedly disproved by advances in quantum chemistry in the mid-twentieth century. *Physicists themselves* were turning to emergence as a conceptual framework for understanding the relationship between CMP and "fundamental physics."

One can see why condensed-matter physicists moved in this direction. CMP is a science of the interaction both of particles and of properties that can be ascribed to different levels or scales, and the formation of new domains decoupled from specific properties of other domains. For instance, as the physicists Feng Duan and Jin Guojin describe CMP, "it is in the subtle interplay of the coupling and decoupling of different strata in the structure of matter that the gist of emergent phenomena lies" (Feng and Jin 2005, p. 6). We will be arguing that the coupling and decoupling is contextual in two senses. First, it depends on multiscale contexts to be produced. Second, such coupling/decoupling brings about changes to those contexts themselves leading to new contexts. A CMP system requires a concrete context to constrain the system in a particular way to allow changes to that system to manifest. In turn, those changes themselves can interact with the context

[1] See Franklin (2019) for a discussion of whether universality in RG explanations is reducible.

to produce a new context for the system, thereby producing a new emergent system/context state.

The circumstantial evidence for emergence in CMP comes in three forms that will be familiar from the contextual emergence pattern that we have encountered in previous chapters:

1. As P. W. Anderson and Laughlin observed, knowing the underlying laws of physics does not allow one to write down—or even predict the general form of—the laws of the condensed-matter system in advance of instigating the system in question.
2. The properties of large collections of particles in a condensed-matter system are radically different from the properties of the constituent particles in small numbers.
3. Condensed-matter systems can be highly sensitive to changes. Simply changing the concentration of a minority element in a system can so profoundly change the context that radically different properties manifest.

One may be tempted to view this evidence as an indication of the methodological limitations of condensed-matter physicists: "If only they had greater calculational tools they would be able to predict the laws of CMP systems ...," as is often mistakenly claimed for quantum chemistry. Yet as we have seen throughout this book, this is a failed response. The laws of CMP systems are very often concerned with properties that simply have no well-defined meaning outside a system with large numbers of particles. This is consistent with the contextual emergence pattern.[2] One must already have the larger-scale CMP system displaying its strange new set of aggregate properties to then reconstruct how the underlying particle interactions might give rise to those properties. Because nature itself requires specific contextually produced larger-scale structures to manifest specific properties in CMP, it is no wonder that one could not epistemically derive those properties without also knowing those contextual structures.

As we have indicated, we believe contextual emergence, in some sense, can be seen as a unifying framework for many of the accounts of emergence in CMP extant in the literature. We will argue that failures of mereological supervenience (Silberstein 1999, 2002, 2017a; Lancaster and Pexton 2015), symmetry breaking (Morrison 2012), universality and RG methods (Rueger 2000a,b; Batterman 2002), elimination of degrees of freedom (DoF) (Wilson 2010; Bain 2013), and fusion of properties (Humphreys 1997) are all different reflections of the way contextual features interact to allow the manifestation of new properties, and the ways the compatibility relations of those

[2] The inability to predict CMP laws is, as P. W. Anderson (1972) has argued, an in-principle, not an in-practice, epistemic failure.

new properties together with the compatibility relations of the contextual properties define the new modal space of an emergent CMP system. This claim is not meant to trivialize the distinctiveness of any of these authors' accounts of emergence; rather, it is to say that these accounts can be reframed in contextual emergence terms.

Although we will argue that CMP is a prime example of contextual emergence, it is not our claim that emergence in CMP is somehow *more contextual* than other cases of emergence. After all, our argument has been that contextual emergence itself is fundamental—context is as important as any laws one can point to. Rather, it is that CMP wears its contextuality on its sleeve in a particularly illustrative way at particularly advantageous scales and intervals for humans to study.

The contextuality of CMP on which we will focus is where physical collections of particles create new environments—new internal or endogenous contexts—as well as change the properties manifested by the elements making up those collections. These new environments create new stability conditions, which, together with the new manifested properties, define new domain boundaries—new object/environment distinctions. Thereby, a new external context for the system arises. These two contextual processes are linked and feed back on each other.

The new object/environment distinctions could be the defining of a new energy regime that the system occupies, for instance. These new contexts produced by new object/environment distinctions are not merely epistemic. They have unique ontological distinction and implications, reflecting genuine structures in the world. Once a system has isolated itself in such a way that a new object/environment context has manifested, that new context can itself dialectically influence the physical conditions of the system in the future by providing new stability conditions.[3]

These new environments, constraints, and boundaries are produced between feedback from the initial collection of particles, their first context, and the emerging context. The particles collectively change that first context rather like an organism manifests its properties in one ecological environment, but by doing so may change that very environment, which, in turn, changes the context of the organism, allowing it to subsequently manifest new properties.

This dynamic and dialectical interaction between object/context and context/object is both the product of the way that property compatibility relations and constraints restrict the available possibility space, and a product of new properties whose compatibility relations and constraints further shape the possibility landscape of the system.

[3] Recall how we saw similar behavior in Bénard cells providing stability conditions changing/constraining access to fluid parcels' states of motion in possibility space.

It is because of the highly contextual and dialectical way in which the possibility space is constrained that results in CMP being so nomically inhibited in a way not dictated by "fundamental" physics. The particles in small numbers have one well-defined context (boundary conditions, constraints, etc.), a robust set of manifested properties, and, hence, a stable possibility landscape reflected in relatively simple and reliable law-like regularities. Once the number of those particles increases to the large numbers seen in CMP systems, the context itself becomes radically different, with the compatibility relations of the new properties of the collective being an important new stability condition. It is no wonder that the shape of the possibility landscape of a complex CMP system is radically different from that of the parts outside of that collective system.

"Fundamental" physical laws and properties only contribute some necessary conditions for the initial context for the collective CMP systems—the starting point. The end point involves new stability conditions not given by the microscale fundamental laws and properties. These stability conditions contribute the remaining necessary and sufficient conditions for system behavior and properties. From this vantage point, P. W. Anderson's observation that more is different applies so strikingly to CMP: *contexts can be frequency dependent; having more fundamental particles changes the contextual framework of those particles, which changes the properties they manifest.*

Note here an important general point. *What matters is how radically the context is changed by interactions, not strictly speaking the number of particles involved.* Hence a relatively small change, such as to the number of defects in a system, can have a huge impact because the context is so sensitive to those changes. By contrast, in another system, adding significantly more particles—adding more air to a room, say—does not necessarily alter the context.

To understand how contexts play an important role in CMP, it is useful to first look at the historical development of CMP along with some of the core concepts in wide use. Then we will look in detail at some CMP case studies highlighting the role of contexts.

5.3 Basic concepts in condensed-matter physics

There are many core concepts in CMP that will be useful to understand for the purposes of discussion. Those curious for more technical details about the physics of the issues discussed are encouraged to consult several good CMP books: Anderson (1997), Lancaster and Blundell (2012), Feng and Jin (2005).

CMP is fundamentally concerned with the state of a collective system known as the phase, and how such a system changes between phases. Each phase is identified by qualitatively very different properties that are manifested only in that particular phase. For instance, water flows only in the liquid phase, whereas water as ice is rigid only in the solid phase. CMP as a discipline covers phenomena over a vast range of scales: time scales from femtoseconds to tens of years, length scales from tenths of nanometers to several meters, and temperature scales from nano-Kelvin to thousands of Kelvin. CMP deals with solids and liquids, metals and insulators, and many strange states difficult to imagine. Generally, heavy particles, such as atoms, are regarded as particles, while electrons are treated as waves. Hence, modeling in CMP combines both classical and quantum descriptions; that is to say, that CMP deals with multiscale models. This creative combination of different conceptual frameworks reflects facts about the domain of CMP such as the huge difference in the temperatures below which ions and electrons respectively become dominated by quantum, rather than classical, effects (the degeneracy temperature).[4]

5.3.1 *Compartmentalization of phase space and contextual boundaries*

In CMP it is conventional to call the state space for condensed-matter systems phase space since the phase is the system state most of interest. A system is said to explore phase space if its salient parameters change to be many different sets of the values corresponding to a large chunk of the representation in phase space.

One important and regularly used concept is that of compartmentalization of phase space. The basic phase space of the system is divided into zones, only some of which the system can explore, but with others it cannot. Take, for example, a gas condensing into a liquid. In this process, a barrier forms between the liquid and the surrounding air, meaning that liquid particles can only occupy spaces inside the barrier, and gas outside it. This is represented by the phase space being divided into two regions representing the liquid and gas phases. When the system is in the liquid phase it cannot freely explore the region representing the gas phase. Similarly, when a liquid is frozen to form a solid, many cells are formed. Particles cannot move outside of their cell, so phase space becomes even more compartmentalized than in a liquid.

Compartmentalization of phase space is an abstract way of talking about the role that constraints play on a system. A particular constraint can prevent a system doing something (e.g., occupying a region of phase space), or it can open an otherwise forbidden part of phase space. One way of imagining phase space is as a house. Some of the spaces are open plan, others are rooms

[4] This situation is similar to quantum chemistry, where scientists work with a quantum/classical algebra of observables and span several length and time scales.

connected only by particular doors. Some of those doors are locked and can only be opened by the right key. Each of the rooms has unique facilities in them, so some things can only be done in one specific room. For instance, cooking can only be done in the kitchen, and if you do not have the key to get into the kitchen you cannot do any cooking. CMP systems are like this, where only in specific regions of phase space can specific properties manifest. Whether it is possible to get to those regions in part depends on the constraints acting on the system—constraints never given by the microscale or "fundamental" laws.

The boundaries that compartmentalize phase space are *contextual boundaries*. They are produced by the stability conditions defining the context of the system. These conditions can define both exogenous and endogenous contexts (a subject we will return to); they reflect contextual conditions in the environment outside the system, and the internal environment of the system, respectively.

5.3.2 *Order*

Much of the behavior observed in CMP reflects some specific kind of ordering of the system. As with phase space, there are several ways that a system can be ordered. For atoms, one important and simple kind of order is positional order, such that the atoms are not randomly positioned. Order reflects correlations between parts of the system.[5] As we saw in Chapter 2, if the correlations cover the whole of the system we have long-range order; if the correlations only cover nearby parts we have short-range order. An ideal gas has nearly perfect disorder—the positions are random. Liquids have short-range order without any periodicity (though when organizing into convection cells, the order becomes long-range). In contrast, a crystalline solid will have long-range order with a periodic pattern to that ordering. The correlations associated with order essentially reflect the system's ability to transmit information to itself. If a solid feels a force because of the order associated with rigidity, every part of the solid will feel the force to some degree very quickly; whereas in a gas, one could disturb a part of the gas, and the rest of the gas would either not feel that disturbance or it would take a relatively long time for the information about that disturbance to be transmitted to through the system. For quantum particles, "order" typically refers to an ordering with respect to momentum space. The same can be said of nonequilibrium classical systems such as hydrodynamic systems.[6]

[5] Recall that Rayleigh–Bénard convection exhibits this same kind of order in terms of the fluid parcels.

[6] For hydrodynamic systems, the important correlations are often in the velocity field, (e.g., Rayleigh–Bénard convection).

The crucial point to stress is that order is produced in a particular environment external to the system (an exogenous context), as well as in the environment within the system (endogenous context). If a particle is displaced, the effect of this is very different, depending on the local environmental context (e.g., if in a rigid ordering the information is transmitted, if in a random gas no information will be transmitted). The same displacement of the individual particle has occurred, but the effect which that same event has depends on the environmental context, both internal and external.

5.3.3 *Symmetry*

The landscape of material physics in the 1930s and 1940s was dominated by work in solid-state physics, leading to spectacular breakthroughs, such as the invention of the transistor. Solid-state physics was principally concerned with the properties and structures of periodic crystals modeled by considering the way in which waves move through a periodic lattice. By the 1970s, solid-state physics had been supplanted by CMP. This change occurred because of a widening of the range of phenomena investigated, but also, and more interestingly, because of a shift in the conceptual framework physicists used. The midwives of this conceptual change were Lev Landau and P. W. Anderson.

The shift in subject matter involved extending the field to areas where many-body interactions are important such as phase transitions. Landau's theory of second-order phase transitions introduced the concepts of the order parameter and emphasized the crucial importance of broken symmetry (P. W. Anderson 1972). Just as it is impossible to be a little bit pregnant, in CMP it is impossible for symmetry to be only partially broken. Either a system has a symmetry, or it does not. The transition from a symmetrical state to one without that symmetry correlates with a phase transition along with the appearance of a lower symmetry state of the system.

One way of thinking about symmetry in CMP is to ask how a system can be changed arbitrarily without noticing a difference. For example, a gas is in a high symmetry state. If we imagined rotating all the gas in a room by 180 degrees, we would not notice any difference in the properties of the gas—there is no preferred direction of orientation. In contrast, solids can have many preferred directions of orientation and, hence, have fewer symmetries. Consequently, they have many more interesting properties than do gases. This is because stability not only cuts off access to various regions of probability space, but can also open otherwise inaccessible regions, as we saw with Rayleigh-Bénard convection. By reducing the symmetry of aggregate systems, we often see novel and unexpected properties appear. "[In CMP] *broken symmetry is closely connected with a change in structure and the emergence of new physical properties*" (Feng and Jin 2005, p. 17, emphasis added). The breaking of these symmetries is related to different CMP properties, and some examples are shown in Table 5.1.

One standard recipe for CMP is to ask what the naturally lowest energy state (or ground state) is for a system that is in an ordered phase. To answer this question we must consider the type of interactions between particles and look for the lowest-level excitations above the ground state. We can then simplify the modeling of the problem by treating the system as if these lowest-level excitations are the particles that make up the system. These excitations are called *quasiparticles*. This is the recipe for finding the quasiparticles of an ordered phase. By doing this, we take a difficult-to-model problem based on the actual particles in the system, which are interacting in complicated ways, and turn it into a simpler description in terms of quasiparticles which are not interacting. The quasiparticles more simply capture the dynamics of the system. Which quasiparticles are the appropriate ones to describe the system depends on the internal environment of the system, what the particles are and their interactions, what the order is, the external environment, constraints, dimensionality, energy, and so forth, and how these determine the systemic properties in question.

To foreshadow the contextual emergence discussion: the interactions of particles create with the wider context a new collective environment. The possibility space associated with that new environment is captured by the properties of the quasiparticles, which is why the quasiparticles have the epistemic utility they do as a preferred set of "good coordinates" to model the system. That is, the interactions of the basic particles compartmentalize the property space of those particles that manifest the properties of the

Table 5.1 Examples of symmetry.

Phenomena	Broken symmetry	Elementary excitations	Topological defects
Ferroelectricity Antiferroelectricity	Space inversion	Optical phonon	Domain wall
Ferromagnetism Antimagnetism	Time reversal	Spin wave	Domain wall
Superconductivity	Gauge invariance	Electron	Vortex line
Liquid crystal	Rotation Translation	Nematic states Smectic states	Disinclination
Crystallization	Translation	Phonon	Dislocation
Electronic crystallization	Translation	Ripplon phason Amplitudon phason Amplitudon	Discommensuration

quasiparticles. It is the compatibility/incompatibility relations of the quasiparticles which define the modal relations salient to the system. It is not that the basal particles cease to exist; rather, it is that the modal space is only mapped out by those basal particles interacting and manifesting the properties of the quasiparticles. We will have more to say about this when we look in detail at the fractional quantum Hall effect.

5.3.4 *Generalized rigidity*

Another crucial concept in CMP related to broken symmetry we already met in Chapter 2: generalized rigidity. In a crystal, translational symmetry is broken, so it is rigid: atoms are fixed in position. Hence, if force is applied, then the atoms transmit that force throughout the crystal without dissipation. P. W. Anderson's great insight was to see that there could be a similar effect due to other broken symmetries, each defining a different sense of "rigidity." For instance, in a superconductor the rigidity is a phase coherence between Cooper pairs, and can transmit current without dissipation.

Anderson's ability to generalize a notion, such as rigidity, that would at first glance seem so materialistically specific (e.g., it occurs in Rayleigh–Bénard convection) indicates the importance of higher-level structure in CMP. As we have seen several times in this book, it is the structural patterns that provide the context for very disparate microscale phenomena to manifest sets of properties having similar compatibility/incompatibility patterns. Larger-scale physical structure defines a context, leading to the manifestation of properties, and generating a structure of compatibility/incompatibility relations defining system possibilities.

Microscale or elementary excitations are too weak to influence generalized rigidity. Sitting above the energy level of elementary excitations are what are known as *topological defects*. These are things, such as domain walls in ferromagnets, which lie between zones of magnetism or dislocations in crystal plasticity. One way of thinking about them is that they are remnants from the higher-energy symmetry phase. As the symmetry is broken, there are pockets that "remember" the symmetry of the previous phase. Indeed, both elementary excitations and defects remember the symmetry that was broken. Topological defects can partly disrupt the generalized rigidity of a broken-symmetry system. Note that a topological defect should not be confused with a topological state of matter, such as the fractional quantum Hall (FQH) state discussed later. Topological defects are present in states of matter described by symmetry, whereas topological states of matter are not described by symmetry but by topological properties.

The notion of a topological defect is contextual by definition. Defects are possible only if there is a minimal environment of order to begin with. A defect cannot exist in the high-symmetry phase. Only once the symmetry is broken and there is some sort of order can there be a deviation from that order which is a defect. Consider a periodic crystal structure. To have an atom out of place there must be a context of a pattern to begin with. The context must be changed to have a defect. Then, the interaction of the defect with the new lower-symmetry context itself manifests new properties of the state as a whole. In CMP it is not just a matter of moving between symmetry states, but of excitations and defects which remember the symmetry of one state interacting with the new lower-symmetry context. In a way, the excitations and defects are where the system puts the lost symmetries.

Also note that the fact that defects remember lost symmetries is consistent with the link between contexts in contextual emergence. In contextual emergence, new contexts are forged out of interactions with old contexts. To capture the ontology of the system, we must ask which properties and their required contexts ground the appropriate possibilities. The properties and contexts may come out of more elementary properties and contexts, but those properties and contexts contribute only part of the necessary conditions for all the salient modal relations, as we have seen time and again in previous chapters.

The notion of broken symmetry (and remodeling systems in terms of elementary excitations and defects) forms the conceptual shift from solid-state physics to CMP. Thinking in terms of broken symmetry was required by expanding the empirical range of the discipline. Nevertheless, not all phase transitions are related to broken symmetry. The gas/liquid transition does not involve broken symmetry, nor does the metal/non-metal. Instead, these changes are due to the idea of breaking *ergodicity*. Ergodicity is an idea of Ludwig Boltzmann, the father of statistical physics in the nineteenth century. Ergodicity, non-technically speaking, is the idea that systems left to their own devices will explore all of their phase space. It can be broken by the compartmentalization of phase space, so that specific sectors of the phase space become inaccessible to constituent particles. Liquids are examples of broken ergodic phases in phase space; they can only explore a subset of sectors such as a particular region of a container. In contrast, gas particles can explore all the positions in a room.

Broken symmetry always implies broken ergodicity, but broken ergodicity does not require broken symmetry. The notion of broken symmetry can be generalized to include broken ergodicity. Finally, there are other states of matter, such as the FQH state, not determined by their symmetry but instead by their topology; moving from one topological arrangement to another determines their phase.

Note the general principle: In all three—broken symmetry, broken ergodicity, and topological states—the difference between phases is a different set of stability conditions on the salient phase space. The nature of the stability conditions at play is unavoidably contextual. Both the interactions of the constituent particles, and other outside factors, such as the dimensionality of the system, combine to produce an overall context for that system defined by the stability conditions on its phase space. It is these conditions that are required to manifest the unique properties of the phase.

All phases share something in common with the properties of other phases. Liquids, solids, and gases have mass, for instance. But the identity of a phase—why we can pick it out and give it a name, which reliably empirically tracks with the world and a specific object/environment distinction—depends on the unique combination of properties. Only in a particular context defined by the stability condition making those properties possible will that particular confluence of properties manifest.

For example, when a non-conductor changes into a conductor, both are linked by a continuum—they share many properties that vary smoothly. But the salient property for the aspect of identity—whether it conducts electricity or not—changes rapidly and flips between radically different qualitative states. Here there is no sense of entirely separate domains of reality existing in parallel. Rather, it is simply that changing the context can sometimes rapidly and radically change particular properties (but not others) salient to object/environment distinctions.

This is an emphasis on, as it were, *meta-context*. The contextual nature of whether a change in a context produces a large or small change stands in stark contrast to the pure structuralism of the "rainforest realism" of Ladyman and Ross' (2007). In their account there are simply parallel tracks, each one defined by modal relations, each one as worthy of ontological status as the next. A set of turtles on each other's backs, independent but equally valid. Such a picture correctly emphasizes the centrality of modal notions, but because it leaves modality unexamined it overlooks much. In contextual emergence, those modal relations are linked to properties through the compatibility/incompatibility relations. The properties are related to other properties in different domains. The properties from one domain play a crucial role in defining, in part, the context for the production of the properties manifest in another domain. In contextual emergence, the contexts are cross-domain and multiscale, as we have seen. There is no ontology of separate domains that float free from each other. Domains are only epistemically distinct because, with respect to specific aspects, there are particular natural object/environment distinctions. The separate parallel tracks of Ladyman and Ross (2007) are all implicitly relying on contextual features provided by other tracks. Only by understanding the dialectic among domains and scales can we account for the world around us. The world is simultaneously continuous and discontinuous—there is no contradiction here; rather, there is a necessary interdependence. Discontinuity in property manifestation and modal relations results, in part, from how different contexts interact with systems. Yet changing contexts requires a continuity: new contexts manifest only out of old contexts, and very often this change of contexts is itself continuous.

5.4 Contextual aggregate environments

One of the ways that CMP displays contextual emergence is the way in which aggregating microscopic particles changes the environment of those particles. A large collection of particles is not a simple summation of their properties in isolation; instead, the aggregate itself creates a new context for the microscopic constituents. In a particular sense, an electron is no longer the same bundle of properties in a complex aggregate environment as it is in isolation, as we saw with Debye shielding in Chapter 4.[7] The aggregation of particles creates new contexts, some of which create constraints that feed back on the aggregate itself (e.g., picture how Bénard cells constrain the fluid parcels). In general, a complex aggregate will have exogenous contextual features from outside the aggregate (such as the dimensionality of the system), and endogenous contextual features from the environment created by the particles for themselves. We will now look at two case studies of how aggregation changes important contextual features of a system: superconductivity, and the FQH effect.

5.4.1 *Superconductivity and gauge symmetry breaking*

Superconductivity is a state in which there is no resistance to current flow. It was discovered in 1911 by Heike Kamerlingh Onnes after testing the electrical resistance of a very cold sample of mercury. Metal atoms are a fixed lattice around which their loosely bound electrons can move around when an electric field is applied. In ordinary metals, thermal fluctuations randomize the electrons and interrupt this flow, causing the electrons to lose energy. This is the nature of electrical resistance. For ordinary metals, this process occurs no matter how cold it becomes, but for superconductors, if the metal is cooled below a critical temperature (which is different for each metal), T_c, then the electrons suddenly shift into a highly ordered state and travel collectively. The thermal fluctuations no longer disrupt the flow, and resistance

[7] This will clearly be seen in the case of the FQH effect, in which electrons become tethered to quanta of flux. In a very real sense, the system of an FQH state is no longer composed of electrons in the standard sense.

drops to zero. It is important to note that this effect is not a classical linear extrapolation. The resistance does not become progressively smaller until it seems so small that it appears to be zero. Rather, superconductivity is a purely quantum phenomenon. At the transition marked by T_c, a phase change occurs, and the superconductor suddenly has *literally zero* resistance.

It was not until 1957 that a widely accepted theory of superconductivity was suggested by John Bardeen, Leon Cooper, and Robert Schrieffer (1957): the BCS theory. In this theory, as an electron moves through the lattice of atoms it deforms the surrounding environment, leaving a trail. This trail pulls in another electron, and the two electrons form a linked state known as a Cooper pair. As the temperature lowers more and more, Cooper pairs form until at the critical temperature all of them form a collective state called a condensate. In this macroscopic quantum state, all the wavefunction phases are aligned.

There are some important things to note here. First, intuition might suggest that Cooper pairing involves electrons that are near each other, but this is not the case. In fact, the pairs can be quite spatially separated. Second, it is only when the phase transition occurs, and a global alignment of the wavefunctions happens, that a superconducting state occurs. In this case, thermal fluctuations cannot disrupt the flow of electrons because to disrupt one Cooper pair would require the whole system to be changed (generalized rigidity). The collective state provides a new environment for the electrons (or Cooper pairs in this case) until the point where the collective environment has changed so much that a new context is formed, and there is a new object/environment distinction—now the system acts as a whole.[8]

Superconductivity has attracted philosophical interest as a case of emergence. The arguments usually involve those who suggest that the BCS theory is indicative of the fundamental microscopic resultant nature of superconductivity (Howard 2007), while others argue that superconductivity involves some top-down form of determination. We will now briefly examine the best-articulated version of the latter argument, due to Margaret Morrison, before suggesting that this distinction between bottom-up and top-down has missed something important: whichever way the phenomenon is viewed, it requires contextual emergence. Viewed as a bottom-up phenomenon,

[8] BCS theory is not the end of the superconducting story, however. In the 1980s, high-temperature superconductors were discovered. The term "high-temperature" is relative here, as these materials still need to be very cold to be superconductors. Explaining high-temperature superconductivity is still an area of active research. There are currently two competing accounts: valence-bond theory due to P. W. Anderson (1987), and spin-fluctuation theory by a group led by Philippe Monthoux (Monthoux, Balatsky, and Pines 1991). The details of these accounts are not relevant here, as neither theory is adequate to account for the range of superconducting states observed. Suffice to say, they both involve mechanisms that allow Cooper pairs to form more easily at higher temperatures by using structural features of the materials involved.

superconductivity still requires exogenous contextual features to intersect with the interactions of electrons changing the collective environment internally (endogenous context). Similarly, for Morrison's top-down emergence to occur, this also requires contextual features. As is usually the case in the literature, the role of stability conditions defining the contexts has been neglected.

For some systems, the aggregate is a necessary gateway to allow a higher-level stability condition to operate. *It is only the aggregate itself that can be subjected to a particular higher-level stability condition—only the aggregate can "see" the context, as in superconductivity and the breaking of gauge symmetry.* Although we will argue that gauge symmetry is unclear (instead, phase coherence or whole-system topological entanglement is more likely the key constraint), to even apply gauge invariance (or these other descriptively similar but ontologically radically distinct constraints) the aggregate is required to be the subject of the relevant stability condition. A contextual disposition, if you will, is only activated by an aggregate. Here we see the *implicit relationality* of contexts. Relevant contextual constraints such as gauge are contextually dependent on the aggregation of microscopic "parts." The lower level contributes some necessary conditions for the larger-scale property to manifest, while the larger-scale property acts as a constraint on the microscopic particles forming the aggregate.

5.4.1.1 MORRISON'S ARGUMENT FOR EMERGENT SUPERCONDUCTIVITY

Morrison's argument rests on the distinction between acknowledging an ontological relation between an emergent phenomenon and its microscopic base, and the notion that the microscopic base constitutes the phenomenon. In the case of superconductivity, Morrison's claim is that the identity of the phenomenon—superconductivity as a natural kind—is properly identified only by the presence of larger-scale gauge symmetry breaking, not by the microphysical base.

Morrison argues that the most indicative properties of superconductivity—and most importantly, exact properties of superconductors—are not derived from microphysical models alone (e.g., BCS theory). Instead, key properties of superconductors are produced by imposing the SSB of gauge on the system. As Morrison puts it:

> One needs detailed models like BCS to explain the specifics of how the SSB occurs, at what temperature superconductivity is produced, and as a basis for approximate quantitative calculations, but not to derive the most important exact consequences of this breakdown—infinite conductivity, flux quantization, and the Meissner effect—properties that define superconductors
>
> (2012, pp. 151–152)

As we saw with the renormalization group in Chapter 4, the microscale details—properties and laws—about how Cooper pairing takes place are not relevant to explaining superconductivity. In other words, "the characteristics that define the superconducting state are not explained or predicted from [microscale] processes and are independent of them" (p. 152). Change the microscale base, and the superconducting properties still emerge. "Although the breakdown of gauge invariance involves the formation of Cooper pairs—a dynamical process—the micro story figures simply as the foundation from which superconductivity emerges (p. 152).

Morrison takes seriously the central paradigm of CMP—that symmetry breaking is related to phase transitions. Superconductivity is a phase transition associated with broken gauge symmetry. Hence, analogous to a ferromagnet gaining its properties by breaking symmetry and having magnetic moments align to create an overall magnetization, so too a superconductor gains its properties by breaking gauge symmetry.

Furthermore, one can mathematically derive these exact properties of superconductors without any reference to microphysical details, such as Cooper pairs. One can simply describe the system in terms of the gauge symmetry breaking—an associated order parameter that allocates a value to the symmetry breaking, and what are called Goldstone modes (mathematical ways of describing fluctuations of the order parameter). By considering the symmetry properties alone, one can derive the properties which constitute the definition of a superconducting state. The insensitivity of such a derivation to microphysical details is so great that it does not even matter if one is dealing with a system of fermions or bosons. As Morrison puts it: "Although superconductors are constituted by their microscopic properties, their defining features (infinite conductivity, flux quantization, the Meissner effect) are immune to changes in those properties (e.g., replacing fermions with bosons)." The breakdown of electromagnetic gauge invariance "provides the dynamical explanation of emergent phenomena, but the specific microphysical details are irrelevant; how the symmetry is broken is not part of the account" (p. 156).

This is the sense in which superconductivity as an "emergent phenomenon is not reducible to its microphysical constituents, yet both retain full physical status" (p. 156). There is no determination relation between the microscale constituents and macroscale phenomena. Hence, Morrison notes that supervenience "understood in terms of a dependency relation, is inapplicable in explaining the part-whole aspects of" superconductivity as an emergent phenomenon (p. 156).

The crux of Morrison's argument is not that we can conveniently ignore some microphysical details; rather, it is that we must refer to larger-scale features such as the symmetry breaking of gauge to account for superconductivity. It does not really matter how gauge symmetry is broken so long as there is an associated phase change to a superconducting state. Superconductivity is multiply realizable with respect to specific microscale details.

There is much merit to Morrison's argument. Clearly there are resonances with previous chapters in this book. When large-scale features are the properties that encompass the relevant possibilities for a phenomenon, it is not just convenience to refer to them. Such references are irremovable. Nevertheless, applied to the SSB of gauge, there are subtleties and potential difficulties in both the standard CMP textbook treatment and Morrison's argument that need elaboration. There are two points to discuss. First, gauge is an unlikely candidate to be the right larger-scale stability condition providing the crucial constraint. Second, even if it were, or some other stability condition is at work, this is not context-free.

5.4.1.2 THE PHYSICAL SIGNIFICANCE OF GAUGE

Despite the often-complicated mathematical descriptions of gauge theories in physics textbooks, the basic idea behind gauge symmetry is relatively simple. When we have a redundant degree of freedom in a system, a gauge invariant description should be one that preserves the fact that this degree of freedom is ours to play with. As a simple example, in an electrical circuit one point is designated as "ground" and given the potential zero. Yet in a circuit we only ever measure potential differences—the change in potential between two points—so we could have ground be any value of potential we like, and ground and its relationship to all the points in the circuit would be exactly the same. This is gauge invariance or "symmetry."

Note that this kind of symmetry is not really a symmetry in the usual physical sense. It is an invariance to changes in descriptions. There has not been much philosophical discussion of gauge symmetry in the context of superconductors; such is the originality of Morrison's account. Fortunately, physics has provided us with a case very similar to superconductivity in the realm of elementary particle physics: the Higgs mechanism. Here, gauge, and its physical significance, has generated much interest.

In some sense, the Higgs mechanism is the application of ideas developed for superconductors to the whole universe. It was P. W. Anderson who first made such an intellectual connection, and crucial steps were provided by many others, including Peter Higgs, after whom the mechanism is named. Essentially, in Higgs' theory the early universe undergoes a phase change in which a gauge symmetry is spontaneously broken. This breaking of symmetry attaches the property of mass to some elementary particles (depending on the specifics of how they interact with the Higgs field). There are many parallels between Morrison's account of what is the agent of determination in superconductivity (SSB of gauge), and the Higgs mechanism, where a similar gauge symmetry is spontaneously broken.

Steven Weinberg notes:

> Higgs, Kibble, and others ... showed that if the broken symmetry is a local, gauge symmetry, like electromagnetic gauge invariance, then although the Goldstone bosons exist formally, and are in some sense real, they can be eliminated by a gauge transformation, so that they do not appear as physical particles. The missing Goldstone bosons appear instead as helicity zero states of the vector particles, which thereby acquire a mass.
>
> (Quoted in Lundqvist 1992, p. 545)

This certainly appears to be very similar in ontology to Morrison's claims for superconductivity. The SSB of gauge, order parameter, and Goldstone modes are invoked by Morrison as the key parts of the essential derivation of superconductors' unique properties.

Nonetheless, in the Higgs case at least, such claims have some philosophers concerned that too quick a jump to ontology has been made. For instance, Holger Lyre writes:

> [T]he status of the symmetries in question, gauge symmetries, is in fact a non-empirical or merely conventional one precisely in the sense that neither global nor local gauge transformations possess any real instantiations (i.e. realizations in the world). Rather their status is comparable to the status of coordinate transformations ... How is it then possible to instantiate a mechanism, let alone a dynamics of mass generation, in the breaking of such a kind of symmetry? Suspicions like this should raise philosophical worries about the true ontological and explanatory story behind the Higgs mechanism.
>
> (2008, pp. 2–3)

John Earman perhaps expresses the same thought even more directly: "A genuine physical property like mass cannot be gained by eating descriptive fluff, which is just what gauge is" (2004, p. 1239). And Christopher Smeenk similarly writes that "if gauge symmetry merely indicates descriptive redundancy in the mathematical formalism, it is not clear how spontaneously breaking a gauge symmetry could have any physical consequences" (2006, p. 488).

The essential complaint of Earman, Lyre, Smeenk, and others is that a gauge symmetry itself is not the same thing as a physical symmetry. This view is also expressed by many physicists. Xiao-Gang Wen has described gauge symmetry transformations as "do-nothing" transformations. In quantum systems, a gauge symmetry is not a symmetry, in the sense that a gauge transformation does not alter any quantum state. Quantum gauge theory is not associated with a Noether theorem, for instance, the way physical symmetries are such as rotating a system invariant under rotations. Indeed, there is a lot of descriptive convention associated with gauge—two different gauge groups can sometimes describe the same physical theory. So, when we describe SSB of gauge, *no physical symmetry breaks*. A phase transition occurs that can be tracked by the associated order parameter (and Goldstone boson), but the gauge itself is just a kind of redundancy in the description of a system. Whereas in a ferromagnet the breaking of a symmetry corresponds to a real physical symmetry, breaking of gauge does not.

There is a further complication, as has been pointed out by Simon Friederich (2013). Recall that there are two forms of gauge theory: local and global. According to quantum gauge theories, *local* SSB of gauge is impossible. This result is known as Elitzur's theorem. In lattice gauge theory there is no local SSB. Now, the spontaneous breaking of *global* gauge symmetries is not forbidden by Elitzur's theorem. The left-over global symmetries are called remnant symmetries, but the breaking of these remnant gauge symmetries (RSSB) in general does not correspond to phase transitions (Caudy and Greensite 2008). Both Coulomb and Landau gauges can have RSSB without phase transitions. Morrison associates breaking of gauge with phase changes, and in superconductors this happens to be true. But, in general, *there is no link between gauge breaking and phase changes*. Moreover, the way these remnants break is dependent on the way one represents the system. According to William Caudy and Jeff Greensite, the parameter values for which there is a transition between broken and unbroken gauge symmetry depends on the (physically arbitrary) choice of remnant subgroup. For example, whether the Higgs field has RSSB depends on a choice of gauge fixing.

Morrison has suggested that SSB is unphysical for some systems: "It would be folly to accept a robust physical interpretation of the SSB story [of electroweak theory]" (2003, p. 348). She contends that this is because of ambiguities in what the vacuum state is. Friederich argues that no physical interpretation of SSB in electroweak theory can be given because "the breaking of the remnant subgroups depends on the gauge fixing" (2013, p. 25).

In general, there is a mixing of descriptive ability with physical determination in the discussion of gauge symmetry. For example, Weinberg argues that phase transitions imply the reality of gauge symmetries: "If a gauge symmetry becomes unbroken for sufficiently high temperatures, it becomes difficult to doubt its reality" (1974, p. 3359). If he is correct, then, in general, phase transitions should align with broken gauge symmetry, but this is not the case. There are good reasons to think that local gauge symmetry is never broken and that global symmetry breaking is physically arbitrary. Thus, phase transitions do not undermine the view that gauge symmetries are merely descriptive formal symmetries with no physical meaning, since whether those phase transitions correspond to a remnant gauge symmetry breaking depends entirely on free parameters.

The status of gauge, then, is philosophically controversial. At the very least it becomes an open question if SSB is explaining anything in superconductivity rather than describing it. Morrison offers no argument for how SSB brings about any changes in the system other than as a form of constraint. Morrison says it does not matter why SSB breaks as long as the microphysical state is such that it does. But this constraint could simply be descriptive. Once we have the microphysical conditions which can be described by SSB, we have superconductivity, but the bringing about appears to be all microphysical.

SSB turns out to be merely a handy descriptive summary, which is perhaps ineliminable epistemically/pragmatically because the microscale conditions that bring about flux quantization, say, are so heterogeneous that we cannot produce an exhaustive list of them. Morrison is quite aware that not having to invoke micro details is not enough for emergence. Rather, we must *not be able to use* micro details to explain phenomena. Hence, implicitly explanatory indispensability is central to Morrison's argument: "What is truly significant about emergent phenomena is that we cannot appeal to microstructures in explaining or predicting these phenomena even though they are constituted by them" (2012, p. 143).

The idea of explanatory indispensability is that if something must be referred to in an explanation, then it is worthy of ontic status. Yet one must be careful in tricky cases relying exclusively on such an indispensability principle. Gauge is surplus mathematical structure which may be representationally essential, but (since we have wider reasons for thinking it does not correspond to anything physical) it does not license an explanatory indispensability argument.

Cases of apparently ineliminable surplus mathematical structure appear in explanations of other physical phenomena. One example is the infinitely self-repeating on all scales of strange attractors in state space of dissipative chaotic systems (Smith 1998, chapter 3; Koperski 2001; Bishop 2016, section 5.1). Our models of strange attractors have infinitely intricate mathematical structure, but actual-world attractors lack infinitely self-repeating structure (Bishop 2016, section 5.1, and references therein). Alisa Bokulich (2008) discusses another example: the use of classical orbits in semiclassical models of quantum chaotic systems. In such chaotic systems, such as a quantum billiard, the wavefunction is "scarred" or altered at positions where the corresponding classical orbits of the system would have been. Nonetheless, the systems in question are purely quantum. There are no classical particles or orbits.[9] To describe the wavefunction scarring, researchers use semiclassical models which use the classical orbits as "explanatory fictions." The classical orbits may be indispensable, but wider considerations block the

[9] At least on most interpretations of QM.

usual move from indispensability to reification of terms. We *know* that the purely quantum system does not have classical orbits. In the same way, if SSB of gauge does not correspond to anything physical, then for wider reasons we know that it is not responsible for anything in the system.

There is another possibility, however. A global gauge symmetry, and its SSB in superconductivity, actually is a description tracking a genuine physical symmetry. Gauge, then, would not be responsible for anything in the system, but this other global symmetry would be. This is quite plausible in the case of superconductivity, especially if we compare it to a related phenomenon: superfluidity. In superfluidity there is no SSB of gauge, but there is a phase transition relating to a global phase coherence being established by the particles of the system. Superconductivity, in essence, is superfluidity, but with charged-particle microphysics. As such, the best way of viewing it may be that the gauge symmetry is actually coincident with a global wavefunction phase symmetry. The superconducting phase change, then, is associated with the formation of a larger-scale wavefunction, the key property of which is that it is impossible to change the relative phase of any Cooper pairing without changing the phase of the whole system at once (generalized rigidity).

We believe that this is more plausibly the sense in which a superconducting state is emergent. It is not that some magic gauge is broken, forcing the system to behave qualitatively very differently. It is that the environment for the microscale states becomes radically different. The way in which the collective set of particles' correlates creates a new internal context for each electron (again, picture Bénard cells creating a new internal context for individual fluid parcels). Instead of elaborating too much on this idea here, we will save discussion of this point for another case where the physics demonstrates this contextual emergence very clearly: the topologically ordered FQH effect. This digression will be philosophically useful, but also directly related to superconducting states, since some of the latest physics is reformulating superconductors as previously unrecognized topologically entangled collectives (see Hansson, Oganesyan, and Sondhi 2004).

Thus there are reasons to be suspicious of SSB gauge as forcing a system to do something. Nonetheless, whether we have SSB of gauge, or gauge is tracking some other global stability condition, there is a second point on which to focus: the application of such a global stability condition itself is contextually dependent on changes in the environment of the parts of the aggregate. That is, it is only once the local environment is changed for the electrons through their interactions that a property, such as gauge, could be broken. Morrison states that the microphysics is irrelevant because it does not matter how we break gauge, only that we do. In response to reductionist accounts such as Don Howard's (2007), this is a perfectly valid point.

But Morrison's claim has the unfortunate consequence of overlooking the role of context. For electrons to form Cooper pairs, particular exogenous stability conditions are needed. This should be obvious, since we cannot make a random collection of electrons form Cooper pairs. Such stability conditions compartmentalize the phase space ordering or structuring of the internal environment—an example of the compatibility relations mentioned above. The electrons form Cooper pairs (and possibly topological entanglements), and in this process the Cooper pairs radically alter the local environment of other electrons. They change the context of those electrons so that they manifest different properties.[10] This leads to a rapid and complete reformulation of the local environment of the particles, so they order and form a condensate with phase coherence. In turn, it is this condensate state that can break gauge and have gauge (if it does) impose a structural constraint leading to the exact natural kind of properties of a superconducting state.

Now, Morrison is correct that the above story could be altered in many microscale respects, but it could not be altered in all of them. There can be multiple realizability with respect to the contexts that produce a feature, but this does not undermine the requirement of a context. Gauge can break only if specific contextual aggregate microscale features are present. These features may not be the narrow set identified in reductionist accounts of superconductivity, but they are present nevertheless.

5.4.2 *The fractional quantum Hall effect and topological entanglement*

The FQH effect is a special type of collective behavior that appears in a two-dimensional aggregate of electrons.[11] When the temperature is low enough, the system becomes like a fluid of electrons that responds to an applied magnetic field in a very specific way. The resistance of the fluid lengthwise becomes vanishingly small. Meanwhile, instead of being a smooth linear relationship, as one would expect, the transverse Hall resistance shows unusual plateaus.

These plateaus occur at specific values of the so-called filling factor ν, which describes a ratio of filled to vacant electronic states. In cases when ν takes an integer number, we observe the integer quantum Hall (IQH) effect in which the Hall resistivity is found to be quantized in units of h/e^2. However, when ν is a ratio of integers (frequently one with an odd denominator),

[10] Again, think of Debye shielding as a simple example of such a collective environmental change.

[11] The FQH effect is at the forefront of putative cases of emergence in physics. Our discussion follows the philosophical analysis in Lancaster and Pexton (2015) and the physics analysis in Lancaster and Blundell (2012). We refer readers interested in (even more!) technical detail to the latter paper. See also Guay and Sartenaer (2016) and McLeish, Pexton, and Lancaster (2019).

then we observe the FQH effect, which has some very unusual properties. It displays a version of generalized rigidity; even though it is a liquid, it will not deform as a liquid. Furthermore, the excitations of the system carry fractional charges. Usually, charges are measured in integer units of the electron charge e; in other words, e, $2e$, and so on. Instead, in the FQH effect, the charges are not integers but fractions: $\frac{1}{3}e$, say. This is extremely unusual. Since an electron is a fundamental particle, it is not physically possible to chop it up into another particle with a fraction of e. This means that the excitations in the FQH effect are very unusual particles, neither bosons or fermions, but instead an exotic particle type only theorized in quantum electrodynamics: anyons.

Despite their similar names, the FQH effect represents a phase of condensed matter different from the IQH effect. This difference arises owing to the role of the interactions between electrons in the former case. Although most phases of condensed matter can be characterized by symmetry considerations, the FQH state is instead characterized by topological order.[12]

Many phases of matter show some form of order allowing distinct phases to be distinguished by the nature of that ordering. Order usually accompanies some form of rigidity, reflecting the fact that the deformation of an ordered state is usually resisted by forces that appear upon ordering, but which are absent in the disordered state. From this point of view, each FQH state (distinguished by different fractional values of ν) is a distinct phase of matter. This is to say, the FQH effect with $\nu = \frac{1}{3}$ is a distinct phase of matter than that with $\nu = \frac{1}{5}$, and so on. The FQH state is, however, a form of order different from that usually found at temperatures below a thermal phase transition. The FQH state is often called "topologically ordered." Topological orders are general properties of many-body quantum states at zero temperature, which have a non-zero energy gap between the ground state and excited states (Wen 2004).

Most ordered phases occur on lowering the temperature of a system, and thus forcing it through a thermal phase transition occurring at T_C. In contrast to a thermal phase transition, a quantum phase transition (Sachdev 2011) takes place at $T = 0$ and is caused by the change of another parameter, such as applied magnetic field or pressure. It is a quantum phase transition that separates each different FQH state. Moreover, although most phase transitions involve a lowering of the symmetry of a state, the FQH states all have the same symmetry, despite having different values of ν (Wen 2004).[13]

[12] The discovery and theoretical elucidation of the FQH effect won Robert Laughlin (1983), Daniel Tsui, and Horst Störmer (1982) the 1998 Nobel Prize in Physics.

[13] Another way of viewing the order in a specific FQH state (Wen 2004) is to consider a picture of semiclassical electron motion in a magnetic field. This reveals that interactions force the many electrons in the fluid to adopt patterns of motion that characterize the FQH state. This picture follows from the Laughlin wavefunctions (Laughlin 1983), which encode the dynamics of the

5.4.2.1 TOPOLOGY AND FQH STATES

So far, we have described the FQH fluid as "topological," without providing a link between the FQH effect and mathematical topology. In this subsection we discuss this link. Topology characterizes the overall shapes of objects and spaces, irrespective of the details of their exact geometry. In topological terms, a doughnut is identical to a coffee mug in that both may be continuously deformed into an identical torus.[14] Therefore, topology is blind to the details of lengths and time-interval which are mathematically described by a spacetime metric tensor $g_{\mu,\nu}$. Topology enters our discussion of the FQH effect in two places. The first is in the determination of the exchange statistics of identical particles in two dimensions. This relies on considering the overall shapes of paths in a two-dimensional plane. The second is in the field-theoretical formulation of the FQH effect—a mathematical theory that does not involve the components of $g_{\mu,\nu}$, and therefore is blind to geometry and is consequently described as topological.

The bosonic and fermionic statistics of identical particles follow from their exchange properties. The topology of the two-dimensional space in which the FQH effect occurs requires that particles are exchanged by moving them around within a plane. We will see that this feature allows more possibilities over and above the existence of Bose and Fermi particles. The crux of the matter is that the way in which particles must slide around the two-dimensional plane may be described by a path in space, and that these paths have topological properties. Although the details of the exact route that the particle paths have taken are unimportant (since these can be altered by smoothly and continuously changing them, as topology allows), there are some key features of the paths which cannot be removed without introducing singularities into the paths (a move that is strictly forbidden by topology). Specifically, moving one particle around another allows us to label a path with a winding number, which counts how many times one particle has passed around the other. Such a number is topologically protected in 2D, meaning that the winding number cannot be changed by any smooth, continuous deformations of the topology. Interestingly, it is not protected in this way in 3D, where all paths can be made equivalent because the extra dimension allows them to be disentangled and then contracted into a point. Here we see an example

states mathematically. We stress that these wavefunctions are not the result of an exact diagonalization of the microphysical Hamiltonian; rather, they were discovered as an ansatz, intended to capture some of the observed behavior of the FQH states. It is also worth noting that they are many-body wavefunctions describing correlations of all the electrons in the material in one wavefunction.

[14] This equivalence under smooth deformations is known as a "diffeomorphism."

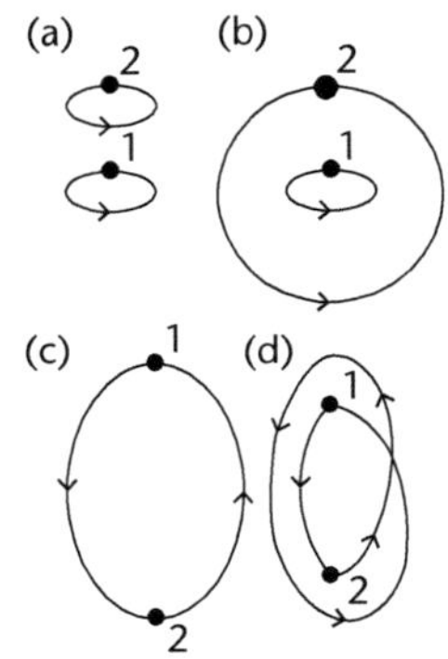

Figure 5.1 Particle exchange of identical particles in two dimensions. (Figure reprinted from Lancaster and Blundell 2012.)

of spatial dimension functioning as an acausal, global stability condition for the existence of FQH effects in such systems.

To exchange two identical particles in 2D, first one particle can simply loop around the other by an angle φ of π (and then translate the entire system if necessary), so we end up with a system that is the same as the one with which we started (Fig. 5.1(a)). In this process, the particles have swapped $x_1 \to x_2$ and $x_2 \to x_1$. Second, one particle could loop around the other by an angle $\varphi = 2\pi$, such that $x_1 \to x_1$ and $x_2 \to x_2$ (Fig. 5.1(b)). In fact, the first process could include winding the first particle around the second by an angle $\varphi = (2p + 1)\pi$, where $p = 0,1,2,3...$, (Figure 5.1(c) and (d)), while the second involves an angle $2p\pi$ (also represented by Fig. 5.1(a) and (b)). Each value of the winding number p describes a topologically distinct process.

This process of winding particles around each other is known technically as "braiding." In terms of QM, we note that during the exchange process, when one particle moves around the other by some angle φ, the contribution this rotation makes to the wave function $\Phi(\varphi)$ should be pure phase and have the property $\Phi(\varphi_1+\varphi_2) = \Phi(\varphi_1)\Phi(\varphi_2)$. This implies $\Phi(\varphi) = e^{i\eta\varphi}$, where η is some numerical parameter. As described previously, exchanging two particles can be achieved by rotating one about the other by an angle π, and then, translating the system, resulting in a multiplicative contribution to the wavefunction of $e^{i\eta\varphi}$. To recover Bose statistics we need only assume that η is an even integer, while for Fermi statistics we take η to be an odd integer. Nonetheless, there is nothing necessitating η being an integer, opening the possibility of anyon statistics described by fractional values of η. The prediction of a fractional value of η falls neatly out of the mathematical formulation of the FQH effect.

5.4.2.2 THE PROBLEM WITH A MICROSCALE DESCRIPTION OF CORRELATED ELECTRON PHYSICS

It might be naively expected that a description of the FQH effect would follow from writing down a Hamiltonian for the electrons in a 2D plane subject to a magnetic field. Although the standard textbook approach assumes that this Hamiltonian should contain all the physics, it is in-principle impossible to deduce FQH states using only the information in this Hamiltonian describing microscale states and observables. It is well known that a theory of a macroscopic number of strongly interacting electrons is a very difficult one, which appears not to admit an exact solution. However, the FQH fluid has special properties preventing, in principle, even those approximate methods used for conventional many-electron physics which have proved useful in elucidating the behavior of metals.

The problem is seen if we imagine being able to (temporarily) turn off electron–electron interactions. Then, for $\nu < 1$, there are more states in the lowest Landau level than particles, and, crucially, these states are degenerate in energy. This causes the system as a whole to have a macroscopic degeneracy. For the $\nu = \frac{1}{5}$ state, for instance, we expect five states in the lowest Landau level per electron. The interactions between electrons, when turned back on, will break this vast degeneracy, but from the microscale properties alone it is impossible to track *how* the degeneracy will be broken.

In fact, it is this degeneracy that prevents our treating the interaction as a perturbation (Murthy and Shankar 2003), which we would gradually turn back on, as one would do for a calculation in the conventional many-body problem using perturbation theory (Lancaster and Blundell 2012). Furthermore, since perturbation analysis is impossible, we cannot look to asymptotic behavior to pick out macroscale states and observables. The other conventional many-body route to treating the interacting problem—Hartree–Fock mean-field theory—also fails rather dramatically for the FQH fluid. It predicts that the 2D electron gas in a strong magnetic field leads to the Wigner crystal state with broken translational symmetry.

5.4.2.3 MULTIPLE REALIZABILITY, STABILITY AND TOPOLOGICAL COMPUTATION

Since it is the topological properties of the FQH system that characterize it (such as the winding numbers for the particle paths), FQH systems are formally stable against perturbations, such as those introduced by impurities in the conductor. This stability reflects the extreme multiple realizability of the states. This multiple realizability goes far beyond many standard examples from physics, such as superconductivity or phase changes. This is because the FQH effect is not only multiply realizable with respect to changes in the microscale makeup (such as the material and its individual level of disorder), but also with respect to changes in the spacetime metric.

It is well known that in QM we can describe the evolution of the system by acting on the wavefunction with operators representing physical interactions. In the case of a topological phase of matter, invariance under a diffeomorphism means that the only locally defined nonzero operator is the identity matrix, which leaves the wavefunction unchanged. This means that the only thing a local interaction of the system can do is keep the system the same. Any local perturbation, such as a photon striking an electron, couples to local operators and cannot change the state of the system. Topological order functions as a stability condition for the FQH effects.

This stability is indicative of the system-wide dynamics which prevent local perturbations destroying the system. Again, it is important to stress that this is not the usual insensitivity to perturbations encountered in standard physics. In these standard cases, a local perturbation is irrelevant for some phenomenon defined on a particular length scale (e.g., small perturbations leaving convection cells unchanged). In topological phases of matter, this stability is much stronger and is grounded in the non-local topological order of the system, not simply the irrelevance of perturbations over some length scale.

This difference can be illustrated by thinking about a potential application of the FQH effect for topological quantum computation (Nayak et al. 2008). Topological computers build logic gates out of the different possible braiding relations of quasiparticles, using those gates to execute programs. If realizable, topological computers would represent a significant improvement over standard classical and even quantum computing. Classical computers have definite states, whereas a quantum computer can use the indeterminacy of states to parallel perform calculations. One significant challenge to realizing quantum computers is the problem of error correction. As a quantum computer performs operations, small errors creep in due to quantum decoherence, and these must be corrected if the calculation is to be performed accurately. A topological computer, by contrast, has no such problems. Since we are braiding quasiparticles around one another to perform operations, the system depends only on properties affecting the topology of space, and no local perturbation can change this. The system is immune to decoherence because no local unitary operation can change the state of the system. In this way, topological computing is protected from errors, with no need for the software error corrections of classical or quantum computers. This stability of topological computers reflects the extreme physical multiple realizability of the system.

Finally, at this point it is worth addressing a possible worry concerning our twin claims of multiple realizability and a failure of supervenience. One may think that since the state is realizable, and realizable in many different ways, then there cannot be a failure of supervenience. However, recall that supervenience is not about realization! It is about the way that changes affect a system. FQH states are realized by having particular microscale entities correlate in specific ways, yet do not supervene on the mereological sum of those entities, because in principle it is not possible to make particular changes to those microscale entities in isolation. For instance, consider a simple example of a single pair of maximally entangled particles, such as in an EPR experiment. This state can be formed by many different types of particles; therefore, an EPR entangled state is multiply realizable. Nevertheless, there is still a failure of mereological supervenience of each concrete particular realizing base (chapter 4).

5.4.2.4 LONG-RANGE ENTANGLEMENT

We can now explore the microscale dynamics that lead to FQH states being holistic and ontologically emergent. The FQH effect can be described by an effective topological quantum field theory (QFT): the Chern–Simons field. This can be thought of as tying a quantum property—quanta of vorticity—to electrons to produce what are known as "composite fermions" (Lancaster and Pexton 2015). The composite fermion is a type of quasiparticle. Just as a crystal might be described in terms of the quasiparticle phonons,[15] the FQH effect has composite electrons. The microscale basis for this is to be found in the electron interactions themselves (Wen 2013; Chen, Gu, and Wen 2010). In the FQH effect, the correlation of local operators is short-sighted. They cannot know the topology of space, which is a global property. Nonetheless, the ground-state degeneracy depends on the topology of space, meaning that the electrons in FQH effects states cannot just be short-sighted; they must have a mechanism of achieving long-range correlations. This mechanism is long-range entanglement.

A system has short-range entanglement if it can be transformed into an unentangled state through local unitary evolution (Chen, Gu, and Wen 2010). Mathematically, we define a local unitary (LU) transformation as

$$U = We^{-i\int H'(g)dg},$$

where W is a path-ordering operator, $H'(g) = \sum_i O_i(g)$ is a sum of local Hermitian operators, and g parameterizes the path between states. All short-range entangled (SRE) states can be transformed into each other via LU transformations. Long-range entangled (LRE) states cannot be transformed into product states via a LU transformation. It can be shown that gapped ground

[15] See Franklin and Knox (2018) for a discussion of phonons.

states achieved via Landau symmetry breaking have only short-range entanglement, while topologically ordered phases of matter have long-range entanglement.

Topological order, then, is underpinned by a pattern of long-range entanglements. It is this that allows the system to know the topology of the space. Because of these long-range entanglements, alternative descriptions of the system, such as in terms of composite fermions (discussed subsequently), cannot ignore the whole system dynamics. This entanglement underpins the radical multiple realizability of the system and its immunity to local perturbations. Any local perturbation (which is not large enough to destroy the system) is suppressed by non-local correlations between electrons.

5.4.2.5 SUMMARY OF THE KEY PHYSICS

We can summarize the important physical features of the FQH effect from the preceding discussion as follows:

- The FQH state is characterized by plateaus in Hall resistivity at fractional filling factors.
- Each state is associated with a family of anyonic quasiparticles which possess a fraction of an electron charge. Anyons are neither fermions nor bosons and obey fractional statistics.
- The fractional statistics reflect the topological constraint (stability condition) on the FQH state. These topological features are also seen in an effective field-theory description of the state in terms of a topological Chern–Simons theory.
- FQH states are topologically ordered. This is quite unlike most condensed-matter states, which are distinguished by their symmetries.
- Particular topological orderings reflect particular patterns of LREs, which knit together the system. Such LREs cannot be removed under application of a renormalization group, so that in principle the FQH state is insensitive to local perturbations.

At this point it is worth clarifying the contrast between the FQH effect and critical phenomena in phase transitions (where renormalization group techniques are vital). During a phase transition a system is typically described as approaching a state with an infinite correlation length, so that the system's fluctuations show long-range correlations over a divergent length scale. This causes the system to be insensitive to perturbations. However, we should not let the similarity of the language here disguise profound differences with the FQH state. For the phase transition, the infinite correlation length is a descriptive term, not one that should be taken realistically (Chapter 4, Section 4.8). It applies only at the critical point (at which there is a singularity in the thermodynamic free energy), not to the system either side of that singularity. This is different compared to the long-range behavior of the FQH state. Here, long-range correlations do not belong to a theoretical description of a never precisely observed transitory point. Instead, they describe a physical state of matter that persists (measurably) over time.[16] This is why it is a genuine possibility to build a computer that will be robust against environmental perturbations from a topological state, but not from a system at a phase transition.

5.4.2.6 DIFFERENT WAYS OF ARTICULATING EMERGENCE IN THE FRACTIONAL QUANTUM HALL EFFECT

The following are different approaches to articulating emergence in the FQHE in the literature.

5.4.2.6.1 Quasiparticle realism

One obvious way of articulating the emergence of the FQH state is by regarding FQH effect quasiparticles realistically. The FQH effect is not the only condensed-matter system with quasiparticles; they are ubiquitous in CMP. In general, the term "quasiparticle" refers to the single-particle-like states in interacting systems. These may be single particles (such as electrons) dressed quantum mechanically (or renormalized) by interactions. Alternatively, quasiparticles may be collective excitations that may be treated as single-particle-like, but whose dynamics involve a change in the coordinates of all particles in the system. For example, a simple crystal lattice possesses particle-like vibrations known as phonons, which have this latter property.

Wallace says of quasiparticles:

> [A]re quasiparticles real? Well they can be created and annihilated; they can be detected (by scattering them off "real" particles like neutrons); in some cases (such as so-called ballistic phonons) their time-of-flight can be measured and they play a crucial explanatory role in solid-state theories. We have no more evidence than this that real particles exist, so it seems absurd to deny the existence of quasipartic les.
>
> (Wallace 2010, p. 59)

This picture seems even more compelling, depending on how one views elementary particles such as electrons. For example, Michael Levin and Xiao-Gang Wen (2005) have proposed a model where photons and electrons are emergent phenomena. The idea is that both types of particle can be modeled as different excitation modes of what is known as "string-net condensation."

[16] Recall that an important role of stability conditions is in guaranteeing the existence and persistence of states and observables over time.

Just as the quasiparticles of the FQH effect are a collective excitation out of the sea of electrons, so the electron itself would merely be an excitation of a deeper unifying physical reality. Less exotically, it is consistent to regard 1) electrons in a metal as quasiparticle excitations from the sea of electrons that form the ground state, and 2) electrons in the vacuum as quasiparticles excited from the vacuum and dressed by quantum fluctuations from the matter fields of the universe.

Despite the ubiquity of the quasiparticle interpretation, there are reasons to be cautious about such sentiments. Quasiparticle realism is undermined in the FQH case by a topological feature that causes an ambiguity in representing quasiparticles, which is not present in non-topological phases of matter (Nayak et al. 2008). For starters, any anyonic system must have multiple types of anyons. Thus, for a state that has anyons of statistics η, that state also has anyons of statistics 4η, and so forth. We can always approximate any two anyons by a single particle whose quantum numbers are obtained by combining quantum numbers (including topological quantum numbers). This formation of a different type of anyon by bringing together two anyons is called "fusion." A classic example of this is when we bring a statistic $\eta = \gamma$ particle together with a statistic $\eta = -\gamma$ particle to form a boson with statistics $\eta = 0$, dubbed the "trivial" particle.

The situation is even more complicated since in addition to the different ways by which we can represent the system in terms of different fusions of anyons, we can also represent the system with a completely different single-particle-like state which is not an anyon: the composite fermion. The composite fermion model of the FQH effect was proposed by Jain (1989) and extended by Halperin, Lee, and Read (2013). The idea is that we can redescribe FQH effect systems not in terms of quasiparticles with fractional statistics, but instead in terms of electrons that have quanta of vorticity attached to them. This compound entity—an electron plus quanta of vorticity—is known as a composite fermion.[17] Composite fermions, unlike quasiparticles, carry integer charge, since the only element of charge is the electron. They are also characterized by fermion statistics. The advantage of this description is that it allows the FQH states to be mapped onto the IQH states. So, the filling factor $p/(2p+1)$ for a quasiparticle becomes a filling factor of p for the composite fermions, and the FQH effect for quasiparticles is an IQH effect for composite fermions.

This might lead one to think that the FQH effect is not so different from the IQH effect, and that there is nothing particularly emergent about it.

[17] Sometimes the composite fermion is described as an electron that has captured a quantum of magnetic flux, but this is only an analogy; the magnetic field is uniform everywhere.

However, the composite fermion picture is deceptive: composite fermions are not merely individual electrons interacting with the magnetic field (Stern 2008). For instance, if we attempt to remove a composite fermion by adiabatically turning on a flux tube, then we alter the charge of the system by a factor $e/(2p+1)$. Even though the composite fermion only has charge e, by annihilating it we remove a fractional charge. The reason for this discrepancy is that composite fermions are *topological entities only defined relative to the whole system*. Therefore, if we remove a composite fermion in one Landau level, we also affect the composite fermions *in all other Landau levels*. Even a single composite fermion is really a collective entity of the whole system. This is because the quantized vortex term is derived from the quantum-mechanical interaction of all the electrons. An FQH state may appear simple when translated into an IQH state in terms of composite fermions, but this is illusionary. A composite fermion is a quasi-entity determined by the highly complex, strongly correlated long-range correlations of the electrons making up the system. A lesson to be drawn from this is that even if we describe the FQH system in microscale terms that appear to remove interactions (i.e., in terms of non-interacting composite particles), this is an illusion. Any way that we describe the system, either in terms of anyons, composite fermions, or Chern–Simons field terms, there is a holistic character to FQH states that is not present for IQH states.

5.4.2.6.2 Good coordinates

Mark Wilson (1993) follows others, such as David Lewis (1999), in concluding that without differentiation, properties in and of themselves are problematic. Too many properties are as bad as no properties at all for delineating the structures of the world. What is required is a way to privilege a set of properties. For example, Lewis privileges the intrinsic properties of point-like objects and the spatiotemporal relations between them. He suggests constructing a "natural" sparse language with reference to only these "fundamental" properties. Any other property can be ranked in unnaturalness by considering how complex and long its definition is in the sparse language.

Wilson's critique of Lewis is that such a way of viewing properties pays almost no attention to what are deemed the physically significant properties by physics. The sets of properties which are privileged by physics (either nomologically, predictively, or causally) will often not be Lewis's natural properties. That this is the case is one of the motivations of P. W. Anderson's critique of privileging high-energy particle physics. In Wilson's terminology, the privileged set of properties are dubbed "intensional." Importantly for him, intensional properties can vary from system to system because they are defined as the properties determining the physical evolution of a system. For instance, a rigid body has its evolutionary dynamics set by its mass, velocity, angular momentum, and moment of inertia.

In general, in physics we can define a system as exploring a state space. Sometimes this state space can be as simple as a phase space, in which the system is described in terms of its position and momentum coordinates. But other state spaces are possible to construct in which a system is described as exploring an abstract space defined by a particular set of coordinates. The nature of those coordinates is defined by the degrees of freedom of the system. If a system is subject to a specific constraint, then this can reduce the dimensionality of the state space that the system can explore. The "natural" or "good" set of coordinates of any system for Wilson are those for which the state-space description is optimally simple. Because these good coordinates pick out the physically significant variables of a system, they also pick out the intensional properties of that system. Since the nature of the constraints that reduce the state space will vary between condensed-matter systems, the intensional properties will too.

The question, then, is what determines a set of good coordinates? In general, it seems to be constraints of some sort (hence the reduction in state-space dimensionality). Specifically, Wilson suggests that order in the form of SSB is a mechanism by which the state space is reduced, and good coordinates are frozen into a system. Since the broken symmetries are specific to each system, so too are the good coordinates. In the case of SSB, the coordinates of interest, presumably, might include the so-called order parameter of the system. This is a field whose vacuum expectation value takes on a nonzero value only when symmetry is broken, whose variation in space determines the emergence of general rigidity, and whose excitations constitute the emergent quasiparticles of the broken symmetry state. Another possibility is to identify the "good coordinates" with the normal coordinates found via the diagonalization of the interacting Hamiltonian of the broken symmetry state. The constraint that leads to either of these sets of coordinates is that a particular symmetry is broken, and a symmetry element, therefore, is missing in the description of the system.

It is clearly not possible that the mechanism of emergence of FQH states is SSB, because all topologically ordered states of matter share the same symmetry. Nevertheless, some argue that SSB is the only possible mechanism by which emergence can occur. Accordingly, topological constraints are just another type of constraint that can pick out intensional properties. Another possibility is that the constraint that the FQH system is incompressible at particular filling fractions leads to the Chern–Simons field being identified as a set of "good coordinates." Identifying the stability conditions leading to the emergence of new states and observables seems to be the only solution to identifying the relevant "good coordinates."

5.4.2.6.3 Failure of mereological supervenience

Lancaster and Pexton (2015) suggest that the basis for the emergent properties of the FQH effect is in the long-range entanglements characterizing topological states of matter.[18] It is the long-range order in the FQH state that explains the plateaus at fractional filling factors. This long-range order is, in turn, explained in terms of topological entanglement entropy. That is, the long-range order reflects the fact that the system is criss-crossed by a context of LREs. So, everything we said about the nature of quantum entanglement in Chapter 4 works for explicating the FQH effect.

For example, the FQH quasiparticles are collective excitations definable only within a system obeying fractional statistics that depend on the topological properties of the whole system. Even though quasiparticles are localized to some extent they cannot be treated in isolation. The same is true of the Chern–Simons field approach. In this representation there is a field which pervades all space and is emergent from the microscale details. This field ties quanta of vorticity to electrons, making them composite fermions. Yet composite fermions are topological entities dependent on the whole system to determine their properties. Finally, at the level of electrons we know that FQH states are those for which there are non-local entanglements. In all three representations, the FQH state is one in which some of the individual properties of free electrons are determined by the strong system-wide correlations which coordinate the FQH states, hence the FQH states are collective.

In the FQH case, we may provide an example via the properties of quasiparticles. The consideration of quasiparticle properties does not commit us to a claim about the reality of quasiparticles. Rather, since the quasiparticles constitute the collective motion of all constituent electrons in the system, quasiparticle properties are a means of indirectly considering the way in which the properties of those electrons are interacting. The quasiparticles of a given system may be just a representational tool, but that representation is still tied to the underlying physical, collective features of the system in question, and can therefore reveal features of it. Specifically, recall that the quasiparticles of the FQH state bear a fractional charge. Now the charge of an isolated electron is as good a candidate for a nonrelational property as any. But when a system of many electrons condenses into an FQH state, the fractional charge on a quasiparticle excitation of that state now arises only by virtue of the entanglements existing across the whole system. More dramatically, we can consider adding an extra electron (carrying its single charge) to the FQH system. If we try to keep track of the extra charge, it will appear to decay into several fractional charges. Of course, the electrons do still carry charge, but it is no longer a charge intrinsic to them as individual particles.

[18] For a contrasting view, see Fletcher 2020.

The electron gives up its charge property to the entanglements. Thus, it is not that the basal properties of electrons (charge, mass, etc.) disappear in the transition to an FQH state; rather, it is that (some at least) of those basal properties *are now inherently relational* for the FQH state.

Any nonentangled relatively "basal property" of an isolated electron, that in a FQH state becomes entwined in the web of LREs, becomes a part of an inherently relational structure. One way in which this relationality can be seen is in the system's robustness to unitary operations/local perturbations described previously. This reflects the fact that the individual elements making up the FQH state cannot respond as individuals to interactions. The FQH state responds as a whole or not at all (recall the unique robustness of a topological computer in comparison with any other type of quantum computer). Furthermore, as well as the transformation of basal properties to relational, the web of correlations of the electrons in an FQH state allow new properties to manifest which can only be associated with a collective state, and hence are inherently relational.

For instance, the properties of the Chern–Simons field cannot be considered intrinsic properties associated with a particular component of the FQH system. These properties are not straightforwardly assignable as nonrelational basal properties of the electrons outside the FQH state. Thus, the FQH state has inherently relational features derived from the conversion of relative basal features of the components of the FQH state and from the production of new features definable only for the collective state. Note that not all relatively basal properties of the electrons need to become inherently relational. Nonetheless, the properties that are responsible for determining the FQH state are those that are contextually emergent given the larger-scale stability conditions.

It is in this sense that the FQH state cannot be reduced to its parts, so must be regarded as a unified whole. Moreover, in the case of the FQH effect, it is the inherently relational properties that determine the physical state of the system. One way of thinking about this is that the physical features of the FQH state picking out the good coordinates are inherently relational that are enabled by the stability conditions. Even if one thinks that the notion of good coordinates in general is merely an epistemic category, the fact that any plausible representation of the FQH state (any set of good coordinates) must include, explicitly or implicitly, these relational aspects, suggests that an inherent relationality of some sort is crucial to the ontology of the FQH state. Using the notion of good coordinates, these inherently relational physical features are the intensional properties.

Another way of thinking about this is that what we have is a contextually emergent "supervenience base," which is not the mereological sum of the individual parts and their intrinsic properties in relation to one another.

Instead, in the relational state—the "supervenience base"—is the "parts" and their now inherently relational properties. There is no causal exclusion possible, since this new "supervenience base," although microscale, is not reductive, as it results from the larger-scale stability conditions. We cannot say that the causal powers of the FQH state are merely a summation of causal powers of the electrons; rather, the casual powers of the FQH state are the collective casual powers of the now inherently relational electrons fused together. This does not necessitate basal loss, however, as the properties of the electrons are no longer their own alone.[19]

5.4.3 *Contextual emergence and the fractional quantum Hall effect*

Recall that in our section on SSB and superconductivity, we stated that some of the latest thinking in CMP is that the superconducting state itself has parallels with the kinds of holism associated with the FQH efect. For instance, Hansson, Oganesyan, and Sondhi write:

> [W]hat is the nature of the ordering in a superconductor? We find that the answer is properly that the superconducting state exhibits topological order in the sense of Wen, i.e. that while it lacks a local order parameter, it is sensitive to the global topology of the underlying manifold and exhibits an associated fractionalization of quantum numbers.
>
> (Hansson, Oganesyan, and Sondhi, 2004, p. 1)

It is a matter for physicists to decide which phases of matter are topologically ordered. Our motivation here is not to rest any significant philosophical conclusion on the possibility that superconductors are. Rather, it is to use the FQH effect as a canary in the mine for the kind of features that are prevalent in CMP in general, regardless of specific mechanisms. The FQH effect is particularly good at revealing these features, since it is topological, and that mechanism stands out with respect to these features. Yet even non-topological states of matter in CMP display the same general aspects.

The crucial point to stress is that phenomena such as the FQH effect and superconductivity (whatever the ultimate mechanism) depend on contexts defined by stability conditions in two senses.

The first sense is in the exogenous stability conditions making such CMP states possible. In the FQH effect, for instance, the system depends crucially on spatial dimension. The spatial dimensionality of the system is a gatekeeper providing access to the possibility space for topological ordering. Spatial dimensional is not part of microscale elements, but a global stability

[19] This contrasts with "fusion" *à la* Humphreys (1997).

condition providing access to regions of possibility space that would otherwise be blocked by compartmentalizing the phase space available to the electrons. It does this by creating conditions allowing a particular topology to be important (the topology of winding). The dimensionality of the FQH effect is a property which is compatible with some topological states and not others. Dimensionality is but one such feature; there are many "external" environmental factors, such as temperature, which play the role both of stabilization and constraint, allowing for the manifestation of FQH states.[20]

The second sense is the internal or endogenous stability conditions for CMP systems. The exogenous stability condition of two spatial dimensions places constraints on, and facilitates, the system of electrons. In combination, the endogenous stability conditions also facilitate the system of electrons. This means that if the interactions are strong enough, as in the FQH effect or superconductivity, they bring about a radical shift in the internal environment of the electrons. This is to say, that the very environment of the electrons and defects changes qualitatively in a radical way. This change in the endogenous context interacts with the exogenous contexts. Specifically, exogenous and endogenous stability conditions give rise to the emergence of a new set of properties associated with the new endogenous context—a form of dynamically generated larger-scale stability and constraint.[21] The states, observables, and laws at the microscale contribute some of the necessary conditions for CMP phenomena, while the exogenous and endogenous stability conditions not given by those microscale properties provide the complete set of conditions for the manifestation of CMP states.

In the specific case of the FQH effect, the endogenous context of the environment electrons is radically altered in terms that can be spelled out in (formally at least) equivalent ways. We can say that the electrons become a network of long-range topologically entangled electrons, or that the electrons become fused with quanta of vorticity to form a composite electron (where the vorticity is a global property of the system), or that a topological Chern–Simons field is formed that changes the dynamics available to the system. Whichever way it is described, ontologically, we move from a set of properties produced by an exogenous stability condition establishing contexts modulating a series of electron objects in an endogenous context of defects, to an exogenous set of contexts modulating (and being modulated by) a whole object which behaves accordingly with a new endogenous context provided by the aggregate environment of interactions. In other

[20] We remind the reader that temperature is a contextually emergent property (Chapter 4, Section 4.3).

[21] Recall that we saw an example of dynamically generated larger-scale stability and constraint in Rayleigh–Bénard convection.

words, there is dynamic adjustment/feedback between two contexts at different scales that each are acting as stability conditions of electron behavior in FQH systems.

The collective environment of electrons and interactions (with the endogenous context) participates in a feedback loop on itself, changing that very environment of electrons so profoundly that the system can no longer be adequately described as electrons in a background environment, but instead is a whole FQH liquid. This shift in endogenous context is the shift to inherently relational properties between the electrons. The context that each electron sees as provided by other electrons shifts so radically that the boundary between an electron and the context of other electrons has vanished. It is now a whole that interacts with an external context. The same is true for superconductivity. The overall phase coherence of a superconducting state creates a new macroscopic environment. The internal environment of the system is so radically qualitatively different that the distinction between what the system and environment is has shifted dramatically.

In general, then, external contexts constrain and stabilize electrons. This allows electrons to interact in just the right ways to radically alter the internal collective environment such that new properties manifest. Those properties have a new set of compatibility/incompatibility relations with the exogenous context properties. The combination of the external context defined by its stability conditions, along with endogenously produced properties, collectively produce the unique set of properties characteristic of a particular CMP system. Neither the exogenous nor endogenous context is a function of elementary particles and their forces alone.

This fits the pattern of contextual emergence. Elementary particles and their forces at the microscale contribute a necessary part of a contingent sufficient condition for the existence of the properties and behaviors of the FQH effect and superconductivity. Yet the existence of as context-free-as-possible (ACFAP) elementary particles and forces at the microscale do not guarantee that either FQH or superconducting states and observables will exist. The total set of necessary and sufficient conditions for these states and observables is contingent on the larger-scale stability conditions and the endogenous and exogenous contexts that they establish. None of the stability conditions are given by the ACFAP elementary particles and forces alone, nor are they reducible to such particles and forces in any sense.

Now we can see that even if SSB works in superconductivity as Morrison describes, it is not as straightforward as SSB of gauge bringing about the characteristic properties of superconductors. This is empirically verified by the general lack of associated SSB of gauge with such phase transitions. Instead, it would be the stability conditions establishing contexts for electrons to interact and change their internal environment such that SSB of gauge occurs, enabling such SSB to interact with the endogenous context and producing the characteristic set of superconducting properties.[22]

The "good coordinates," as described by Mark Wilson, pick out the unique contextual topology for which the salient observables are manifested, and combine with the external contextual properties. By this means, the good coordinates provide an epistemic way of tracking an ontological shift that has taken place in the world. By the change in collective environment (endogenous context) and how this combines with the previous exogenous context, a new object/environment distinction is produced. This is to say, that such a set of feedbacks between objects and their contexts produces a new class of object and a new context, hence new states and observables absent from the microscale simpliciter. We have a new object/environment boundary tracked by the new set of good coordinates. There is a new contextual topology representing the partitioned space of what is intrinsic versus extrinsic to the object differently.

This change in context is in spirit with Paul Humphrey's (1997) fusion account of properties, though there is no requirement that one set of object/environment distinctions cease to manifest a new such object/environment distinction. Instead, we have a hierarchy of multiscale properties, contexts, and objects combining to produce new modal relations at multiple partitions of object/environment that can coexist. In the FQH effect, for instance, although electrons obviously exist, once the FQH state properties emerge, the multiscale contexts for electrons have altered profoundly.

Here we can contrast contextual emergence with two other accounts: 1) in Humphrey's (1997) fusion version of emergence, "lower-level" properties cease to be, and that is crucial to bringing about "higher-level" properties, and 2) in Ladyman and Ross's (2007) ontic structural realism and rainforest realism, "higher-level" and "lower-level" properties exist separately and in parallel. In contextual emergence, the ACFAP microscale properties do not vanish, but are contextually altered to become new contextual features allowing the manifestation of larger-scale properties. The modal spaces mapped out between "levels" can be described as separated, since features can be associated with implicit contexts; ontologically, however, the levels are not genuinely separable, because there is a feedback loop across the multiscale contexts.

Here is another way to think about this shifting object/environment distinction. Given contextual emergence, there is no possible world in which an entity or object could exist by itself alone in that world, let alone have states and observables. The electrons in FQH states have a series of contextually

[22] Although we suspect that SSB of gauge is not the relevant property manifested by the change of endogenous context, the physics of superconductivity still has many unanswered questions.

emergent properties (e.g., fractional charge). Those powers are directed to manifest in a particular way by the constraints and opportunities afforded by the stability conditions defining the exogenous context. Specifically, the system is constrained to have the correct topological properties, which opens access to a portion of possibility space previously closed to the electrons. The standard properties of the electrons act in this context in a way that alters their own local environment. At a specific point, the environment is so changed that the standard contextual properties expressed by the electron are changed (a network of topological entanglements forms). This, in turn, means that a new set of contextually emergent states and observables are manifested by the whole network of electrons in their new contextual environment (a combination of endogenous and exogenous contexts). This new set of contextually emergent properties is associated with a new demarcation between object and environment. In this case, the new object is the FQH liquid taken as a whole. That is the new "atomic" unit—the new ontological break between object and environment relevant to the new contextual states and observables along with their manifested effects. The good coordinates epistemically pick out the new set of contextually emergent observables because they track the modalities associated with them. The new states and observables shape the object/environment distinction, which distinguishes a new context.[23]

Although we have focused largely on the FQH effect, the general contextual emergence framework applies across CMP to aggregate systems. The mechanism of what the initial contexts are, what the interactions are, new stability conditions, new collective environments, and hence what the new states and observables along with the object/environment distinction are, will all be specific to each system. For some systems, changes in symmetry contexts will be crucial; for others, the compartmentalization of phase space by specific contextually generated boundaries will be important; while for others, such as the FQH state, topological constraints will be of most salience. Of course, in most cases a combination of many of these will be operating. Superconductivity is a prime example in which the compartmentalization of phase space, symmetry, and subtle topological constraints probably all play a role.

Nature is complex in the way in which contexts interact with and in systems. Fortunately, our epistemology naturally tracks contextually emergent states and observables, and defines object/environment distinctions so that

[23] We saw something similar in Rayleigh–Bénard convection, where the Bénard cells provide a new context-determining behavior of the fluid parcels (e.g., determining to which cell any individual fluid parcel would migrate).

most often in scientific inquiry it is easy to miss stability conditions and shifts of contexts or to let them slide into the background. Indeed, this is one reason why it is so easy to miss the importance of stability conditions and the contexts which they define. Our epistemic tools, along with the good coordinates, sciences generally speaking naturally hit upon the object that bears contextually emergent states and observables of interest. As we have said previously, contextual emergence makes explicit what scientists often leave implicit. And it is the work performed by the implicit that renders faulty the judgment that CMP systems, say, are ontologically reductionistic but epistemically emergent.

Of course, one may be tempted to think that this shifting of object/environment distinctions, and what is part of the exogenous context of a system itself, suggests that such distinctions are arbitrary or at mostly epistemic. They are convenient to partition the world to follow and talk about it, but they are not genuinely part of the world. Such a conclusion is mistaken, principally because there is no scenario, no matter how supposedly fundamental, that does not rely on some contextual features. If contexts and object/environment distinctions were just epistemic notions, then this should imply that there is a context-free and genuine "real ontology." The contention of contextual emergence is that there is no such thing, and no such explanatory situation.

We will examine that issue in the next section by looking at another feature of CMP, the use of effective field theories, and how these compare to the theories of elementary particle physics. Such a comparison leaves the notion of a context-free fundamental theory in great difficulty.

5.5 Contextual object/environment distinctions and effective field theories

Effective field theories (EFTs) are widespread in CMP. In an EFT, a kind of "approximation" to a lower-level or underlying field theory is used in a very specific sense. This approximation includes the salient degrees of freedom (DOFs) to describe a particular system, usually by imposing a cut-off in energy scale or length scale getting rid of irrelevant DOFs. The EFT rewrites the Lagrangian describing a system, thereby ignoring DOFs that do not fall within the cut-off. EFTs are most obviously useful where there is a large separation between phenomena based on scales, and a decoupling of DOFs such that if we are interested in low-energy phenomena in CMP, then we can ignore much of the high-energy dynamics; these even out and are not salient to the low-energy phenomena.

The use of EFTs is intimately connected with the RG and symmetries. Recall that in RG methods we integrate out degrees of freedom that are irrelevant, changing the scale upon which we view a system until we hit upon the relevant scale for which the phenomena are sensitive to changes. Typically, the construction of EFTs relies on assessing the symmetries at work in a system, and using these to construct the theory.

For many philosophers and physicists, the notion of high-energy particle physics as "fundamental" rests implicitly or explicitly on the assumption that the theories of fundamental physics are, well, "fundamental." Yet this notion of fundamentality is notoriously difficult to spell out in a non-question-begging way (as we have seen in previous chapters). These are thought to be context-free or first-principles theories that describe reality in the raw; in contrast, the theories used in CMP are thought to be approximate and apply only to particular cases or work only heuristically because of the intractability of applying the fundamental theories to large particle number systems. Paradigm examples of fundamental theories include the standard model of particle physics and general relativity (or whatever the presumed form of quantum gravity is). Nonetheless, there has been a major shift in how such fundamental theories are viewed, with many physicists now regarding them as EFTs; that is, these theories are low-energy versions of a deeper underlying field theory.

Typically, in fundamental field theories, one worries about renormalizability—the property that a perturbation series integral of the theory does not blow up to infinity, but instead provides a nice well-behaved answer. EFTs do not need to be formally renormalizable, but fundamental theories do. EFTs are well-behaved if all the tricky terms that would blow up to infinity are outside the scale of interest. But this raises an intriguing possibility of underdetermination. It may well be that rather than our fundamental theories being renormalizable because they are fundamental and special, instead it is that they are actually EFTs.[24] Theories usually thought to be renormalizable, such as quantum electrodynamics and electroweak theories, may include non-renormalizable terms in their true Lagrangians, but these terms do not manifest since they occur at relatively high energies compared to the energy scale at which we apply those theories. So, all our "fundamental" theories very well may be EFTs that break down at sufficiently high energies. Indeed, this view has become common among physicists. As Howard Georgi summarizes, "the most appropriate description of particle interactions in the language of QFT depends on the energy at which the interactions are studied" (1989, p. 448).

[24] See Rosaler and Harlander (2019) for a discussion of the "naturalness" or not of renormalization parameters.

This immediately suggests two possibilities: 1) that there is a genuinely fundamental theory underlying all our current EFTs, or 2) that there is an endless series of EFTs sitting atop each other, and that the notion of fundamentality is relative to a particular set of constraints and cut-offs. One should view this situation not as "turtles all the way down," neither in terms of axiomatic set-theory, nor in terms of foundationalism with the foundation infinitely deferred. Rather, view it as contextually nested, recursive, and self-referential. After all, given contextual emergence, we simply make explicit that nothing exists without multiscale contexts, and the very same "multiscale chunk of reality" can simultaneously count as a system to be explained and the context for another multiscale system.[25] We think that this view shares many features of a similar account articulated by Tian Yu Cao and Sam Schweber (Cao and Schweber 1993; Cao 1999). EFTs are always contextual, and they always require input from outside contextual information to formulate them, such as stability conditions and symmetry considerations. Even the most fundamental parts of fundamental theories in physics require contextual features for their construction which reside outside that domain. Moreover, elementary particles and forces cannot do anything but come to expression in a concrete context.

The philosophy of EFTs therefore recreates structural features of the debate around RG methods to describe physical systems (Chapter 4, Section 4.8). As we coarse-grain to view a system at a different scale, say, are we approximating a fundamental ultimate view, or are we throwing away irrelevant details to isolate the salient states and observables? In the latter, as we have argued, contexts matter in two ways: 1) as stability conditions that the "approximation" procedure reveals as crucial to the existence and persistence of new states and observables, and 2) as partly determining what is salient for a relevant description.

For instance, context always plays a role in determining what the object/environment distinction is. In EFTs, this is the taking of a potentially infinite space of theories and producing a particular EFT by imposing an energy cut-off. This cut is conceptually similar to defining the boundary of an object of study to separate it from its environment. In cellular biology, the boundary between a cell and its external environment is implicitly suggested by the presence of a membrane. In the construction of an EFT, the separation requires a subtler evaluation of the domain of properties of interest. Once this domain of properties, and associated modal relations, are defined, a set of symmetry and other constraints immediately suggest the DOFs of interest—the division of the object (the EFT) from its environment (the other theories it "approximates").

[25] This is related to the idea of relative onticity (Atmanspacher and Kronz 1999).

In set theoretic terms, this contextual splitting into object/environment distinctions is the imposition of a contextual topology on the system (Chapter 3, Section 3.2). Once a specific contextual topology is imposed, the world has been divided. The reductionist response is typically to view such divisions as epistemic only, with the promise of the ontology of an undivided nature at the bottom. But the relations between theories suggested by EFTs do not support the reductionist intuition. Our best "fundamental" theories are no less "effective" than any theory in CMP.

The key misconception to remove is the misleading textbook description of EFTs as "approximations" of other theories. To say that an EFT, such as QFT, approximates some deeper yet undiscovered genuinely fundamental theory, is a misnomer. The relations between these theories is not approximation in any philosophically robust sense. A given EFT is not encapsulated by another EFT with a different cut-off. There is no idealization[26] going on, nor is there a Boolean hierarchy of EFTs. If we change the cut-off, we produce a different theory entirely. These theories are related mathematically in particular ways, but one is not just a simpler version of the other *contra* Elena Castellani (2002), for instance.

5.5.1 *Objections to giving up on fundamental theories*

Even if current theories are EFTs, does this license the more radical claims about the nature of fundamental physics presented here and by other authors? There are certainly those who think not. For example, Castellani (2002) states:

> The fact that current QFTs are now seen as EFTs does not imply any specific thesis about the existence of a final theory. The high-energy theory could be another EFT, a local QFT, or something completely different (in this sense the picture of a tower of EFTs is in some ways misleading).

She continues maintaining that "The EFT approach does not imply that the idea of a theory being more fundamental than another is meaningless." She thinks this because the belief that a low-energy EFT is equally fundamental to an original high-energy formulation relative to a scientist's concerns "does not entail that the high-energy theory could not be seen as more fundamental from a different perspective" (Castellani 2002, p. 15). As far as it goes, Castellani is right about the lack of entailment of this belief.

Michael Redhead (1999) defends the notion that fundamental theories should not be abandoned on methodological grounds: "[F]rom a point of view of methodology of science a recurring theme has been the search for an

[26] See Liu (2019) for a discussion of idealization and infinite limits for contrast.

ultimate underlying order characterized by simplicity and symmetry that lies behind and explains the confusing complexity of the phenomenal world." If this is the case, then "To subscribe to the new EFT programme is to give up on this endeavour and retreat to a position that is admittedly more cautious and pragmatic and closer to experimental practice, but is somehow less intellectually exciting." Here he is appealing to "the regulative ideal of an ultimate theory of everything" as "a powerful aesthetic ingredient motivating the existence of the greatest intellectual ingenuity in the pursuit of what may admittedly, in itself, be an illusory goal" (Redhead 1999, p. 40). This is how regulative ideals function, to be sure.

Meanwhile, Stefan Hartmann sees EFTs as not really theories at all, but rather as some methodological chimera between theories and models. As such, the usage of EFTs as tools is quite different, and EFTs can only really be understood on a case-by-case basis rather than in terms of general principles:

> The relation between some underlying theory and an EFT can be reconstructed along the lines of Rohrlich's (2000) two-step model: first, a deductive relation between the two formalisms has to be obtained. This typically involves approximations and a limiting process. Second, ontological bridges between the incommensurable concepts of both theories have to be established. This model, designed to account for the relation between mature theories, does not help to analyze the relation between a model and a theory, which is typically much more involved. This calls for detailed case studies.
>
> (2001, p. 36)

Yet he also notes that "EFTs are only applicable if the energy scales of a system separate well. That is why EFTs work well in particle physics, but do not work so well in the physics of complex systems" (2001, p. 32). Taken together, these statements suggest that Hartmann has a view of EFTs as specialist tools, which captures much of their methodology but misses some of the general importance.

Granted, most CMP researchers would categorize every field theory used to account for a condensed-matter system of complex interactions as an EFT. Furthermore, if *every single physical theory science has ever discovered* is an EFT, then what does the regulative ideal of a "theory of everything" actually contribute, or in what sense are EFTs methodologically distinct from "real" theories?[27] Leaving aside how much of science is model-building or theory generation, is it the case that EFTs represent some form of practice distinct from theory generation? The lessons of EFTs are not limited in scope, because they apply only to select branches of particle physics, or because EFTs represent their own heterogeneous distinctive practice from which no general considerations can be learned.

[27] And what are the non-question-begging criteria distinguishing "real" theories from EFTs?

Another criticism that Hartmann raises against taking EFTs as evidence of the lack of fundamental theories is the decoupling theorem of Appelquist and Carazzone (1975). For a simple example, consider two coupled systems with different energy scales E_1 and E_2, where $E_2 > E_1$, and suppose that this coupled system is described by a renormalizable theory. The theorem implies that there always exists a renormalization condition such that the effects of the physics at the E_2 scale can be effectively included in the theory with the smaller E_2 scale by changing parameters of the corresponding theory. The decoupling theorem also implies the existence of an EFT at the E_1 scale which will cease being applicable as the energy approaches E_2. Hartmann asks:

> Does the decoupling theorem imply that empirical reality is, as a whole, layered into quasi-autonomous domains, as suggested by some authors? No, since the decoupling theorem is based on assumptions which may not always be fulfilled. Most importantly, the decoupling theorem ... presupposes that there is a renormalizable theory of the composite system which is the starting point of the decoupling procedure. Without such a theory, which is supposed to be valid on all energy scales, the decoupling theorem cannot be applied.
>
> (2001, p. 20)

This critique in defense of an ultimate theory X, which is renormalizable, is too hasty though. All the decoupling theorem shows is that an EFT can be decoupled from a renormalizable starting theory—a "local" rather than "global" result. As previously stated, theories such as the electroweak theory are perfectly renormalizable and can generate decoupled EFTs. Yet the underdetermination point remains that electroweak theory, say, itself may be an EFT of a higher-energy theory, and electroweak theory is renormalizable because it has cut-off degrees of freedom. The coupling theorem does not imply that there must be an ultimate fundamental theory, and the practical efficacy of EFTs speaks to their ability to decouple from higher-energy theories even when we know that those high-energy theories are themselves EFTs.

What of Redhead's criticism of abandoning the search for fundamental theories? Of course, Redhead's definition of the methodology of theory construction can be questioned. Perhaps researchers have been constructing theories, effective or not, to account for the observed phenomena in a domain. It is just that some of those domains are broader than others. In any case, we offer no argument here against reductionist methodologies in the history of science and their utility—merely an argument against a metaphysical thesis being supported by them.

In short, this is due to two features: 1) the role contexts play in the ontology of the world allowing it to form domains that are decoupled sufficiently for research, and 2) the role that implicit contextual features play in epistemic constructions. One cannot model or theorize about anything without constraining the parameter space beforehand, and these constraints are tied to contexts, as we argue and illustrate throughout this book. Hence we do not refute the utility of particular reductionist approaches at particular times; we merely claim that the choice of reductionist methodologies is always epistemically contextual.

Just because our current fundamental theories are EFTs, does that imply that there is no such thing as a fundamental theory? Modulo defining "fundamental," there are several points to be made here. First, one could argue that there is a burden of proof on the reductionist side. Why should we believe in fundamental theories if science has never produced one? Moreover, given the radical underdetermination of any such suggested fundamental theory, how could we ever have good reasons for thinking we have struck gold? Second, those who suggest that EFTs are leading us towards discovering the fundamental theory usually invoke the notion of approximation at some point in their argument. Supposedly, EFTs are approximations of the more fundamental theories, hence are taking us one step closer to the ultimate fundamental truth as we move up the energy scale. Third, there is the open question of what it means to be fundamental in the first place. We will now address these last two points in turn.

5.5.2 *Effective field theories and approximations*

EFTs are not approximations of other theories. They are formally distinct subsets of the set of the higher-energy theory. In general, EFTs bear a non-Boolean relation to each other and are not linked by approximation or idealization.[28] The EFT that describes the FQH effect—the Chern–Simons theory—is not "approximately true" with respect to the in-principle possibility that a better ACFAP higher-energy field theory could exist that would capture more of the dynamics. In fact, quite the reverse is true. A higher-energy theory with a different cut-off—a different set of constraints—would wholly fail to account for the exact set of properties and modalities manifested by the FQH state.

Generally, a smaller degree of approximation implies greater accuracy, and this is the intuition behind interpreting EFTs as being irrelevant evidence regarding the existence of some "final" ultimate theory. In contrast to this intuition, imposing a cut-off that includes a larger energy range does not make an EFT more accurate. Which cut-off and degrees of freedom are relevant depends entirely on the context of the system under investigation, its symmetry conditions, and the constraints on the parameter space, none of which are provided by some higher-energy or more "fundamental" theory.

[28] This situation is similar for theories related by RG methods, as we saw in Chapters 3 and 4.

This is exactly the kind of situation where we have the contextual emergence pattern.

Moreover, even when we have a fundamental theory, or aim to construct one as in QFT, we implicitly use contexts outside that theory to construct it, as noted previously. We always use empirical information, symmetries, and other constraints, invoking stability conditions defining the context for the construction. We never epistemically start with a bottom-up eye ACFAP view of nature; rather, we have our "higher-level" considerations, and these implicitly inform the construction of "lower-level" fundamental theories. Our claim is that we see no reason for believing in a fundamental theory of physics in the traditional—and largely unreflective—sense. Yet even if one were to be defined by imposing another criterion as the measure of fundamentality (as per Castellani), that would not absolve such a theory from the stability conditions making the contextual construction and application of the theory possible. Whatever fundamental phenomena are, they still must come to expression in a concrete context that constrains such phenomena.

To see this, it is useful to look in detail at Jonathan Bain's (2013) analysis of EFTs in terms of DOFs, which suggests a similar conclusion to ours with respect to EFTs as being distinct from mere approximate theories. It focuses on the role that elimination of DOFs plays, and as such it has similarities with Jessica Wilson's (2010) account of emergence in terms of DOFs.

The key point is that by imposing a cut-off, surplus DOFs are removed.[29] This procedure produces a new Lagrangian density that depends only on the new set of DOFs. The effective Lagrangian density is formally distinct from the original high-energy Lagrangian density. Indeed, when one generates the equations of motion for the system—the Euler–Lagrange equations—the set of equations produced by the effective Lagrangian density is not the same as the set from the high-energy Lagrangian density. *More importantly for us, the Euler–Lagrange equations produced from one are not an approximation of the Euler–Lagrange equations for the other* (Bain 2013, p. 3). This is not a scenario where the equations describing the system derived from one cut-off are related to the equations derived from another cut-off (or no cut-off) by a parameter that one takes to be small or large.

This lack of approximation between the two theories is suggested by the fact that to construct an EFT we do not need to know the more "fundamental" high-energy theory at all. In such cases of bottom-up construction of EFTs, the EFT is constructed not by considering a higher-energy theory, but instead by considering the symmetries and constraints appropriate for an energy scale of interest. In other words, to construct an EFT we need only to pay attention to the contexts that apply to the domain of interest, not to

[29] Technically they are integrated out of the Lagrangian density.

any alleged ultimate "fundamental" ACFAP set of considerations. Furthermore, knowing the EFT does not tell us much about the high-energy theory. Nor does knowing the high-energy theory tell us anything about what the dynamics of equations describing a system will be once we eliminate the appropriate DOFs.

All of this suggests that the notion of approximation used in textbook discussions of EFTs is not what amounts to approximation in the usual philosophical or scientific sense. Bain neatly summarizes the three key points given the working assumption that we do not know the form of the high-energy theory and the fact that "for a given high-energy theory, more than one top-down EFT can be constructed":

(a) Failure of law-like deducibility. If we understand the laws of a theory encoded in a Lagrangian density to be its Euler–Lagrange equations of motion, then the phenomena described by an EFT are not deducible consequences of the laws of a high-energy theory.
(b) Ontological distinctness. The degrees of freedom of an EFT characterize physical systems that are ontologically distinct from physical systems characterized by the degrees of freedom of a high-energy theory.
(c) Ontological dependence. Physical systems described by an EFT are ontologically dependent on physical systems described by a high-energy theory.

As we noted previously, Bain's (a) and (b) follow from the formal distinction between "an effective Lagrangian density and a high-energy Lagrangian density, and their corresponding Euler–Lagrange equations of motion." The DOFs of an EFT are "dynamically distinct from those of a high-energy theory; moreover, the former are typically encoded in field variables that are formally distinct from those that encode the latter (i.e., different field variables appear in the Lagrangian densities of an EFT and a high-energy theory)" (2013, p. 4).

Of course, the DOFs of the low-energy theory are a subset of the high-energy theory, so the theories are connected. But note that in set-theoretical terms a subset is not the same as an approximation of a larger set. One way of thinking about this is that by imposing a contextual topology on a set, DOFs are removed. But there are many different contextual topologies that can be imposed on the same underlying set producing radically different sets of reduced DOFs. Depending on the contextual topology, the properties of these sets will be very different.

Therefore, the relationship between the "effective" sets and the "fundamental" set is not one of mere approximation. The main set does not have property P, and a given subset property P* which approximates P. The property P*, as a product of the unique contextual topology/removal of DOFs, can be entirely distinct and unrelated to the property P.

Note that *this is quite different from how approximations usually work*. In approximation, one element corresponds (at least in some lose way) to another element; it is a stand-in for it if we are happy with a degree of imprecision. Therefore, in the literature there are two distinct notions of approximation that must be distinguished:

1. The use standardly in scientific models is to coarse grain, but to be connected to, and represent, the approximated to some degree. For instance, a frictionless plane approximates a low-friction plane. If we add friction back into the model we obtain a more accurate result, but it will have a family resemblance to the approximating model.
2. Approximation as nearness in a set. This is the sense in which EFTs approximate higher-energy theories. Under this notion, to approximate is merely to have conceptual proximity. In the set of field theories, theory A is a subset of the DOFs of theory B. This ordering does not imply all the properties of approximations of notion 1, however. Simply being nearby in the set does not imply that theory A is a rougher version of theory B, or that theory B uniquely determines the form of theory A.

By being a subset, EFTs are constrained by higher-energy theories, but the form of them depends on the nature of the kinds of symmetries, constraints, and contextual topologies that go into making a choice of cut-off, and which DOFs to abandon as irrelevant. There is nothing intrinsic in the higher-energy theory indicating which contextual topology is applicable, the pattern of contextual emergence. The higher-level theory only contributes some necessary conditions to a contingent sufficient condition for the EFT in question.

Recall that deriving an EFT is not a purely mathematical affair. The construction of an EFT requires choosing a cut-off, and the low-energy DOFs. These choices are determined by the context of the system being investigated. There is no automatic way of generating an EFT, and for any high-energy theory there are many EFTs that could be generated from it.[30]

This last point is important if we think of our current best fundamental theories as EFTs related to an as yet unknown high-energy theory X. Why do we have general relativity as our EFT rather than one of the many other EFTs one could generate from X? The reason is that general relativity reflects the DOFs and constraints salient to the domain of the world we have evidence of and wish to account for. It is the contexts of the world that determine

[30] This situation is very similar to the relationship between QM and the classical world discussed in Chapter 4, Section 4.5.

the correct contextual topology, and therefore pick out the correct EFT even when we do not know the high-energy theory. This is because the EFT is a set of good coordinates. It hits upon just the right combination of factors to account for the relevant states and observables and all the modal relations in a domain, because those good coordinates track the properties manifested in the relevant context. Properties manifest in a concrete context; they are described by the Euler–Lagrange equations of an EFT with a contextual topology, specific set of DOFs, and cut-off constraint.

5.5.3 *Effective field theories and fundamentality*

As we mentioned previously, explicating fundamentality is notoriously difficult, and we will not seek to settle the debates. We have already seen one operative sense of fundamentality in the discussion of physical possibility space in Chapter 2, Section 2.3.3: nothing can happen outside the boundaries of possibility space delineated by the most "fundamental" or ACFAP laws and the dimensionality of the universe. This is the most basic sense of fundamentality that can be defined in a non-uestion-begging way. Yet as we have seen in this book, the most basic microscale laws do not fully determine all the actual concrete outcomes within physical possibility space. That is the roll of contexts.

For example, the resources of elementary particle physics at the microscale do not fully determine actual chemical outcomes at the macroscale, though the former resources do bound the space of possibilities for chemical systems such that they cannot behave in physically impossible ways. Additional stability conditions in the form of broken entanglements, broken symmetries, molecular structure, and so forth create the context in which concrete chemical phenomena are realized. Even the "most fundamental laws" contribute only some of the necessary conditions for chemical phenomena.

This contrasts with the typical ontological reductionist intuition that the fundamental laws, whatever those turn out to be, do all the work of bringing about all phenomena in the universe. The contextual emergence pattern that we have been exploring in this book calls into question the idea of a fundamental theory of everything from which EFTs are derived out of pragmatic convenience.

Another problematic ontological reductionist intuition is the notion that fundamentality is linked to composition: the smallest parts we find in physics, characterized in terms of length and time scales, are fundamental, or the theories with widest applicability are, as in unification. Everything else in the universe is built out of fundamental constituents, so everything else is derivative. Recently, Elizabeth Barnes (2012; also see Pexton 2015) has proposed a separation of the notions of composition and fundamentality that challenges this ontological reductionist intuition in a way that we can make better sense of through the contextual emergence pattern that we have been exploring in this book.

In Barnes' terminology, the categories derivative/fundamental are conceptually delineated from the categories ontologically independent/ontologically dependent. Ontologically independent entities are simply those entities which do not rely on any other entities for their continuing existence, such as mereological simples. But given contextual emergence, no such entities exist because everything is enmeshed in multiscale relations. Ontologically independent entities could also be thought of as the basic entities of fundamental physics (e.g., the particles and fields of high-energy physics). This second interpretation, however, is not straightforward. Although physics has "basic" particles, such as electrons and quarks, they are not ontologically independent; they are not even independent of phenomena existing at other larger and smaller scales.

Ontologically dependent entities are those that require other entities for their existence, and ontological dependence is defined by a strong sense of dependence. This means that a dependent entity is sustained in its existence by other entities. Barnes contends that the categories fundamental and derivative are distinct from independent/dependent.

We see that the fundamental/derivative and independent/dependent distinctions can come apart, as Barnes argues. Nonetheless, given contextual emergence, there are no absolutely independent entities; instead, entities always emerge from a web of multiscale relations forming and induced by contexts. In other words, the ACFAP entities at the smallest length and time scales are never independent of context. Whether an entity is fundamental or derivative is going to be relative to ontological and explanatory contexts in a contextually emergent world.

Barnes' notion of fundamentality, however, is relevant when thinking about EFTs. If what is fundamental is whatever must be put in by hand, then clearly, at least *in an epistemic sense*, EFTs in CMP are as fundamental for those systems as any theory of particle physics.[31] As noted previously, the Euler–Lagrange equations for an EFT do not come from the Euler–Lagrange equations of a higher-energy theory. To derive the equations that describe the dynamics of the system, the right cut-off must be imposed and the right DOFs must be thrown away. Nothing in the "fundamental theory" indicates how to make these choices. Even the most fundamental ACFAP theory must come to expression in a concrete context—a context not given by such a theory. In terms of modality, the ACFAP theory tells us only about the possibilities

[31] This, we argue, is the best way to make sense of the fundamentality debates between P. W. Anderson and Weinberg.

included in the possibility space, not how particular regions of possibility are accessed or actualized. Contexts have as much claim to fundamentality as the ACFAP theory as explanatory of our world.

Speaking ontologically, the properties of CMP systems are produced only when nature itself interacts with the relatively structural properties of a system or the aggregate environment created by complex interactions. To manifest the properties of CMP systems, those systems must themselves be interacting with their ontological contexts. It is only in the ontological context of a breaking of symmetry that a ferromagnet manifests magnetization. It is only in the ontological collective environments produced by long-range topological entanglement that the FQH state (and possible superconducting state) is produced.

Nature does not get any of these properties for free simply by posing the properties and modal relations of higher-energy regimes or particles in isolation or different aggregates. Only with the combination of those properties and contexts to modulate them are the precise set of conditions to manifest new properties produced, and hence new modal relations based on the compatibility relations of those properties.

5.6 Conclusion

The detailed elaboration of CMP exhibits the contextual emergence pattern that we have been explicating and illustrating in this book. Interestingly, concepts and techniques from the study of soft condensed-matter systems have been successfully applied to biology, so taking a detailed journey through CMP prepares us for exploring contextual emergence in biological systems. In Chapter 6 we will see that the contextual emergence pattern is the norm in biology and neuroscience as well.

6

Contextual Emergence in Biology and Neuroscience

> When you think about the complexity of our natural world—plants using quantum mechanics for photosynthesis, for example—a smartphone begins to look like a pretty dumb object.
>
> Jeff VanderMeer

6.1 Introduction

In Chapter 1 we introduced contextual emergence with a biological example of patterned follicle formation (Shyer et al. 2017). The underlying genetic machinery provides some of the conditions necessary for follicles. Yet it is the larger-scale mechanical forces due to dermal contraction that provide the crucial stability condition for patterned follicle formation. This example can be extended: smooth muscle contractions are also vital to the cycle of hair-follicle apoptosis and regeneration (Heitman et al. 2020). Again, genes and stem cells provide only some of the conditions necessary for follicle apoptosis and regeneration. The rest of the necessary and sufficient conditions for these processes are provided by the larger-scale dermal sheath contraction due to the stability condition provided by smooth muscle behavior actualizing the genetic and stem-cell machinery through large-scale mechanical stresses.

These are biological examples of the contextual emergence pattern that we have seen throughout this book, particularly in physics—the supposed king of reductionism. Along the way we have hinted that the contextual emergence pattern is broader than physics and chemistry. In this chapter we turn to biology and neuroscience, and we will see that the same pattern arises in these domains. Perhaps this is not too surprising. If physics—the most basic science—is not an exemplar of the reductionist vision, then it is less plausible that biology would really fit the vision.

Emergence in Context. Robert Bishop, Michael Silberstein, and Mark Pexton, Oxford University Press.
 DOI: 10.1093/oso/9780192849786.003.0006

On the other hand, one might argue that biology has been successfully reduced to cellular mechanisms and biomolecules such as DNA which, in turn, are aggregates of chemical molecules. Maybe one cannot reduce biology to physics, but have not molecular biologists shown that biological phenomena are successfully reduced to biochemistry? The short answer we elaborate in this chapter is "No!"

Historically, the biological sciences have played a very important role in discussions of emergence.[1] Massimo Pigliucci (2014) has argued that one can understand the history of biology as a struggle between reductionism and holism. The reductionist side can be illustrated by Richard Dawkins (1976) arguing that evolution is based on genetic changes and that all other larger-scale phenomena in evolution are derivative from such changes in bottom-up fashion. This yields a picture of organisms as "just 'lumbering robots' in the service of their genetic masters" (Pigliucci 2014, p. 261). In contrast,

> there is a long holistic tradition that focuses on the complexity of developmental systems, on the non-linearity of gene–environment interactions, and on multilevel selective processes to argue that the full story of biology is a bit more complicated than that.
>
> (p. 261)

It is true that the successes of genetics and molecular biology over the past decades might appear to fuel the reductionist intuitions of Dawkins and others. But this is far from the whole story, Pigliucci maintains: "Holism has built on the development of entirely new disciplines and conceptual frameworks over the past few decades, including evo-devo [evolutionary development] and phenotypic plasticity." He further observes that "a number of biologists are still actively looking for a way out of the reductionism–holism counterposition, often mentioning the word 'emergence' as a way to deal with the conundrum" (p. 261).

As formulated, the reductionism–holism dilemma looks to be structured as another false forced choice. One impediment to a way out of this dilemma is the hold that the mechanistic paradigm has throughout biology and philosophy, including, of course, neuroscience. This sort of explanation in the literature is known as localization and decomposition (L&D). The core idea of L&D is to break down a mechanism or whole into operations of interrelated parts and organize them into modules, which, when properly ordered, explain the workings of larger mechanisms or submechanisms that they make up. According to this line of thinking, interacting and hierarchically organized parts causally produce the phenomenon in question (Bechtel 2011; Bechtel and Abrahamsen 2005; Machamer, Darden, and Craver 2000).

[1] Indeed, we could have started our tour of contextual emergence with biology and worked our way towards physics as a viable path.

Perhaps the most well-known example is the attempt to localize specific cognitive functions, or even aspects of conscious experience, to specific regions in the brain. The contextual emergence pattern challenges the reductionism of the L&D strategy.

While many defenders of neuromechanistic explanation and reductionism now grant that simplistic forms of L&D often fail, and also grant that mechanisms are often multiscale and multilevel, they still often want to defend neomechanistic explanation, mechanistic reductionism, and even L&D in *some* form.[2] However, even most "liberal" neomechanists would like to restrict the explanation of cognition, behavior, and conscious experience to the brain and central nervous system as much possible, and not beyond. In short, most neomechanists are methodological individualists who still regard "emergence" as almost as dirty a word as "holism" (Silberstein 2021)—at least, "emergence" in any but the merely epistemic or weak sense. In part, this is because they can conceive of only two kinds of emergence—the mysterious strong or epistemic variety. As we argued in Chapter 1, if one must make a false forced ontological choice between mysterious strong emergence and ontological reductionism, the latter wins hands down if our goal is a unified, coherent understanding of the world.

While this ancient debate rages on in the literature on neural and biological mechanisms, what many often fail to notice is that contextual emergence in various forms is not only well confirmed by the biological, cognitive, and neural sciences, but even constitutes textbook science. Granted there still is a lot of room for debate about what the data entails; nevertheless, we shall argue that contextual emergence is the best and perhaps even most common interpretation of the data. Not only that, but in cases where it looks like L&D is a successful strategy, this is only because the role of context has been underestimated. The spatial ordering, growth, apoptosis, and regrowth of hair/feather follicles is an example: neither ordering, growth, nor apoptosis can be localized or decomposed into "underlying" genetic mechanisms.

In summary, we will see that L&D fails because of the following:

1. Multiscale co-dependent, co-determination relations ranging from the molecular to the social (Bassett and Sporns 2017).
2. Global, higher/larger scale, stability conditions (e.g., "small-worldness" in multiscale topological explanations).
3. Sensitivity to contextual changes at multiple scales, both "internal" to and "external" to "the system" in question (e.g., robustness and plasticity).

[2] See Silberstein (2021) and Silberstein, and Chemero (2013) for details.

4. Multiple realizability and overall partial autonomy of "higher-level" explanations. For example, the difference-maker in topological explanations is the topology of the network in question (e.g., being a small-world network). Such networks are ubiquitous across species, come online early in the development of individual organisms, and are instantiated by very diverse structural components and interactions (the "universality" of such networks). Thus, what accounts for the explanatory power and autonomy of such networks is the topology itself.
5. The fact that relevant stability conditions cannot be explained mechanistically or reductively in principle (i.e., in terms of L&D).
6. Such stability conditions explain the behavior of the system in question at multiple scales, as well as the emergence of new behaviors, entities, processes, and so forth.
7. These explanations show why integrative and explanatory pluralism of various sorts (e.g., Mitchell 2002) are essential, such as the necessity of multiple disciplines in understanding the target explanandum and the need for different types of explanation with different focuses for the same system.

We well begin by digging into systems biology (Section 6.2), and then turn to neuroscience and cognitive science (Section 6.3). In these areas, multiscale contexts play key roles in all the phenomena. Consciousness and mental causation will receive some attention in these discussions, though we have more to say about the latter in the Chapter 7. Finally, we will discuss some briefer examples from biology and ecology that illustrate the broadness of the contextual emergence pattern.

6.2 Systems biology

Let us begin by reminding ourselves of a recent episode in the history of molecular and developmental biology. The Human Genome Project (HGP) grew out of the 1980s and 1990s in which genetic determinism was a belief that many biologists held in common. The HGP was motivated in large part by its own version of L&D and modularist thinking, for example, in the form of the gene doctrine (one gene for one protein). Upon completion of sequencing the human genome, many scientists were shocked to discover that humans have only 30,000 genes as opposed to the predicted 100,000. They were also surprised to discover that we humans have only 300 unique genes distinguishing us from a mouse. And again, surprised to learn that genes (more accurately gene networks) can give rise to many different proteins. Genes are pleiotropic, and phenotypic traits are mostly multigenic.

In reaction to learning all this, famed Harvard biologist Stephen Jay Gould wrote in the *New York Times* (19 February 2001): "The collapse of the doctrine of one gene for one protein, and one direction of causal flow from basic codes to elaborate totality, marks the failure of [genetic] reductionism for the complex system we call cell biology."[3]

In the years since this statement was published, more and more evidence has accumulated to show that Gould was right. As Luciano Boi puts it: "There is unidirectional flow of information from one class of biological molecule to another. Genomic functions are inherently interactive, isolated DNA is 'virtually inert', cells have many mechanisms to 'complexify, modify and change' the structure and function of DNA" (2018, p. 196). Hence the rise of epigenomics and the choir of people claiming that genetic determinism is dead. As Carl Zimmer notes: "Yet the epigenome is not simply a rigid program for turning genes on and off in a developing embryo. It is also sensitive to the outside world" (2018, p. 160). Examples include methylation, diet, and all kinds of stress. Other key variables include chromatin configurations, chromosome configurations, organelles, membranes, hormones, morphogenesis, and so on. The very meaning of epigenetics is that the genetic "code" by itself is not sufficient to determine what is happening in development or afterwards. For instance, epigenetic modification to histone proteins that wrap around genes change gene expression. It is now well known that there are environmentally induced epigenetic alterations that arise early in development, that happen throughout our lives and can be inherited via transgenerational epigenetic effects (so called "epigenetic inheritance"). Such extragenetic inheritances can outlive actual mutations and can be reversible.

Molecular biologists have discovered that the relationship between genes and proteins is many–many, and we now must think in terms of regulatory gene networks (genomics), regulatory RNA networks, protein networks (proteome), and the complex non-linear interactions among them. Furthermore, these relationships are also affected by several multiscale contexts, including the cellular environment, the wider organismic environment, the ecological niche of the organism's population, and various features of the larger environment in which this niche is situated. These interactions are obviously multiscale, multilevel or organization, inextricably interrelated, and interdependent (Noble 2006, p. 105). In the developmental process, what any key biological player does, such as genes and proteins, and what it results in as output, is a function of multiscale contexts (Bechtel 2019, pp. 461, 488;

[3] This is not to say that people were not already working on gene regulatory networks before the HGP. For instance, the recently deceased Eric Davidson pioneered the field decades ago (e.g., 1968, 1994; Britten and Davidson 1969).

Francis 2011, p. 159; Noble 2006, pp. 17, 34). For example, since at least the 1980s, biologists have been discovering how much so-called junk DNA contains switches that regulate protein coding genes and body-plan genes that orchestrate these switches. This is all part of gene regulatory networks that also produce many different types of RNA, which play key regulatory roles in their own right, such as regulating chemical modifications of the DNA and associated proteins in response to environmental signals.

Similarly, protein folding and protein function are embedded in multiscale contexts. Amino-acid sequences alone determine neither three-dimensional structure nor function. Other factors include many features of the cellular environment and cellular activities such as various properties of water, lipids, and the interactions of "many other molecules that are not coded for by genes" (Noble 2006, p. 35). Nothing in the genome determines the topology of proteins. As Boi observes, "the biological information of proteins does not derive only from structural information, but also from the complex functional networks that connect specific binding sites at the molecular level to the cell's activity and to the more global organismic level of organization and functioning" (2017, 195).

Likewise, embryogenesis is not determined by the genome in bottom-up fashion. Early on, in the embryo a ball of identical pluripotent cells becomes differentiated into various cell types and organs because of a network of physical and chemical environmental gradients and signals. It is because such biological processes are so contextual and not driven by instructions from the genome, that many biologists call them self-assembling, self-organizing and self-maintaining regulatory networks, with interdependent interactions at all scales and "levels."

Mark Povich and Carl Craver (2018, p. 193) have expressed skepticism about any departure from modularity in complex biological systems on the grounds that modularity is a clear evolutionary advantage, since such modular networks will be better able to survive environmental changes, damage, and so forth. However, DNA–RNA–protein networks are modular in the sense that they retain a kind of autonomy across several different contexts that is highly conserved across evolutionary history. There are two things to note here. First, the reason that the modularity regarding such networks obtains is because the relevant stability conditions and context for the networks conserved by evolution come along for the ride when transplanted into a new environment. Second, even given this sort of modularity it does not follow that the modular network in question will produce the same output/effect given *any* change in its context. The very same network, as defined structurally, can produce different effects, operations, or products in the context of different networks. The point is that one must think of modularity and context together—which L&D approaches, even at their most liberal, do not do.

As we will show in the next section, people often talk about plasticity, robustness, reuse, and the like, in the context of the brain, yet such features are common in complex biological systems more generally. In addition to the brain, one can find multiscale networks, reuse, plasticity, robustness (invariance in the face of environmental and *contextual* changes), and autonomy (adaptability and flexibility in the face of environmental and *contextual* changes) in many complex biological systems (Silberstein 2016, 2021). Thus it is no surprise to see that these same features exist in neuroscience. This is to say, that scientists find phenomena such as plasticity of all varieties operating at multiple scales, such as behavioral plasticity, neural plasticity, molecular plasticity, and so forth (Bateson and Gluckman 2011). The point here being that brains inherit or exemplify their multiscale network properties, organizational principles, and other global organizing constraints because there are basic biological processes and principles applying across multiple biological scales.

Perhaps all of this is best illustrated by the relationship between plasticity, robustness, and autonomy. There are many different forms of robustness and plasticity, such as developmental, phenotypic, and a variety of neural, behavioral, immunological, among others. Let us take phenotypic plasticity and robustness as an example. This is the phenomenon in which genetically identical individuals will develop different phenotypic traits in different environmental conditions (Kaplan 2005, 2008). Because of phenotypic plasticity, a single genotype or genome can produce many different phenotypes, depending on environmental and developmental contingencies (Gilbert and Epel 2009). Phenotypic plasticity is just one example of epigenomic processes in which various mechanisms create phenotypic variation without altering base-pair nucleotide gene sequences, altering the expression of genes but not the gene sequences.

In contrast, there are cases in which genetic or environmental changes have no phenotypic effect. This persistence of a particular organism's traits across environmental or genetic changes is called "robustness." Robustness is illustrated by various knock-out experiments in synthetic biology whereby a particular gene (or group of genes) known to be involved in the development of some protein or phenotypic trait is disabled without disturbing the presence or production of the developmental end product in question (Jablonka and Lamb 2005).

As we will see, to account for and model plasticity and robustness, developmental biologists have used network and dynamical systems forms of explanations. The ongoing development of an organism acts as a global stability condition that constrains or "enslaves" the components necessary to maintain its dynamics. Because of this, a developing system will have highly plastic boundaries and will be composed of different enslaved components over time. This plasticity serves the autonomy and robustness of the developing organism, making it more likely to be viable and adaptive. The development of brains is no exception.

Robustness is closely related to autonomy—another key concept in evolutionary developmental biology. Autonomy is the property of living systems making use of their environments to maintain themselves. It is sometimes explained in terms of recursive self-maintenance. Some systems are plastic such that they can maintain stability not only within a range of conditions, but also within a range of changes of conditions: they can switch to deploying different processes, depending on conditions that they detect in the environment.

Robustness and plasticity are interdependent. For instance, "plasticity is often regulated by robust mechanisms and robustness is often generated by plastic mechanisms" (Bateson and Gluckman 2011, p. 46), in an interplay of evolutionary and developmental processes. The ever-growing varieties of robustness and plasticity are co-creating and co-maintaining, allowing a complex biological system to have autonomy: "Development involves both internal regulation and reciprocity with the environment. Careful analysis of what happens during development suggests that it is no longer helpful to retain a hard and fast distinction between robustness and plasticity" (2011, p. 62).

As the previous remarks indicate, plasticity, robustness, and autonomy are global or universal features of complex biological systems. They apply or are exemplified at multiple scales, and function as organizing principles in a variety of different fields of biology. Such principles have been well known for a long time, and they are well confirmed without any appeal to abstract mathematical models (Koonin 2011, pp. viii–ix; Jaeger and Calkins 2012,p. 27). It is also well known that such global conditions can impose the same function even across different species, using different structural components. One can demonstrate equivalence classes of networks across species, including network function: "Given that function is conserved and the mode [i.e., specific structural mechanism] isn't, it suggests regulation by the organism as a whole" (Jaeger and Calkins 2012, p. 27). This is an actual-world example of multiple realizability. Plasticity, robustness, and autonomy, then, function as stability conditions. They exert what is often called "top-down information control," by which the working "parts" are constrained in their behavior in service to the larger function, the functional network.[4] Nonetheless, the explanation for plasticity, robustness, and autonomy is not found in the genes of organisms, nor in the proteins alone that genes produce. Plasticity, robustness, and autonomy are conditions that

[4] Compare with how individual fluid parcels are constrained to serve the larger Bénard cells in convection (Chapter 2, Section 2.3).

make functional networks what they are, and genes behave as they do in the context of the networks that they inhabit. This is the contextual emergence pattern.

Take gene-environment interactions, for example. It is well known that cultural and social processes constrain such interactions by reshaping the environment and population structure. Genetic, cognitive, and cultural evolution all go hand-in-hand as one multiscale process once the conditions enabling genes, forms of cognition, and culture manifest. There are many epigenomic studies revealing correlations between cultural changes and changes in genetic activity, which genes are turned off and on, what various proteins those genes together are contributing to, and so on. Cultural changes happen more rapidly than genomic changes, thereby creating new environmental constraints that impose new selective pressures on the genome (e.g., changes in brain size, brain complexity, neural plasticity due to tool-use, and language).

Indeed, the development of agriculture may have been the most influential cultural change in hominin evolution: "Large-scale scans using patterns of linkage disequilibrium to detect recent selection suggest that many more genes evolved in response to agriculture" (Richerson, Body, and Henrich 2010, p. 8985). A simple example is the rise of lactose tolerance in the eighth millennium BCE in response to cultural changes in agriculture. Think also of the various effects of controlled fire and cooking—not only on the human genome but on all the other genes in our gut microbiome.[5] There are neurons in the heart and the gut that communicate with the brain, and there is growing evidence that the networks of trillions of bacteria in the gut's microbiome communicate with various neural networks in the brain. There is evidence from mouse studies and some human studies that imbalances in the gut's microbiome and failure of communication between microbiome networks in the gut and neural networks could be causal factors in neuropsychiatric disorders, including disorders that adversely affect sociality itself (Wu et al. 2021).

This is not so surprising, as new evidence reveals the importance of the human gut's "enteric nervous system," sometimes called the "second brain" (Ahmadzai et al. 2021). This second brain includes a network of various types of neurons and glia—roughly the same number as those found in the brain of several other animals such as cats. Just as with the brain in the head, the glia in the second brain heavily modulate and modify the signals carried by neural circuits (Ahmadzai et al. 2021).

Recent studies indicate that by transplanting the microbiome of young mice into older mice, many of the effects of aging on the older mice as relates

[5] The microbiome itself is modeled with network analysis and dynamical systems theory due to its complexity.

to memory, learning, and the like, were greatly reduced or reversed. Microglia are thought to be a key part of this process (Boehme et al. 2021, p. 666). Thus, not only did cooking outsource and externalize digestion, but it allowed us to consume and conserve more calories and get better nutrition, which in turn increased the size of Homo genus bodies and brains, and so on. We can already see that what we think of as the human cognitive system, nervous system, digestive system, and immune system are inherently multiscale and inextricably intertwined. And we can see that cooking and other cultural factors played a big role in this coming about and developing in the way that it did.

This is all to say that genetic heritability is not only or even primarily a function of genes, traits, and ecology, but of cultural factors once culture exists in some form. Cultural changes drive much of our within-species phenotypic variation. At best, talk of genetic effects makes sense only with respect to specific ever-changing cultural contexts: "The question is not whether genes or culture contribute more to a behavioral trait, as behavioral traits can only be understood as emergent products of our dual inheritance, genetic and cultural. Nothing in behavioral genetics makes sense except in the light of cultural evolution" (Richerson, Boyd, and Henrich 2016, p. 23).

What we see is that human evolution is a process illustrating contextual emergence. Development of agricultural practices, creation of cities, and other forms of culture and technology are stability conditions defining contexts full of multiscale constraints that limit genetic evolution in some directions while enabling it in other directions. Culture changes the accessibility conditions in possibility space—genetic possibility space, if you will. So, while genes contribute some of the necessary conditions for human evolution, the full set of necessary and sufficient conditions for human evolution—and many of the changes in genes that become stable parts of human populations—are contributed by the stability conditions at the larger scale of culture.

One implication is that neither structural and functional modularity, nor L&D, are the key players in complex biological systems. And even when there is modularity and some form of L&D, these always function in a larger context, as language illustrates. Language processing is enabled by highly distributed multiscale neural networks composed of very many smaller brain regions or nodes that themselves are multifunctional and domain-general (Zerilli 2021, chapter 7). Nevertheless, language capacities do not develop properly in individuals unless they are fully exposed to natural languages within a particular window or developmental time frame, and this is shaped by cultural practice, not neural modularity. As we will see in the next section, language processing is no exception in this regard, but rather exemplary.

6.3 Contextual emergence in neuroscience and cognitive science

The context-dependent nature of language processing opens on to the fact that the brain's structure and functions are inherently multiscale. Molecular neuroscience, systems neuroscience, and computational neuroscience can and do use a variety of different tools to explore all kinds of activity at multiple interacting scales. These tools include molecular techniques and dynamical systems theory, as well as different neuroimaging modalities and computer simulations involving network or graphical models. All the tools together are used to create mechanistic (structural) and statistical (functional) models, as well as other formal models. One goal is to relate more deeply structural and functional analyses, as they both enlighten and constrain the exploration of each other.

Suppose we try to start with a single neuron and a single neurotransmitter system such as serotonin. One cannot properly model ionic currents without bringing in the background condition of the inherently larger-scale property of cell voltage. In turn, the biophysical properties of neurons affect the propagation of electrical currents (action potentials). Modeling this most basic process in the brain already requires interactions among the scales of molecular, subcellular, cellular, and neural circuits.

Action potentials cause the release of neurotransmitters to their postsynaptic target. Neurotransmitters are varying chemicals that travel across neurons and bind to specific receptor sites, depending on the type of neurotransmitter. Accordingly, the neurotransmitters in turn affect the behavior of the cells with which they bind. Action potentials and neurotransmitters regulate one another in all kinds of complex ways. Stimulation of excitatory receptors by neurotransmitter binding causes depolarization of the postsynaptic plasma membrane, promoting generation of an action potential. Conversely, stimulation of inhibitory receptors causes hyperpolarization of the postsynaptic membrane, repressing generation of an action potential.

While we often think of neural signaling via action potentials as being strictly digital (all-or-nothing) and being the central mode of information transmission, there is now evidence that things are much more complicated than that in both respects. For instance, subthreshold changes in presynaptic membrane potential before triggering the spike also determines spike-evoked release of neurotransmitter. The changes in presynaptic voltage that regulate spike-evoked release of neurotransmitter (through the modulation of biophysical state of voltage-gated potassium, calcium, and sodium channels in the presynaptic compartment) are themselves analog (Zbili et al. 2016).

Even at the molecular scale, neurons and neurotransmitters are not anywhere near the whole story. For example, regulatory RNA is generated at the synapses and in turn controls the production of various proteins at these sites. The latter proteins are gene-regulatory proteins that can travel to cell nuclei and affect gene expression therein. Thus the local activity at different synapses can have a profound effect on the long-term genetic behavior of the neurons involved.

Some of the regulatory RNA produced by neurons is packaged in membrane-bound vesicles called exosomes, which can carry information to distant cells, affecting their function and behavior. For instance, traveling from a postsynaptic neuron to presynaptic ones, exosomes can strengthen and enhance connections related to learning. Exosomes communicate with other neurons in brain and glia (or "neuroglia") cells, helping to activate and inhibit both, as well as helping to synchronize gene expression in nearby neurons (Parrington 2021, p. 114).

Glia cells—once thought to be nothing but support structures and maintenance for neurons—do not generate action potentials. However, they do communicate chemically with many of the same surface receptors as neurons, as well as other glia cells (Parrington 2021, p. 69). Glia, such as astrocytes, also aid in the formation of synapses. Microglia both help defend against disease as well as help pruning neural connections (p. 70), and are involved in memory formation by activating NDMA receptors related to long-term potentiation via amino-acid D-serine, thereby changing synaptic structure and gene expression, all of which in turn involves regulatory RNA networks (p. 112).

Moving up and away from the molecular and cellular scales, highly coordinated and oscillating neural synchrony affects the action potentials of individual neurons, which in turn affects the overall charge of cells and their background activity. Just as with small-world networks, many different insects and animals have similar endogenous synchronous global brain wave activity associated with sleep, attention, and other processes existing at many different scales. This is thought to be yet another method for coupling and communication in the brain. For instance, there is evidence that primates use beta waves to consciously switch between different pieces of information, while gamma waves are associated with, among other things, retrieval of different pieces of information. Beta waves are thought to act like signals that open access to working memory.

On a more global scale, all this activity generates electrical fields in the brain which can affect spike timing dependence and synchronization of neural activity. Thus the electrical fields are generated by neural activity, while these fields alter that activity in turn, such as ionic flux across cell borders

(Godfrey-Smith 2020, pp. 187–197; Parrington 2021, pp. 73, 122). Such global and multiscale coordination is possible because multiscale networks and their properties enable a brain–body–environment binding exhibiting contextual emergence.

6.3.1 *Systems neuroscience and social neuroscience*

Consider neuroscientist Lisa Feldman Barrett's bestselling book *7½ Lessons About the Brain* (2020). Feldman Barrett's second lesson is that "Your brain is a network—a collection of parts that are connected to function as a single unit" (p. 30). As she notes: "Your brain network is not static—it changes continuously. Some changes are extremely fast ... These network changes happen instantaneously and continually, even as your physical brain structure seems unchanged" (p. 36). We will return to discuss brain networks later, but her key point is that the primary unit for modeling and understanding the brain in systems neuroscience is the functional and structural network. This is on a scale larger than that of neurons, local neural circuits, or even local assemblies of neurons. Such networks often encompass a variety of spatial and temporal scales, entities, and activities, as well as interactions across scales, and can be used to model brain-wide organization, as well as more regional activity (Silberstein 2021; Silberstein and Chemero 2013). Multiscale modeling of such structural and functional networks is more and more the norm in systems neuroscience and computational neuroscience, where the goal is to create sophisticated mathematical tools to model and simulate such networks (Schirner et al. 2018, p. 1).

Feldman Barrett's third lesson is that "little brains wire themselves to their world" (p. 47). Her point is that whether you want to call it embodied, embedded, enactive, or extended cognition (more on those shortly), one must acknowledge that the brain and the world often should be treated as one system or at least as inextricably connected. This includes the brain's connections to both the physical and the social: "Little brains *require* a social world in order to develop typically. For example, certain physical inputs, such as photons of light bombarding their retinas, must be provided or the brain won't develop normal vision" (p. 57, original emphasis). Other examples of essential connections that she presents include the following: the role of caregivers in the social world for tuning and pruning the brain in a newborn's neural development, guided attention, various stable features of the fixed environment, niches, training the senses, exposure to natural language, many different social inputs such as love and affection, education, and various socioeconomic conditions (pp. 57–58). In her own words: "Our three examples of tuning and pruning demonstrate how the social world profoundly shapes the physical reality of the brain's wiring" (p. 57). As she notes, all of this illustrates the limitations of the nature-versus-nurture debate (pp. 62–63).

Echoing the field of social neuroscience (to be discussed shortly), Feldman Barrett's fifth lesson is that "Your brain secretly works with other brains" (p. 83). The fact that we are a social species does not stop at early brain development or even at childhood. "Ultimately, your family, friends, neighbors, and even strangers contribute to your brain's structure and function and help keep your brain humming along" (p. 84). Our social interactions with others coregulate and synchronize several biological and cognitive processes such as breathing, brain waves, motion and bodily movements generally, heart rate, circadian cycles, menstrual cycles, linguistic capacities, and learning, among others (Spivey 2020). Most of us are familiar with all of this, in part because, whether "in the wild" or in the laboratory, we can observe what happens to people when they are removed from their social connections.

In the widely used textbook *Introduction to Social Neuroscience* (2020), Stephanie and John T. Cacioppo note that this relatively nascent field of social neuroscience is based on the following assumptions (Chapters 1–3):

- Human brains are not regarded as isolated computational devices, but like a device networked with other brains and people both physically and socially.
- Evolutionarily speaking, there are conserved neural, hormonal, cellular, and molecular mechanisms involved in social behavior.
- Social connectedness, social complexity, and social/cultural learning are some of the driving forces behind the evolution of the human brain.
- Brains and their evolution underly social processes, but the reverse is true as well. This can be seen on both evolutionary and developmental time scales.
- The social brain hypothesis. Larger and more complex brains enabled more social interaction, and vice versa. As culture developed, many more complex problems were solved by social groups, all leading to positive feedback in the direction of ever-increasing neural and cultural complexity.
- The focus is on connection and coordination in inherently social functions such as communications, social perception, recognition, imitation, empathy, competition, and cooperation. For instance, language is not viewed primarily as an information-processing medium, but as a means of communication.
- Multiple interacting scales and levels of organization from genes on up connect brains and social interactions such that there are multiple

> multiscale avenues of mutual determination and multiple multiscale interacting causal factors. There is reciprocal influence from social to biological and vice versa at multiple scales. Such biopsychosocial dynamical systems are highly complex, often non-linear, and "interaction dominant" (Spivey 2020). Examples include the growing evidence that social environment can modulate gene expression, the severe effects of social isolation and loneliness on neurological, cognitive, and genetic processes (Ingram et al. 2021, p. 935), and that the effects of pharmacological interventions, such as stimulants or even placebos, are partly a function of social hierarchy and other social factors.

Social neuroscience, then, is an entire discipline devoted to studying the way that neural and social processes interact across multiple length and time scales—including both evolutionary and developmental time scales—to constrain and enable one another to emerge. Highly regarded hard-nosed and mechanistically-minded neuroscientists at the top of their profession think that we can now regard it as well confirmed neuroscientific common sense that multiscale contexts are the right way to think about the relationship between brains and their physical and social environments. This has clear parallels with the growing body of evidence from epigenetics and epigenomics surveyed more generally in the previous section.[6]

From the point of view of *everyday* common sense, none of this comes as a surprise. Think of the recent effects of social media on all aspects of human cognition, behavior, politics, mood and affect, and the like. What humans are, how we cognize, how we act, how we feel, what we believe, and so forth, cannot be isolated from our social and technological contexts (e.g., Bail 2021; Spivey 2020).

What we see in our survey is that neurons provide some of the necessary conditions for human brains and behaviors. But it is the larger-scale, sometimes even brain-wide, neural networks, electromagnetic fields, and chemical environment that supply the rest of the necessary and sufficient conditions for individual neuron behavior. The properties of large-scale networks provide the stability conditions for microscale behavior of neurons. Nevertheless, even these networks provide only some of the necessary conditions for human behavior. The cultural—technology, social media, religious traditions, and so forth—provides the rest of the necessary and sufficient conditions for the behavior of large-scale neural networks and brain dynamics. Cultural conditions provide the stability conditions that large-scale neural networks and brain dynamics need to have the properties and behaviors that they exhibit.

[6] For more discussion, see (Silberstein 2021).

This is the contextual emergence pattern like we saw previously with genetics, epigenetics, and culture, and it is the same pattern as we saw in physics and chemistry. Neural and cognitive science is not simply a matter of molecular and cellular properties and modules determining all phenomena at larger length and time scales. Instead, the properties of biomolecules, neural cells, and any modular units these form[7] by themselves are not even sufficient for their own behaviors. The larger-scale contexts in which these microscale properties and modules come to expression are crucial to their behavior.

6.3.2 *Organizational features of the brain and related complex biological systems*

Given what we have discussed so far, we should expect that the explanation for our malleability and adaptability is not a reductionist one as many philosophers and neuroscientists have imagined. Let us begin with a summary of those features of the brain that help explain human general intelligence, the contents of conscious experience, and our overall adaptability:

- Neural plasticity. Generally defined as changes in the structure, activity, or function of the brain on some scale relative to *some change in context* such as injury, stroke, or simply learning. An example would be synaptic plasticity, where experience, learning, and memory formation change the synaptic connections in the brain; cross-modal plasticity, where, for instance, the loss of one sensory modality inducing cortical reorganization that leads to enhanced sensory performance in remaining modalities such as the relocation and transfer of somatosensory and auditory functions to the former visual cortex; intramodal plasticity (plasticity within a modality) such as the expansion of cortical maps to neighboring regions of intact cortex that have been deprived of sensory input from within the same modality as supported by the expanding cortex; and supramodel plasticity (not unlike cross-modal plasticity but need not involve injury, sensory deprivation, or special training) such as occipital cortices not only serving as a basis for non-visual information processing, but also contributing something inherently visual to the non-visual input, as in "non-visual input is being processed visually" (Zerilli 2021, p. 20).[8]

[7] We will forego discussion of how even the modules cognitive and neuroscientists tend to focus on supplying stability conditions for the behaviors of the neural cells composing such units. The contextual emergence pattern repeats on these scales.

[8] See Zerilli (2021, chapter 2) for more details on these types of plasticity.

- As John Zerilli notes, "plasticity is an intrinsic and persistent property of the nervous system" (p. 10) at all scales in the brain, including not only the aforementioned cortical map reorganization, but neurotransmitters, neuromodulators, cellular changes caused by learning and memory consolidation, neuromorphology, neurogenesis, among others. Perhaps the most well-known example of plasticity is sensory substitution, such as converting visual images into soundscapes via a "visual-auditory sensory substitution device" (p. 88). Neural plasticity, as we saw, is joined by a host of other types of plasticity in biology such as phenotypic plasticity, all of which helps explain the robustness and autonomy of complex biological systems in general.
- Neural reuse or *recontextualization*. Each region of the brain ends up participating in many different functional coalitions over time/at a time. In other words, the same neural circuits can end up contributing to different tasks or functions, *depending on context* (Anderson 2010, 2014; Zerilli 2021, chapter 3).
- Neural redundancy. Numerically distinct brain regions have the same structure and function. These regions become active due to *changes in context* such as injury or stroke.
- Neural degeneracy. Different neural structures and mechanisms perform the same function, depending on specific *contextual features* that change over time. As Feldman Barrett puts it: "Degeneracy in the brain means that your actions and experiences can be created in multiple ways. Each time you feel afraid, for example, your brain may construct that feeling with various sets of neurons" (2020, p. 39).

One certainly can debate how all these different context-dependent adaptive features of the brain relate to each other (e.g., whether some are more fundamental than others in some specified sense). In the case of neural reuse, the very same neural circuits—the same mechanisms—are contributing to different functions under different conditions *in different contexts*. Yet, at least in some cases of plasticity, not only do we have different mechanisms subserving the same function, but sometimes the mechanism (what Michael Anderson calls "the working," 2010, p. 297) itself might change because of its new role or *context*. In the case of degeneracy, distinct mechanisms that have always been distinct end up subserving the same function or task at different points in time.

What all these features have in common is the fact that they are triggered by *changes in context* such as environmental changes, changing cognitive tasks, changes at larger and smaller scales within the brain, and so forth. In other words, the neurons and neural circuits do not drive the changes bottom-up, regardless of context. Instead, it is changes in large-scale contexts that trigger the smaller-scale changes at the scale of circuits/mechanisms. Again we see that whatever L&D and modularity might be going on in the brain, none of this is functioning in absence of stability conditions defining the larger-scale contexts and their changes as expected from the contextual emergence pattern.

6.3.3 *Network neuroscience*

The following is a brief overview of network neuroscience and the nature of topological networks.[9] Network analyses of the brain are based on the thought that brain function is not just relegated to individual regions and connections but instead emerges from the topology of the brain's entire network (e.g., the "connectome" of the brain as a whole). In such graphical models of neural activity, the basic units of explanation are not individual neurons, cell groups, or even brain regions, but multiscale networks and their large-scale, often distributed, and nonlocal connections or interactions. The study of this integrative brain function and connectivity is mostly based in topological features or architecture of the network. Such multiply realized networks are partially insensitive to, decoupled from, and have a one-to-many relationship with respect to smaller-scale neurochemical and "wiring" details.

More specifically, a graph in this case is a mathematical representation of some actual many-bodied biological systems. The nodes in such models can represent neurons, cell populations, brain regions, or other relevant objects, and the edges represent connections between the nodes. The edges can represent structural features such as synaptic pathways and other wiring-diagram-type features, or they can represent more topological features such as graphical distance and network types. What matters in such graphical explanations is the topology or pattern of connections. Different geometries or arrangements of nodes and edges can instantiate the same topology.[10]

When mapping the interactions (the edges) between the local neighborhood networks, scientists are interested in global topological features (i.e., the topological architecture of the brain as a whole). While there are local networks within networks, it is the global connection between these that is often of greatest interest in systems neuroscience. Graph theory has many kinds of network topologies, but one of great interest to systems neuroscience are small-world networks. This is because various regions of the brain, as well the brain as a whole, are thought to instantiate such networks. The key topological properties of small-world networks are:

[9] See Silberstein and Chemero (2013) and Silberstein (2021) for more discussion.
[10] See Silberstein and Chemero (2013) and Silberstein (2021) for technical details.

- Sparseness: relatively few edges given the large number of vertices.
- Clustering: edges of the graph tend to form knots, so, if X and Y both know Z, there is a higher-than-normal chance they know each other.
- Small diameter: the length of the most direct route between the most distant vertices, for example, a complete graph, with $\frac{n(n-1)}{2}$ edges, has a diameter of 1, since you can get from any vertex to any other in a single step. Most nodes are not neighbors of one another, yet can be reached through a short sequence of steps.

For small-world networks, there is 1) a much higher clustering coefficient relative to random networks with equal numbers of nodes and edges, and 2) a short topological path length. Thus, small-world networks exhibit a high degree of topological modularity and nonlocal or long-range connectivity. There are many different types of small-world networks and other types of networks with unique topological properties that allow researchers to make predictions about the robustness, plasticity, functionality, health, and so forth, of brains that instantiate these networks (Sporns 2011).

One type of network of particular interest is called the "Rich-Club" network (van den Heuvel and Sporns 2011; Pedersen and Omidvarnia 2016) Such network architectures are called "Rich-Club" based on the analogy with wealthy, well-connected people in society. "Members" of this club constitute a few "rich" brain-regions or central "hubs" that distribute many of the brain's global neural communications. The "Rich-Club" topological brain architecture is instantiated when the hubs of a network tend to be more densely connected among themselves than nodes of a lower degree (see Fig. 6.1).

There is recent evidence that the hubs are "interwoven" so as to *simultaneously* participate in multiple networks and cognitive tasks. Some hubs are more connected than others, both structurally and functionally, while some hubs are more regional versus larger-scale or longer-range connections. There

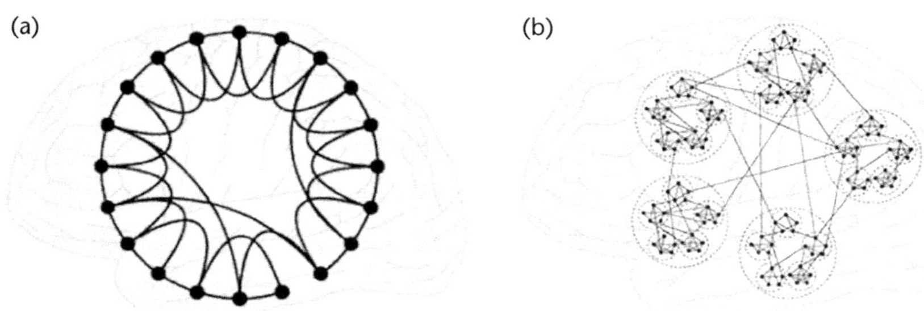

Figure 6.1 Small-world and rich-club networks.

are, no doubt, also "meta-hubs" such as the amygdala, which is part of a "rich club" of various regions such as the prefrontal cortex. According to (Najafi et al. 2016, p. 104) such networks have the following properties:

> Among others, our analysis of rest and task data revealed several properties of overlapping brain networks. 1) Overlapping brain networks exhibited general features that resemble those of standard disjoint clustering; however, community membership values spanned the whole range, from weak (closer to 0) to strong (closer to 1), showing that disjoint clustering discretizes important information regarding the association of brain regions to multiple networks; thus, disjoint communities do not capture the information that is present in mixed communities. 2) Functional diversity of brain regions (that is, the range of functions in which they participate) was linearly associated with membership diversity (that is, the extent to which a brain region participates across multiple networks). 3) Task performance substantially altered the structure of functional connectivity across brain regions. 4) Enhanced communicability across the brain (that is, modularity decreased during tasks relative to rest). 5) We were also able to study the distribution of "bridge" nodes, including bottleneck and hub bridges. Task performance altered the role of regions in important ways.

The structure–function mapping of networks in the brain is an ever-changing dynamical process that changes over time as a function of multiscale changes in context such as cognitive task performed, attention, emotion, social interactions, and what other brain networks are doing. For instance, the amygdala and frontal-parietal cortex affiliates with many different networks in a dynamical and context-dependent manner, depending on various changing needs (Pessoa 2014 p. 19). The complexity of the structure–function networks relationship and the partial autonomy of functional networks from structural networks is exhibited by the fact that the relationship between distinct structural networks and cognitive tasks or behavior can be many-to-one or one-to-many. Furthermore, it is well known that normal functional brain networks can emerge even in cases where the brain has profound structural alterations pertaining to structural connectivity, such as the severing or absence of the corpus callosum (Pessoa 2014, p. 19). There is growing evidence to suggest that functional connectivity in the brain depends more on the network scale than on pairwise structural interactions (Pessoa 2014, p. 20). There is also evidence that emotional salience and motivation, among other things, affect the global functional connectivity (global efficiency) or decomposability of brain networks. All of this illustrates what we have repeatedly seen; namely, that brains are inherently multiscale processes and are highly context-dependent in nature.

The dynamical interactions in such brain networks are recurrent, recursive, and reentrant. Therefore, the arrow of explanation or determination in such systems is both top-down (graphical to structural) and bottom-up (structural to graphical). Global topological features of complex systems are not explicable in principle via L&D. The many-to-one relationship between the structural and the graphical features demonstrates that specific structural features are neither necessary nor sufficient for determining global topological features (i.e., topological features such as the properties of small-world networks exhibit a kind of "universality" with respect to smaller-scale structural details). In the case of random networks, for example, power laws and other scale-invariant relations can be found. These laws, which by definition transcend scale, help to predict and explain the behavior and future time evolution of the global state of the brain, irrespective of its structural implementation. Power laws are explanatory and unifying because they show why the macroscopic dynamics and topological features exist across heterogeneous structural implementations, as we saw in Chapter 4.[11]

What we are seeing in this chapter is that the nature of networks is a unifying factor explaining and binding all the features of complex biological and cognitive systems discussed previously. Indeed, networks have plasticity, reuse, redundancy, and degeneracy built into them. As Olaf Sporns observes, "the same set of network elements can participate in multiple cognitive functions by rapid reconfigurations of network links or functional connections" (2011, pp. 182–183). It is not surprising, then, to see graph theory and network tools being so widely used in systems biology, cognitive science, and neuroscience.

As we noted previously, network analysis links structure and function, revealing the very complex statistical, dynamical, topological, and various causal relationships that can exist between them. These relationships generally do not look anything like the neomechanist's L&D (Silberstein 2021). Daniel Weiskopf (2019) puts it this way: "One upshot of this form of organization is that the neural regions that participate in this assembly may have no identifiable cognitive function outside of their role in the ensemble." The traditional L&D explanatory program "assumed that distinct cognitive systems would have disjoint physical realization bases." What research reveals, as we have observed, is how different cognitive systems or functions are multiply realizable or can have "shared physical structures spread out over a large region of cortex." The idea of distinct, localizable neural mechanisms corresponding to specific functions is oversimplified at best. As Weiskopf summarizes, "the relevant anatomical structures are multifunctional in a highly context-sensitive way" which challenges the L&D program (p. 681).

One essential feature of network analysis is to illuminate the topological structures of brain networks. For instance, small-world networks and "Rich-Club networks" appear over and over in the brains of many organisms. The

[11] See also Silberstein and Chemero (2013).

point for now is that such topological features explain many key capacities and organizational principles of brains just in virtue of the topology type. This is a form of universality in that the topology type is independent of the underlying structural details (e.g., whether the network is at the molecular, cellular, or modular scale). Properties such as topology type are stability conditions defining the networks, which in turn constrain the behavior and dynamics of their ever-changing structural elements.

As noted previously, these networks are highly dynamical, and network neuroscience is melding tools from dynamical systems theory with network theory to better model such dynamical networks (Kaiser 2020). As Bassett and Sporns (2017, p. 356) summarize, "methods from network science are expanding in new directions, going beyond descriptive accounts of network topology and toward addressing network dynamics, generative principles, and higher-order dependencies among nodes."[12] An example would be methods for examining multiscale organization in networks to link microscale properties of network organization to macroscale neurophysiological dynamics. Other areas of research include using network tools to explore processes that generate the topologies of networks observed in empirical studies. "Such generative models can clarify the contributions of spatial embedding and other (non-spatial) wiring rules in shaping the network topology of the connectome, and can also reveal potential factors driving the selection of functionally important network attributes (p. 356). Bassett and Sporns point out that "the first conception of dynamics of networks grew from the recognition that many physical and biological systems display patterns of connections that change over time, in different contexts or in response to varying external demands" (p. 357).

Such modeling of brain networks takes place at molecular and ever-larger spatial and temporal scales. Nevertheless, brain-wide networks harness, recruit, integrate, and unify all these scales and their components in the service of various cognitive functions and to subserve contents of conscious experience. The scales, components, and their complex interactions often include the following: ionic flux, subcellular structures, proteins, genes, RNA, neurons and neural assemblies, glial cells, neurotransmitters and neuromodulators, hormones, large-scale neural synchrony and neural oscillations, and electrical fields (Godfrey-Smith 2020, p. 193; Parrington 2021, p. 114).[13]

[12] The ultimate goal is to meld both dynamical systems-type models along with ever more realistic, complex, and multidimensional graphical models such as those employing hypergraphs and simplicial sets that can represent greater quantities of, and more types of, connections and dependencies (Battison et. al. 2021).

[13] As you will recall, regulatory RNAs are secreted by both neurons and glial cells and travel to other cells in the brain, modulating various functions.

Considering all this, it is a mistake to focus only on networks involving neurons and their action potentials and oscillations as the L&D strategy typically does. Complex networks often involve brain-wide integration at every scale. The brain (and the body and the environment) must manage, modulate, and coordinate these processes that are happening at very different spatial and temporal scales. There are many different heterogenous time scales in the brain ranging from milliseconds to seconds to minutes and beyond. This sort of integration is what networks do (Bassett, and Sporns 2017; Silberstein 2021). It should be immediately obvious from all this that, just like the genes as codes misnomer, brains are not merely or primarily simply computational devices wherein synaptic activity and their connections are all that really matter. Neither single genes nor single neurons are the sole or primary unit of interest and action.

Graph theory and the big-data techniques of network analysis, such as network simulations, time series analysis, various sorts of causal analysis (e.g., Granger causality), dimension reduction, and universality class analysis are perfect for illuminating these multiscale relationships and connections. As we noted earlier: "One example is the joint investigation of gene co-expression patterns and patterns of brain connectivity" (Bassett and Sporns 2017, p. 358).

In every area of cognitive neuroscience—ranging from learning to memory to decision making and action to empathy and mind reading to consciousness and the contents of consciousness—the central theoretical unit of study and analysis is the multiscale network (Bertolero and Bassett 2020; Jansson 2020; Schirner, et al. 2020; Shine et al. 2019). Whether we are talking about task neutral networks such as the Default Mode Network, or task positive networks such as Executive Control Network, the Salience Network, Dorsal Attention Network, Frontoparietal Network, Amygdala Network, Action-Perception Network, Empathy-Network, the thalamus-anterior -posterior cingulate cortex-angular gyri network (whatever it is called), the idea is that various functional brain networks are the key unit of investigation (Huang et. al. 2020; Scheinin 2020).

Complete details of how all these specific networks work need not concern us for the purposes of this discussion. The point is that the explanations involving these networks for various cognitive abilities and various states and contents of consciousness all have to do with the structural, functional, and topological changes in these networks as well as their various multiscale interactions (or lack thereof) over time under different changes in context (e.g., the presence or absence of anesthesia, psychedelics, deep sleep, stress, specific cognitive tasks, social interactions, etc.).

Spirited reductionists might accuse us of "cherry-picking" and "special pleading" with our focus on network neuroscience. We hope all the

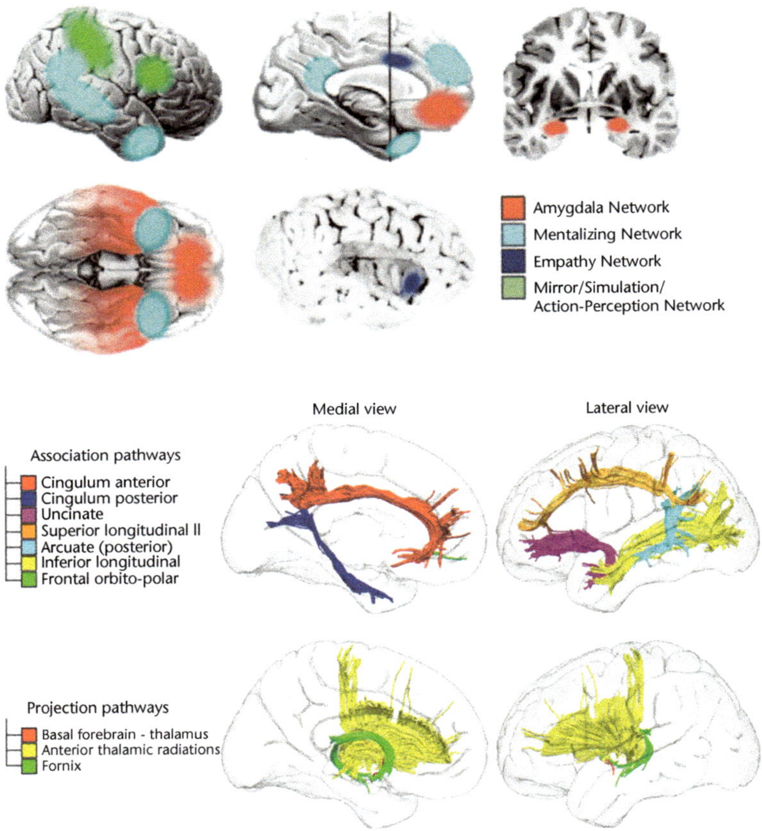

Figure 6.2 Various brain networks.

preceding points make it clear that their charge is empty (see Fig. 6.2). There is a reason that Feldman Barrett's second lesson is that the brain is a network. Bassett and Sporns explain why networks are so central to neuroscience and systems biology: "Network science tools are perfectly suited to accomplish the important goal of crossing levels or scales of organization, integrating diverse datasets, and bridging existing disparate analyses" (2017, p. 368).

As with social neuroscience, many hard-nosed neuroscientists from different subdisciplines now fully acknowledge that the networks in question also encompass and include social networks. Working up to social networks proper, Bassett and Sporns note that "large-scale studies of brain–behavior relations and behavior–behavior dependencies, although still in their infancy, promise to provide a rich database for mapping the relations among brain processes and their contributions to perception, action, and cognition" (2017, p. 354). Such network analysis obviously includes correlates that fall under the category of embodied, embedded, enactive, and extended cognition such as action–perception cycles. They go on to say that

> network neuroscience asks how all of these levels of inquiry help us to understand the interactions between social beings that give rise to ecologies, economies, and cultures. Rather than reducing systems to a list of parts defined at a particular scale, network neuroscience embraces the complexity of the interactions between the parts and acknowledges the dependence of phenomena across scales.
>
> (p. 358)

Such analysis is often called the modeling of various types of "networks-of-networks" or meta-networks (Silberstein 2016). Just as Feldman Barrett and Cacioppo and Cacioppo noted previously, in such meta-networks, brain networks depend on social networks and vice versa. Analysis of such networks can be used to study the complex relationships between ever-changing social conditions, various other environmental features, changes in genomic, RNA, and proteome networks, changes in gene expression and modulation, behavioral patterns, mental and physical health, various types of neural activity, and so on (Bassett and Sporns, p. 361).

This summary of network neuroscience illustrates how central the contextual emergence pattern is to neuroscience. The properties of nested networks at successive larger scales provide stability conditions for the behavior of networks at smaller scales. And all the neurobiological, neural, and cognitive networks are embedded in an ecological and cultural environment, providing additional stability conditions crucial to these networks' behaviors.

6.3.4 *The return of localization and decomposition?*

Sporns might serve as a summarizer of the examples and discussions in this chapter:

> Everywhere in biology and cognitive science we deal with systems made of parts or elements with different functionalities acting in a selective and harmonized way, coordinating themselves at different time scales, interacting hierarchically in local networks, which form, in turn, global networks and, then, meta-networks.
>
> (2011, p. 322)

He goes on to note that "the organization of living systems consists in different nested and interconnected levels which, being somewhat self-organized in their local dynamics, depend globally one upon the others." This certainly sounds like contextual emergence. Sporns concludes: "This means that both the components and the subnetworks *contribute to the existence, maintenance, and propagation* of the global organizations to which they belong. And, in turn, those global organizations *contribute to the production, maintenance, and propagation* of (at least some of) their constitutive components" (p. 322, emphasis added). Basically, this is the contextual emergence pattern, and L&D plays very little role.

Surely the advocates of L&D have a response to the examples that we have offered. One can try arguing that the larger-scale stability conditions and multiscale nature of the properties of networks that we have observed to be so crucial to biological and neural phenomena are just representations, while the real explanatory power lies in the L&D strategy with its modular mechanisms. As Alan Love (2018) argues, this will not do, because "first, reciprocal interaction between genetic and physical causes does not conform to the expectations that mechanism descriptions 'bottom-out' in lower-level activities of molecular entities." The L&D explanatory strategy is successful only if the "bottom-most" mechanisms are the primary players (i.e., deliver the necessary and sufficient conditions for the existence and behavior of larger-scale phenomena such as networks). He continues: "Second, the reciprocal interaction between genetic and physical causes does not require stable, compositional organization, which is a key criterion for mechanisms." Several examples illustrating the failure of this criterion for mechanism have been surveyed in this chapter. An example that Love notes is: "Although compositional differences can alter relationships of physical causation (fluid flow or tension), these relationships do not require the specificity of genetic interaction predominant in most mechanistic explanations from molecular biology." His conclusion is: "Standard mechanistic strategies of representation and explanation appear inadequate to capture these mechanisms" (Love 2018, p. 341).[14]

What we see with the contextual emergence pattern is that stability conditions and their multiscale contexts determine the behavior of the "parts," not primarily or solely the other way around. The L&D strategy appears very convincing in the abstract, but fails in the face of what is being learned in biology and neuroscience. Even William Bechtel (2017)—formerly a champion of the mechanistic approach of L&D—has revised how he thinks of mechanisms in biology: "A concern I raised earlier about Craver's and my treatment of top-down causation is that it rendered all causal relations at the lowest level ... This interpretation [of network representations], however, is mistaken" (pp. 269–270). He gives several reasons for this conclusion, such as the fact that "there are no grounds for treating the nodes in a given graph as at a common level," and nodes "should not be treated as representing entities at some basic level" (p. 270). Bechtel is clear that organism-wide functions and global constraints are mechanisms constraining the behavior of their parts, such that:

[14] For more discussion, see also Love (2012, p. 120) and Love and Huüttemann (2011).

> In biology, the constraints imposed in a mechanism are specific to the conditions in the living system. From this perspective, the physical is far from closed but rather is extremely open-ended. Wherever one finds a set of components organized into a module with sufficient interactions, one will encounter constraints that limit the behavior of the components and how they respond to external inputs. The phenomenon described as top-down causation is not unusual, but common.
>
> (2017, p. 272)

He goes on to acknowledge that network explanations often bear out this "top-down causation" via larger-scale constraints, potentially even operating over the entire network or organism (p. 253). This is a clear rejection of the L&D strategy, at least in spirit. We simply note that although Bechtel is not using the language he is describing the contextual emergence pattern, so we can understand his "top-down causation" as top-down constraint due to larger-scale and global stability conditions defining multiscale contexts constraining the possibilities for behavior of the constituents of mechanisms.

Indeed, Jason Winning and Bechtel have recently been arguing that mechanistic explanation is best conceived in terms of constraints: "We provide a new account on which the causal powers of mechanisms are grounded by time-dependent, variable constraints" (Winning and Bechtel 2018, p. 288).[15] They also note: "The framework of constraints can be applied iteratively—a macroscale object can be further constrained by incorporating it into a yet larger-scale object" (p. 293).

All of this sounds a great deal like contextual emergence, but especially the following characterizations: "Thus, on our view, when constraints enable objects to have novel, emergent behaviors, this is tantamount to the emergence of causal powers ... by means of possessing such emergent powers, mechanisms and components causally produce the effects they do" (p. 294). And finally: "By restricting some degrees of freedom of its components and thereby enabling the whole mechanism to do things that would otherwise not be possible, constraints determine the causal powers of a machine or mechanism. Of particular importance are those constraints that are flexible and time-dependent. These enable machines to operate in different ways on different occasions" (p. 307). Winning and Bechtel argue that mechanisms conceived as constraints solve the "mysteriousness problem," thus grounding the causal powers of mechanisms (p. 292). The idea seems to be that mechanisms just are sets upon sets of constraints. If we understand them correctly, then we agree.[16]

[15] See also Bechtel (2018, p. 574).
[16] See also Silberstein (2021).

One could try a more sophisticated version of L&D; for instance, a kind of "contextual decomposition" (Burnston 2016a, 2016b, 2021). The idea is to relativize L&D to context. Instead of the L&D strategy always driving down to the smallest mechanistic unit—the "atom" of explanation—the explanatory strategy stops at the contextually relevant level in terms of assessing the mechanism responsible for the phenomenon in question and how the mechanism and its components change function or respond differently in different contexts. On this view, a network of mechanisms operating together to achieve an effect different from the effect of each component mechanism would qualify as a contextually relevant L&D, as would the mechanism shifting function in a changed context.

This seems like a fine strategy for analyzing complex biological and neural systems so long as one does not claim that it is a reductive strategy. First, the stability conditions defining the relevant context (e.g., defining the relevant network) is not given by any of the component mechanisms. Indeed, the network of interest for this kind of contextualized L&D is still situated in a multiscale context with larger-scale stability conditions defining that multiscale context which, in turn, shapes the behavior of the network in question. The stability conditions determining the functionality are context-defining, not defined or determined by the components in the way the L&D strategy presupposes. Another way to put this is that the workings of the "parts" at any given time—the particular "contextual decomposition" at that moment—is going to be a function of, or explained by, the multiscale stability conditions that are crucial to the context defining the specific biological or neurological system's integrity and functionality. Hence, even if one can give a contextual decomposition (e.g., Burnston 2021, section 4), this can be done only by explicitly invoking how the conditions associated with the context shape the function of the network and its constituents. This is all grist for the contextual emergence pattern.

Or perhaps one might object that the appeals to networks and larger structures is just the mathematics of network analysis. What really matters is the structural biological details: the structural difference makers. For instance, "biological circuits are composed of a diversity of neural cell types that have different structural and functional properties" (Kaplan 2018, p. 278), and "differences in axonal 'weights' can reflect underlying structural differences in myelination patterns that influence how well the input signal propagates through the axon, etc." (p. 276). The point is that biological (and even artificial) neural networks possess functionally or structurally distinguishable parts, so surely those details matter in a way that mathematical-network and design-based models might hide.

It is true that sometimes the structural details matter the most for the purposes of a particular explanation—they are the difference makers in such cases. Nonetheless, sometimes they do not matter, as with some network-based explanations as we saw previously. And sometimes the structural details are what they are, and the parts do what they do, precisely because of the larger-scale stability conditions defining the multiscale contexts with global and systemic constraints delimiting the functionality of any components in systems. All three of these possibilities are indelibly context-dependent—the opposite of what an L&D approach would predict.[17]

6.3.5 *Broader implications of contextual emergence*

Taken together with Section 6.2, this section presents significant evidence for, and examples of, contextual emergence in systems biology, cognitive science, and neuroscience. But we would be remiss if we did not mention some important broader implications.

First, recall that at the end of Chapter 1 we briefly mentioned the problem of reductionist approaches to psychology. Contextual emergence in cognitive science and neuroscience further reinforces the importance of avoiding reductionism in clinical and counseling psychology if we want to deal with people as they are in context (Bishop 2007, and 2022).

A second implication that we will discuss is the debate between more internalist models versus embodied, enactive, embedded, and extended models of cognitive science (4E) taking place at the intersection of neuro and cognitive sciences. The internalist school of cognitive science gets its name from its orientation that what counts as the cognitive system and where to look for explanation as regards cognition or conscious experience is in the brain. The internalist school historically has two basic flavors: a cognitivist/functionalist one and a brain-centric one. The latter is more common these days. The internalist brain-centric school predicts that in principle, cognitive science (and psychology) should reduce to, or be eliminated by, neuroscience. As our survey of complex biological and neural sciences has indicated, the kind of reductionism of the internalist school is not particularly well supported by the evidence.

The school of 4E cognition eschews this reductionism, and is to some degree a repudiation of both schools of internalism. Historically, many different traditions of thought and many different thinkers feed into and are the foundation for the 4E movement (Chemero 2009; Chemero and Silberstein 2008; Newen, de Bruin, and Gallagher 2018, chapter 1). Furthermore, there are many different brands and varying strengths of 4E cognition.

The big debate in the 4E movement is whether the cognitive system is *constituted* by brain-body processes only (embodied) or by extra-brain-body

[17] See Silberstein (2021) for a lengthy argument to this effect.

processes (extended). The word "enactive" implies that the cognitive system is partly constituted by "active engagement in the agent's environment" (Newen, de Bruin, and Gallagher 2018, p. 6; Silberstein and Chemero 2012). Such engagement is sometimes called "perception-action cycles." The focus here is to see cognition as driven by, and partly constituted by, action. The notion of "environment" here means not just the organism's physical or social environment as defined from a third-person perspective, but their "phenomenological niche" with its first-person affordances (Silberstein, and Chemero 2012). The word "embedded" generally refers to the strong causal dependence or "scaffolding" on the environment of the cognitive system. Most people will grant some degree of embeddedness, but the degree of autonomy varies, depending on how reductionist or internalist one is in their thinking.

So again, regarding the 4Es, the big debate seems to be about where one falls on the spectrum between *constituted by* versus *causally dependent on*. As has been noted throughout this book, in every field in science the lines between what is system and what is environment are constantly redrawn as our pragmatic and explanatory interests change and our knowledge grows (e.g., Bishop and Silberstein 2019). An obvious example to bring to bear here is developmental biology-epigenetics-epigenomics and the quaintness of the nature-versus-nurture debate. Are we partly *constituted* by nonlinear and *causally interdependent* gene networks, RNA networks, proteome networks, human cells, bacteriological cells, a host of chemicals, bodily processes at multiple scales, and so on, all taking place in larger physical and social environments, with which they all share co-determination, co-constitution, and massive causal interdependence? Yes. For all the reasons we illuminated at the beginning of this chapter, one would be hard put to find many educated biologists who still believe in some form of strict bottom-up genetic determinism. Given contextual emergence as the nature of reality and of scientific explanation, depending on our explanatory purpose, needs, and capacity for manipulation and intervention, we change the boundaries as needed (Primas 1983; Silberstein 2021).

The point is that regardless of one's stance on internalism versus externalism or one's position on 4E cognition, the facts speak loudly for themselves. Network neuroscience, social neuroscience, systems biology, 4E cognition, separately and collectively strongly suggest contextual emergence when we dig into the details. Again, part of the point we are trying to make is that *even key areas of neuroscience* exemplify contextual emergence.[18] We are not alone in making this claim. As Michael Spivey notes:

[18] For even more evidence, see Newen, de Bruin, and Gallagher (2018), Spivey (2020), and Silberstein (2021).

> Cognitive scientists Harald Atmanspacher and Peter beim Graben call this *contextual emergence.* They have studied systems like these in intense mathematical detail and designed computational simulations of how a mental state can emerge from a neural network. They find in these mathematical treatments that being in a particular mental state requires more than just for the neurons to be producing a particular pattern of activity. The context of that simulated brain, such as the body and its environment, also need to be accounted for in order to develop a statistically reliable characterization of the mental states that the mind takes on. Hence, based on these computational simulations, the brain by itself is not sufficient to generate a mental state. A mental state, such as a belief or desire, only emerges between a brain and its context, that is, its body and the environment.
> (2020, pp. 142–143, emphasis added)[19]

No matter how one wants to *count systems*, and *wherever* one wants to locate the mind and cognition, ample evidence from across several disciplines, including biology and neuroscience, points to the contextual emergence pattern.

Internalist versus externalist orientations to cognitive science are not simply academic debates. They have actual-world consequences. Although we do not have space for extended discussion, one can see the consequences in how addiction is understood and treated. Based on the L&D strategy, an internalist approach to addiction focuses on the brain, for instance, treating addiction as a brain disease (Levy 2013). In contrast, an externalist approach focuses on brain, body, and material/social environment. For example, recognizing that dopamine itself does not equal pleasure and acknowledging the evidence that having a dopamine-based neural reward system is neither necessary nor sufficient for addiction (Hart 2021, Levy 2013).

Contextual emergence helps us make sense of the relative advantages that an externalist orientation has in dealing with addiction. The mechanisms that neuroscientists have identified provide some of the necessary conditions for addiction. It is not surprising that various features of our brains would be involved in addictive behaviors. But these neural mechanisms do not possess the necessary and sufficient conditions for addiction. The physiology of the body (embodiment), the environment and social contexts in which people are participating (extendedness), their activities in their environments (enactivity), and the ways environments shape people (embeddedness), need to be brought into the explanation of addiction. The key is in discovering what stability conditions at various scales define the nested contexts enabling neural mechanisms to become entrained in addictive behavior. For instance, the strongest correlates and biggest predictors of

[19] See Allefeld, Atmanspacher, and Wackermann (2009) and Atmanspacher, and beim Graben (2007) for discussion.

addiction are socioeconomic, such as poverty, family dysfunction, isolation and loneliness, lack of education, emotional abuse, trauma, and deprivation (Hart 2021).[20]

An exclusively internalist orientation does not stand a chance of making much progress on understanding and treating addiction, so has negative actual-world consequences for dealing with something like the opioid crisis in America. As we suggested at the end of Chapter 1, reductionism is far from an innocent and merely philosophical belief!

6.4 Biological and ecological examples of contextual emergence

So far, we have been focusing on cases from systems biology and systems neuroscience. However, in this section we will discuss some case studies from biology and ecology to illustrate how pervasive the contextual emergence pattern is in the biological realm.

6.4.1 *Microbial community assembly*

Consider how microbial communities assemble and maintain themselves in complex ecological niches. Such communities are found everywhere (e.g., inhabiting organisms, soils, oceans, and surfaces in hospitals and your home). Furthermore, microbial communities play important roles in natural processes, from animal development and host health to biogeochemical cycles (Goldford et al. 2018). There are several empirical patterns that have been observed about microbiome organization, including

> high microbial diversity, the coexistence of multiple closely related species within the same functional group, functional stability despite large species turnover, and different degrees of determinism in the association between nutrient availability and taxonomic composition at different phylogenetic levels.
> (2018, p. 469; see references therein)

The organizing research question for Goldford and collaborators was to discover the common organizational principles in microbial community assembly and how to better connect theory and experiments. Typically, this kind of research focuses on a bottom-up interventionist approach with a

[20] Think about poverty for a moment. Multiscale network analysis indicates "that childhood poverty may result in widespread disruptions of the brain connectome among girls, particularly at the lowest [international normalized ratio] levels, and are differentially expressed in females and males" (Kim et. al. 2019, p. 409). Why is poverty such a predictor of disruptions in neural networks? As with addiction, because poverty is strongly correlated with all the social, emotional, and cognitive predictors.

few species. This is just the approach that Carl Craver and other mechanists think is bread and butter for their L&D explanatory accounts. Although such an approach may yield insights into specific mechanisms and interactions, Goldford and collaborators point out that it is "unclear to what extent findings from these studies scale up to predict the generic properties of large microbial communities or the interactions therein" (p. 469).

Goldford and collaborators used very diverse and taxonomically rich microbial communities from a variety of soils and plant leaf surfaces, and started with "minimal media containing a single externally supplied source of carbon, as well as sources of all of the necessary salts and chemical elements required for microbial life" (p. 469). Glucose was the first energy source used, and they grew the different microbial communities separately under the same conditions. Their initial observations where that the vast majority of communities stabilized after approximately sixty generations, "reaching stable population equilibria in nearly all cases" (p. 469). From this first set of experiments they observed (pp. 469–480):

- The twelve natural communities assembled into highly variable compositions.
- All twelve stabilized communities—with very diverse environmental origins—converged into similar family-level community structures dominated by Enterobacteriaceae and Pseudomonadaceae ... a similar family-level composition arose in all communities despite their very different starting points.
- The same is not true when we compare taxonomic structure at the subfamily (genus) level.
- The reproducibility in population dynamics between replicate communities indicates that experimental error is not the main source of variability in community composition.

These observations display what they call "family-level attractors" resembling what is observed in the wild. Recent studies of natural communities have "consistently found that environmental filtering selects for convergent function across similar habitats, while allowing for taxonomic variability within each functional class" (p. 470). This is a clue that context is important to microbial community assembly.

Next, the team tested whether there was any effect due to differences in externally supplied energy sources, so they repeated the same set of experiments with citrate or leucine as sources. The microbial communities exhibited a similar pattern of family-level convergence with genus or lower variety, though with discernible differences in the specific kinds of genus and species level variations between the communities grown on citrate versus leucine. This indicates that an energy source, holding all other factors equal, is a crucial factor in the kinds of variation with microbiome community assembly (p. 471).

Obviously, a change of externally supplied energy source is a change of context, so there is some change in outcomes for the lower-level variations. Yet the family-level convergences held. To the extent that an L&D approach can explain these observations, the context must be considered as something more than background. Not only are there differences in the energy sources that the microbe communities are consuming, but also the specifics of the "minimal media" and other aspects of the setup are crucial to defining the specific ecological niche that the communities occupy. The observed patterns both reflect these contextual differences as well as transcend them.

Goldford and collaborators also used a machine learning approach to model the communities in their experiments. What the modeling showed for the first two series of experiments was that "rather than selecting for the most fit single species, our environments select complex communities that contain fixed fractions of multiple coexisting families whose identities are determined by the carbon [energy] source in a strong and predictable manner" (p. 471). These modeling results—along with the first sets of observations—suggested the hypothesis "that taxonomic convergence might reflect selection by functions that are conserved at the family level." Further experimentation revealed that the genes and pathways conferring each community's "ability to metabolize the single supplied resource are distributed among multiple taxa in the community, rather than being present only in the best-competitor species" (p. 471).

This led to a third series of experiments in which L&D strategies to understanding biological behavior proved inadequate. An L&D approach to understanding the self-assembly of microbial communities leads to the expectation that if individual species feed on the same resource—whose finite supply naturally limits growth of microbial communities—these individual communities will compete to the exclusion of other communities, as in classic consumer-resource models. But as Goldford and collaborators observed, an L&D strategy did not reflect the behavior of microbes. It turns out that when multiple species of microbial communities occupy the same ecological niche, features such as microbial communities' ability to engineer their environments, their ability to facilitate each other's structuring and growth, and environmental filtering to select for convergent functions across similar ecological niches, allow taxonomic variability within each equivalence class of functionality observed (p. 471).

In their third series, Goldford and collaborators demonstrated the coexistence of competitor species communities due to these different communities being able to consume the byproducts secreted by microbes as a source of energy.[21] The ecological constraints and affordances, dynamically evolving cooperation as multiple communities produce byproducts that feed and support other communities, and environmental filtering for functionality, lead to stabilization of multiple distinct microbial communities in the face of limited resources (pp. 471–472). They also took steps to test for cell death as a competing source of consumable energy, and were able to show that it would only be the case at undetectable levels (p. 472). As well, the spatial structuring of communities played a discernible role in the growth and maintenance of the multiple taxa (p. 472). Moreover, the context shaped by the multiple species even mediates acidification stability for the entire ecosystem—an effect not seen in monocultures (p. 472).

Their empirical observations that "nonspecific metabolic facilitation"—cooperation or affordance—led to stable multiple taxa coexistence were also reproduced in their machine learning simulations (pp. 472–473). Such cooperation is possible only in a context where there are multiple species interacting—the norm outside of pristine laboratory environments.

Goldford and collaborators' result that there are family-level attractors, while at the same time species and genus level diversity related to different energy sources, is an example of universality, robustness, and plasticity. Not only did family-level order appear across multiple microbial taxa, but there was also *functional convergence* at the family level.

> Importantly, this functional convergence was exhibited even when consumers were drawn from uniform distributions, with no enforced family-level consumer structure, suggesting that the emergence of functional structure at the community level is a universal feature of consumer-resource models.
>
> (pp. 473–474)

Their conclusion about the concordance between experimental and model results on functional convergence was that "Our model suggests that species are stabilized by a dense facilitation network ... consistent with observations of widespread metabolic facilitation in experiments" (p. 474).

The results obtained by Goldford and collaborators illustrate the contextual emergence pattern. The larger-scale context due to the concrete ecological niche, as well as the evolving contextual conditions as microbial communities reshape their ecological niche, are not L&D features, even though L&D properties provide some of the necessary conditions for these communities (e.g., the basic workings of a microbial cell as it consumes an energy source). The various kinds of stability forged by multiple microbial

[21] This was done by ensuring that only a limited initial amount of glucose was available and was exhausted after 48 hours. The only remaining energy source for microbes to utilize was the secretions of other microbes (pp. 471–472). The secreted byproducts were more diverse as energy sources than the initially externally supplied glucose.

communities in the same ecological niche is a situation where the smaller scale L&D features contribute only some of the necessary conditions for such stability. Most of the stability is established by "collective rather than pair wise interactions," where "higher-order interactions play a critical stabilizing role in complex microbiomes" (pp. 472). This is exactly what stability conditions do. Goldford and collaborators' summary of their work reads like a rejection of L&D approaches and more like contextual emergence:

> In the absence of a theory of microbiome assembly, it is often difficult to determine whether empirically observed features of natural microbiomes are the result of system-specific determinants, such as evolutionary history and past selective pressures at the host level (10), or whether they are simply generic emergent properties of large self-assembled communities. Our results show that the generic statistical properties of large consumer-resource ecosystems include large taxonomic diversity even in simple environments, a stable community-level function in spite of species turnover, and a mixture of predictability and variability at different taxonomic depths in how nutrients determine community composition. All of these features are not only observed in our experiments, but also have been reported in systems as diverse as the human gut ... plant foliages ... and the oceans.
>
> (p. 474)

Moreover: "Our theoretical results thus provide an explanation for the ubiquity of these empirical findings and suggest that they may reflect universal and generic properties of large self-assembled microbial communities" (p. 474), articulating a common pattern that we have observed among physics, chemistry, and biology systems. Neither the theoretical model nor the empirical results depend on the biochemical details, as is consistent with how universality and multiple realizability show up in various sciences. The network of interacting communities and their dynamic reshaping of, and responding to, their evolving ecological niche are the important ontological factors, as would be expected of larger-scale stability conditions contributing the remaining necessary and sufficient conditions for family-level and functional convergence with species and genus-level diversity:

> We find evidence that densely connected cross-feeding networks may stabilize competition within guilds of highly related species that are all strong competitors for the supplied carbon source. Such cross-feeding networks naturally lead to collective rather than pairwise interactions, supporting the hypothesis that higher-order interactions play a critical stabilizing role in complex microbiomes.
>
> (2018, p. 474)[22]

[22] It will come as little surprise to the reader at this point that microbiomes can be helpfully analyzed by using graphical techniques (e.g., Ha et al. 2020).

6.4.2 *Appendage regeneration*

Some species can regrow lost appendages. For many decades, biologists thought that since such regeneration processes worked with shared tissue stages (e.g., forming a specialized wound epidermis, inducing a blastema) and gene activities, appendage regeneration in tails and limbs, for instance, were the same or highly similar. Recent studies indicate this is not the case (Aztekin 2021 and references therein). Differences in specific kinds of appendage regeneration are observed at the cellular scale and absent at the genetic scale. For example, the SHH gene provides "instructions" for producing the Sonic Hedgehog (*Shh*) protein—a protein playing many roles in embryonic development. During tail regeneration, *Shh* is expressed in the spinal cord and notochord mediating axon growth and extension, while during limb regeneration it is expressed in mesenchymal cells mediating digit development and patterning. The different cells in which the SHH gene is operative lead to different mediating roles for *Shh* in regeneration.[23] Moreover, regeneration involves several coordinated cellular mechanisms such as proliferation, migration, and differentiation. Depending on the cell type, "each of these cellular mechanisms may involve distinct gene activities"(p. 2).

Signaling centers (networks) are required to orchestrate phenotypes leading to morphological changes in appendage regeneration such as forming the wound epidermis at the beginning of regeneration. Nevertheless, the specifics of the wound epidermis formed varies by appendage and across animals (p. 3). Also, differences in the application of other signaling centers in further stages of appendage regeneration exhibit variations by appendage and across animals (p. 4). At the very least, one can see that there are different transcription programs at work for expression of proteins such as *Shh* in these different contexts. Hence, despite involving the same genes, differences are observed at the cellular scale as well as at the signaling-center scale. Aztekin concludes that signaling centers are necessary for appendage regeneration: "Nonetheless, these centres do not need to be the same cell type, be present at the same location, nor exhibit the same cellular mechanisms to instruct the building blocks (stem/progenitor cell types) to reconstitute the lost structure in the correct manner" (p. 4).

Similarly, appendage regeneration requires stem and progenitor cells located in the blastema tissue. It has been discovered recently that blastemas do not contain pluripotent stem/progenitor cell types, but heterogeneous stem/progenitor cell types specific to lineage (meaning that evolutionary

[23] Similar cellular differences account for the differing roles that the *Shh* expression plays in embryonic development.

history plays a role in the blastemas that arise in different animals). Again, this variation is observed by appendage and across animals (p. 5). Furthermore, different cellular mechanisms (e.g., differentiation vs. *Pax7* + satellite cells) are used, depending on appendage, age, and species, meaning that different cells are recruited for regeneration tasks in these different contexts. The signaling centers also orchestrate regeneration involving these different blastema.

All of this can only lead to appendage regeneration if there is a "regeneration-permissive environment," which Aztekin defines as "the cellular composition of the amputation plane, state of the extracellular matrix and available secreted factors enabling signaling centers and stem and progenitor cells to form and interact with each other" (p. 6). The structure or environment remaining after the amputation provides a context for how the regenerative process accomplishes its work. For example, in laboratory studies of salamanders an appendage may be amputated with an almost perfectly vertical cut. In such cases, nearly perfect appendage regeneration occurs. But in the wild, appendage loss may happen at any angle with respect to the plane. Imperfect appendage regeneration occurs in these instances. Also, species with more sophisticated immune systems tend to have lower regenerative abilities, probably due, in part, to the suppression of inflammation. Sustained inflammation is one of the factors promoting regeneration post-amputation. The way in which immune cells are recruited and deployed makes a significant difference to whether and how regeneration takes place (p. 7).

Summarizing, although amputation triggers cellular processes, it does not trigger the same gene expression and cellular processes in different appendages, at different organism age, or in different species. Moreover, transcriptional programs not observed during embryonic development are involved in appendage regeneration.[24] Different cellular processes are operative at different appendages, despite sharing tissue descriptions and gene expressions. Once again, we see the role of contexts. Although genes and expressed proteins, immune cells, and other parts of the biological machinery are involved in regeneration, they contribute only some of the necessary conditions. It is the stability conditions at larger scales (such as angle of the amputation plane) defining the multiscale context that contribute the rest of the necessary and sufficient for appendage regeneration. Regeneration exhibits the contextual emergence pattern.

[24] It is currently unknown whether gene regulatory networks used in embryonic development are rewired for regeneration upon injury/amputation, or if new gene regulatory networks are developed to perform regeneration.

6.4.3 *Bee colonies*

All bees are born genetically identical due to coming from the same queen. It is epigenetic changes that modify each bee's DNA at birth, leading to the division of labor in the colony (e.g., forager, worker). The whole hive provides a context for the activities of any individual bee by regulating how these epigenetic changes take place. The hive provides stabilizing properties, allowing individual bees to manifest specific tendencies. Therefore, there can be new properties manifested by the individual only in the context of the hive, and collective properties of the hive produced which are not possessed by any one individual of that hive—the contextual emergence pattern.

Moreover, bees can switch roles. For instance, if forager bees are removed, the structure or spatial ordering of the bee colony shapes their replacement as well as social cues regarding food resources (Crall et al. 2018). Worker bees at the center of the colony are the most likely to switch to a forager role. The innate genes and instincts of individual bees supply some of the necessary conditions for bee behavior. Yet it is the social structure of the colony that supplies the rest of the necessary and sufficient conditions for this role switch in the face of the contingent loss of foragers.

Before they leave the hive to forage, these bees make a blend of cuticular hydrocarbons specific to the colony to which they belong, so that guard bees will recognize them on return to the colony (Vernier et al. 2019). While the making of cuticular hydrocarbons is instinctual in individual bees, this property supplies only some of the necessary conditions for making the colony-specific blend. The colony's society provides the rest of the necessary and sufficient conditions for a unique identifying blend. This is another way by which the colony determines individual bee behavior.

Roles in the social structure of the colony also determine longevity (Law 2021). Queens long outlive all other bees, and nurse honey-bees long outlive foragers. However, when foragers switch back to nurse roles their longevity returns along with production of different cellular proteins apparently related to longevity. These changes in cellular protein production are not due to the presence of different chemicals or dietary changes. Instead, theye are due to differences in roles and colony interactions (remember that bees in the same colony are genetically identical except for the colony-induced epigenetic changes at birth). Again we see that the larger-scale social context provides stability conditions for the alteration of longevity due to a role that does not exist in the individual bee.

6.4.4 *Ecology*

Ecology provides numerous examples of the contextual emergence pattern. We will sketch two examples as we close this section. The kind of top-down constraint in food webs is an interesting example. In the open ocean, theories focusing only on the body size of marine animals predict that marine ecosystems should be bottom-heavy, with more plants and animals at the lowest levels than at the highest. In contrast, observations reveal that many marine ecosystems are top-heavy. Complex food-webs function as stability conditions for top-heavy marine ecosystems (Woodson, Schramski, and Joye 2020).

Actual-world communities in oceans are comprised of multiple energy pathways that do not balance between bottom-up and top-down processes. Instead, energy pathways range from shared resources with several more or less isolated pathways to complex intertwined webs. It turns out that food-web complexity plays an important role in determining trophic structure—the partitioning of biomass between trophic levels in an ecosystem. In turn, trophic levels are subsets of organisms in an ecosystem that gather energy and nutrients in similar ways (e.g., some are producers, some are carnivores, some are herbivores). Unlike ocean ecosystems with relatively isolated energy pathways, those with complex food webs are unbalanced towards top-heavy. In other words, there is not a bottom-up determination meeting top-down constraints leading to a kind of ecosystem stasis. Rather, the complexity of the food web with intertwined energy pathways creates a stability condition favoring production at higher trophic levels. This leads to an inversion of the trophic structure as top-heavy rather than bottom-heavy.

One implication of this work is that bottom-heavy marine ecosystems appear to have been adversely affected by human activity (e.g., overfishing) which has disturbed the complex food web in these ecosystems.

As a second representative ecological example, ecological constraints of a specific ecosystem delimit the possibilities for behavior. In a recent survey (Barsbai, Lukas, and Pondorfer 2021), many environments that shared ecological constraints across environments delimited behaviors of animal species in these environments. So, it is not just that closely related species engage in similar behaviors; the similarity in local ecological conditions shapes the space of possibilities for behavior by limiting "choice."[25]

What the team found was that foraging, reproduction, and social behaviors among mammals and birds were more like the behavior of other species found in environments subject to the same conditions than they are to the behaviors found in environments subject to different conditions. For instance, in environments having similar ecological constraints, mammals and birds lived in similar densities; in environments having

[25] What "choice" means will be species-dependent. The point is that species subject to local ecological constraints have a limited set of possibilities open to them for making a living in a specific ecological niche.

similar ecological constraints, mammals and birds exhibit social systems with dominant breeders; in environments subject to the same conditions, mammals and birds tended to hoard food; and so forth across several measures. This is highly suggestive of local ecological conditions creating or constraining selection pressure for species to arrive at similar solutions for making a living wherever those conditions have been stable for long periods of time. Here is an example where although genes contribute some of the necessary conditions for behavior, the ecological conditions provide stability conditions defining a context in which evolution can explore delimited possibilities, where the stability conditions are the gatekeepers to access specific regions of possibility space.

6.5 Conclusion

Just as in physics and chemistry, we see that the contextual emergence pattern is part and parcel of biological phenomena. There are many other examples we would love to discuss if time and space permitted: 1) the growing evidence that photosynthesis in some bacteria such as green sulphur bacteria and marine algae requires quantum coherence (Lewton 2021). 2) The growing evidence that in some migratory bird species their navigational prowess depends on quantum entanglement—the claim being that quantum-entangled particles in proteins called cryptochromes are found in some migratory species' eyes. The explanation is that while in flight, small changes in the Earth's magnetic field are registered by these entangled states and then passed on to the brain of the bird (Lewton 2021). 3) The growing evidence that at least in some animals there is communication and essential coregulation between the networks of microbiome in the gut, and brain networks (Lanese 2021). 4) As mentioned in the preface, their is growing evidence that the brains of individuals sync in various ways when people interact in specific ways (Feehly 2021). Along the way, we have seen that the L&D strategy favored by many philosophers and some scientists does not match with the data and results in biology, neuroscience, and cognitive science.

As an ontological form of emergence, contextual emergence has broader implications for metaphysics (some of which we have hinted at from time to time). Many philosophers of science would say that both science and the philosophy of science can be done without ever mentioning or caring about metaphysics. They are simply interested in scientific explanation and methodology. We are interested in explanation and methodology too, but we should not naïvely believe that ignoring metaphysics somehow means that the kinds of explanations scientists give are metaphysics-free.

Indeed, if the contextual emergence patter that keeps showing up in the sciences is ontological, then it will have metaphysical implications. Similarly, if reductionism is ontological, then it would also have metaphysical implications.

Hence, in Chapter 7 we shall pursue some of these broader implications that, we think, help to clarify why a contextually emergent world is an ordered, coherent, stable, and unified world. Of course, as we will see, this all bears on the unity and coherence of the sciences.

7

Ontological and Scientific Implications of Contextual Emergence

Sonnet for The Perch
The spiral staircase to the upper room
delivered us to our longed-for vintage—
'75 Glenrothes, Speyside-aged
single malt, a ripe deep gold perfumed
with candied orange peel, made with honeycomb
and Bergamot marmalade. We engaged
each present friend on the fringe of language,
each blooming late with stories, quirks, and rumors.
If our mouths spoke of fate, we heard it said
with deeper intonations, like the blissful
youth of our host in emerald Oregon's mists.
While darkness fell we toasted to *echad*,
meaning "one," the sort of union that consists
of fellows gathered here, lamplit and glad.

Brett Foster (1973–2015), 1 May 2012

7.1 Introduction

In the Preface to this book we used fireflies and whisky to illustrate the role that contexts play in the order and beauty of worldly phenomena. Brett Foster's poem brings us back to the richness of life and how much context matters to our ordinary lives as well as special occasions such as the one this poem commemorates. Achieving the rich golden perfume, candied orange peel, honeycomb, and Bergamot marmalade character of The Glenrothes 1975 does not happen in a purely bottom-up fashion. Of course, without the grains, fermentation process, and chemical reactions, there would be no scotch. Nevertheless, to achieve this quality of expression of a scotch with

Emergence in Context. Robert Bishop, Michael Silberstein, and Mark Pexton, Oxford University Press.
 DOI: 10.1093/oso/9780192849786.003.0007

these characteristics requires far more: the specific barrels in which the raw spirit is stored with their preparations (e.g., the char after aging American oak barrels), the location of the barrel in a specific warehouse tended by a master distiller exercising special care over the whole process from grain roasting to blending the barrels, and aging the spirit for 18 years, 8 months, and an odd number of days. All these factors span multiple length and time scales, as we pointed out in the Preface. The multiscale stability conditions and the contexts they define make all the difference in how the underlying physics, chemistry, and biology contribute to produce something as rich and complex for the nose and palate as the 1975.

Foster's poem also brings out how the social context makes a crucial difference to life as experienced, including how things taste and are remembered. In this case, the context is that of a small group of good friends gathering in a special nook over the garage known as "The Perch," talking about and sharing their lives together over an exquisite scotch whose phenomenology was heightened and accentuated by the magical quality of friendships lived out that evening. No art of the master distiller could engineer how the 1975 was enjoyed and brought joy to that evening. The context made all the difference to how the underlying physics, chemistry, and biology contained in that bottle played out that evening. Moreover, there was unity of purpose and spirit marking that social context into which the many layers of ordered contexts shaping that Glenrothes worked their magic—a particular magic that could not have happened in any other context.

The final phrase of the poem, "lamplit and glad," captures something of the specialness of that evening, when a set of nested contexts within a larger context work together across time and space to make something special possible. But it also harkens back to the wonder and joy of the patterns of fireflies that the authors imbibe as we enjoy our scotches and bourbons in the evening after a long day of writing. There, we experience the magical effects worked by nested contexts—physical, chemical, biological, social—that neither the biology of the fireflies nor the chemistry of the whiskys can produce on their own. As Foster's poem speaks, and as we have observed throughout this book, it is context up, down, and sideways at every scale.

This is the contextual emergence pattern we experience in life, but it is also the pattern we see throughout all of nature. First we introduced contextual emergence and illustrated how it illuminates some traditional topics in philosophy of science. Then, through many representative case studies, we have observed that the contextual emergence pattern is pervasive in the sciences. We have discussed in detail how contextual emergence illuminates the production of novel properties and why those novel properties matter. Along the way, we have made several metaphysical claims about the nature of such things as properties, laws, causes, and dispositions, and have indicated that contextual emergence tells us something deep about the nature of reality. In this final chapter we turn to these larger implications as well as to objections that have been raised against contextual emergence.

We have argued that contexts play an essential role in the structure of nature and scientific explanations, such as the stability conditions ensuring the existence and persistence of relevant states and observables that define these contexts. Neither the stability conditions nor their associated contexts are given by the most "elementary" or "fundamental" properties and laws at the smallest length and time scales. Indeed, the recurring pattern that we have observed is that multiscale contexts modally constrain systems and their smallest-scale properties and components. It is in contexts where we find multiscale stability conditions representing both the screening off and opening up of access to possibility space at all scales.

For a given domain of reality there can be combinations of conditions and constraints at various interacting scales defining the context for the existence and effects of relevant properties. Evidence for this is found in the epistemology and methodology of science, where modalities are expressible only by using vocabularies tied to specific scales and domains, contextually relevant observables, and contextual topologies. Given the importance of modality to concrete contexts, it is important to say something more about modality and its implications. Our starting assumption is that there are modal facts. Contextual emergence is realist about modality—possibility is not merely epistemic. The question that we will be concerned with is this: what grounds modal facts; in other words, what are the *truth-makers* for modal truths?

We will pursue the claim that modality of all varieties is grounded in dispositions and their relations (Sections 2–10). This will involve reminding readers of what we said about possibility space and modal constraints as they relate to contexts, while extending those discussions to the relationship between dispositions and modality, and how these relate to properties and property incompatibility relations. We shall also discuss how a contextually emergent world affects multiple realizability, laws, causation, causal closure of the physical and microphysical, determinism, the notion of a fundamental ontology, and the hard or generation problem of consciousness. Along the way we will compare contextual emergence with other accounts of emergence in the philosophy and metaphysics of science, answer common questions, and address objections (Sections 11-13). We believe that the prevalence of contextual emergence in scientific explanation is simply a reflection of the contextual nature of reality. The reader should note that many accounts of laws, modalities, dispositions, and so forth are driven by concerns about the nature or essence of laws versus non-laws, the essence of causation, realism about laws and scientific realism more generally, distinguishing nomological claims from causal claims, and so on. Our pursuit of modality is not driven by any of that. For our purposes those are not the most interesting questions to ask, and such questions do not help us to understand what scientific investigation has revealed about the nature of reality.

We think the sciences and our analysis demonstrate that modality is the key to the ontological categorization of contextual emergence. The ultimate reason why we are forced to use contextual observables is that those are the ones that adequately capture the modalities of system states. The reason why those are the observables expressing those modalities uniquely is that they track the contextually produced properties grounding those modalities. As "Sonnet for The Perch" illustrates, contexts are essential to emergent qualities and properties. Likewise, contexts are key to the production of the properties grounding the modalities that are only captured in contextually emergent states and observables. It is this link to the salient modalities that privileges specific, contextually emergent properties as being those that matter physically.

7.2 Context, possibility spaces, and modal constraint

Stability conditions defining contexts play a crucial role in generating properties by modally constraining domains in a multiscale way. Modal constraint involves both screening off and gatekeeping access to new regions of possibility. We remind the reader (chapter 2, Sections 2.3.3–2.3.4) that one can figuratively think of the set of all modalities as a state space, a higher-dimensional space where each combination of the possible and impossible is located. We know from experience that systems are not completely free to explore this space (e.g., your car engine is not free to produce nuclear fusion or create vaccines). To enter specific regions of possibility means to cut off access to other regions of possibility (recall the transition from the initial quiescent state to the convective state in Rayleigh–Bénard convection in Chapter 2, Section 2.4).

As we have observed, stability conditions play the role of gatekeepers to access specific regions of possibility space where the states and observables of concrete systems are actualized. To enter a region, a system might have to plot a particular path through the modal state space; that is, it must possess a set of precursor properties and modalities to even allow it to enter that region. The case of Rayleigh–Bénard convection illustrates this, where possibilities for motion available to fluid parcels in the initial quiescent state are not the same as the possibilities in the convecting state. The stability conditions associated with convection provide modal constraints, opening access to participation in the convective dynamics while foreclosing participation in the dynamics of the initial state. Or think of multicellularity.

Once a biological system develops multicellularity it can do things that were previously impossible (such as develop a differentiated body plan or grow to be a larger size because it is not constrained by the effect size-to-volume ratios have on diffusion). Yet it cannot do other things (e.g., the capacity for regeneration can be extremely limited for specific cell specializations).

We have discussed numerous examples where stability conditions and their associated contexts play these modal gatekeeper roles. Hence, stability conditions and the contexts they define are difference makers which are not given at the smaller length and time scales. We remind the reader that stability conditions need not always be dynamical or causal; they can be adynamical and acausal, and even relatively abstract such as dimensional or topological (Chapter 2, Section 2.3.2).

We rehearse all this to remind the reader of the power of contextual emergence and also to draw the reader's attention to how inherently relational nature is! To be a system is for smaller-scale components of a system to already be enmeshed in multiscale relationships. This relationality goes as far "down" in length and time scales as exist, and up to the largest length and longest time scales in existence. The ontological picture is one of multiscale relations from smallest to largest scales without reified pregiven "levels of reality." There are levels of organization to be sure, but these are always multiscale in nature or dependent on multiscale contexts. In short, ontologically speaking, contexts are an indelible part of nature and are as fundamental on all scales as anything to which philosophers typically point (e.g., "fundamental" laws).

Over the ensuing sections we shall unpack the relationality that we believe the sciences reveal that is part and parcel of the contextual emergence pattern of nature. We simply note for the moment that the relationality of the contextually emergent world has much in common with other recent proposals such as ontological structural realism (Ladyman and Ross 2007), the scale-free universe (Thalos 2013), or the counterfactualism of laws and explanations (Marletto 2021). We will eventually discuss the key differences between some of those views and ours.

7.3 Modalities and dispositions

There is much talk about laws and dispositions these days in both metaphysics and philosophy of science. A central idea is to ground an account of laws in an account of dispositions. As Mauro Dorato puts it: "The manifestation of dispositions codified by scientific laws essentially involves a relationship between different properties" (2005, p. 115). Naturally, there are many different competing accounts of dispositions, properties, and powers, and how those relate to one another. Instead of starting with properties and powers, we want to start with dispositions, because "a disposition, by definition, manifests itself only in appropriate circumstances (as specified in the *ceteris paribus* clause)" (p. 112). That is, a disposition is already a contextual thing! Exactly how to best conceive of contextual dispositions is a matter of debate.

For instance, Dorato points out that debates about dispositions and contextuality have been central in quantum mechanics (QM) since its inception:

> Contextualism is in fact a necessary consequence of a *realistic* interpretation of quantum mechanics, owing to an important mathematical theorem credited to Kochen and Specker, on the basis of which a quantum physical system describable by a space of at least three dimensions can have simultaneously defined properties or magnitudes only if their values holistically depend on the context of measurement.
>
> (p. 116)[1]

Dorato notes that different seminal thinkers in QM disagree about exactly to what degree and in what way quantum systems are dispositional and contextual. Nevertheless, he claims that the belief that quantum systems are "essentially dispositional" represents the "convergence point of a series of interpretations of quantum mechanics as different from each other as instrumentalistic and realistic readings of the same scientific theory can be" (p. 116).

Conventionally, *dispositionalism* is the view that properties are intrinsic dispositions in the sense that objects, entities, and systems have no intrinsic properties except for dispositional properties. Objects, for instance, have intrinsic dispositions to interact with the intrinsic dispositions of other objects in a symmetric fashion such that, for example, "the solution of water and salt is the joint product of the soluble salt and the solvent water, but these substances have many other dispositions, depending on the particular interactions they undergo" (Dorato 2016, pp. 239–240). What makes these dispositions intrinsic is that they could presumably exist in a possible world in which only that object exists, and indeed, such dispositions can exist unmanifested in that world, "in virtue of the fact that dispositions are present even when they do not manifest themselves" (Dorato 2005, p. 113).

We agree that modalities in the broadest possible sense are rooted in dispositions, and that properties (or "powers," if you like) just are dispositions. Nevertheless, contextual emergence implies that dispositions are not intrinsic. What exactly is an intrinsic disposition anyway? Is that just an unanalyzed "fundamental" feature of reality? How exactly does that work?

[1] See our discussion in Chapter 4, Section 4.7.2.

Does every object have an infinite number of such dispositions, and is it like a hidden program or "instruction set" telling the object what to do when it encounters one of an infinite number of things it might interact with? Presumably not, otherwise one is stuck with some metaphysical equivalent of the frame problem. But if not, then what does it even mean to claim that X and Y have an *intrinsic* disposition to yield Z upon interaction? What makes such dispositions *intrinsic?* If the answer is that they exist even when unmanifested, why would one ever be justified in believing that when by hypothesis they only manifest upon interaction? How can there be a property that is both inherently intrinsic and inherently relational? Contextual emergence is a rejection of the very idea of an intrinsic property.

What contextual emergence implies is that dispositions are fully contextual or relational, meaning that there is no possible world which contains only one entity possessing "unmanifested intrinsic" dispositions. Indeed, contextual emergence implies that there are no substances and no entities with "primitive thisness." As we have indicated throughout this book, entities and objects are networks of dispositions. There is absolutely no need to add any second-order metaphysical glue of any kind to the world to explain why it is ordered and stable, and why novelty arises.

The point is that dispositions do not metaphysically *ground* the sciences nor any scientific claim; rather, they just *are* the domain of scientific investigation. Hence, we are equally skeptical about the metaphysical glue implied by intrinsic dispositions in some accounts of properties, as well as that implied by mathematical/modal structures in at least some accounts of ontological structural realism (OSR).[2] We do not believe there is any need for even this much grounding. This is not an antimetaphysical anthem. We are making a bold metaphysical claim about the nature of reality based on the implications of contextual emergence derived from what scientific investigation is revealing about nature.

Regarding "intrinsic" dispositions, they can be only relatively intrinsic or invariant with respect to specific contexts, according to contextual emergence. In the metaphysics community, this contextual nature is not the usual emphasis of dispositionalists. Whether one claims that "system X has the disposition to manifest pre-existing property Y only in context Z," or "acquires property Y only in context Z," seems tangential to the question of the necessity of context Z for property Y to influence the world. After all, for contextual emergence the bottom line is that contexts are necessary to produce the properties that do stuff. An example would be the muon anomalous magnetic moment. It depends on all particles in nature that can interact with the muon (Bennett et al. 2006; Keshavarzi, Nomura, and Teubner 2018).

2 More on all this subsequently.

The muon anomalous magnetic moment manifests only in a context where the muon is interacting with eligible leptons and hadrons.[3] In short, the relationality of contexts is what grounds dispositionalism, not the other way around.

If "intrinsic dispositions" is an oxymoron, as we believe, that does not mean that they are "external" as opposed to intrinsic. Namely, dispositions and everything else are *extrinsic*, not external.[4] We share with conventional dispositional views that no "external" modal glue such as necessitarian laws are needed to ground anything, including nomological and causal claims. However, we go further in claiming that the sciences are showing us that no modal glue is needed at all—not even the oxymoronic intrinsic dispositions.

So, where do order and stability come from, you ask? As we have observed throughout this book, order and stability exist because contextual emergence is the fundamental nature of the universe, not an add-on. In metaphysical parlance, this is a rejection of the hierarchy or foundationalism thesis (Chapter 1, Section 1.2). Relations at various scales are not antisymmetric, transitive, or antireflexive. It is also a rejection of the very idea of intrinsic properties and primitive thisness.[5] Such views ultimately engender a self-referential model of explanation. For contextual emergence, no metaphysical grounding is needed, no modal glue of any sort, and nothing METAphysical need be reified.

It is for these reasons that ontological contextual emergence is common, universal, and non-spooky, and does not defy scientific explanation. It is for these reasons that emergence is not about *essentially* different properties, entities, and the like, "popping" into being in a discontinuous and disunifying fashion as the result of brute bridge-laws, brute causal powers, or brute intrinsic dispositions.

As we said in Chapter 1, we are not alone in adopting such a picture of reality. Carlo Rovelli explains the consequences of such a view as "the impossibility of separating the properties of an object from the interactions in which these properties manifest themselves and the objects to which they are manifested." In much the same fashion as contextual emergence, he goes on to say: "The properties of an object *are* the way in which it acts upon other objects; reality is this web of interactions ... for *there are no properties outside of interactions* ... When the electron does not interact with anything, it has no physical properties. It has no position; it has no velocity" (2021, pp. 78–79).

[3] This includes any particles that may exist which are unaccounted for in the Standard Model. Hence, one potential explanation for the discrepancy between predictions of the muon moment and experiments is the existence of undiscovered elementary particles outside the standard model.

[4] See Silberstein (2017) for a lengthy argument to that effect.

[5] See Bliss and Priest (2018) for more on all this.

As we have observed throughout this book, contextual emergence is not just an artefact of the weirdness of quantum mechanics or relativity. It shows up in a wide variety of physics, such as convection, chemistry, biology, and neuroscience. As Rovelli notes: "The discovery that quantities we had thought of as absolute are, in fact, relative instead is a theme that runs throughout the history of physics. Beyond physics, relational thinking can be found in all the sciences ... we understand things (organisms, chemicals, psychological life) through their being *in relation to* other things" (p. 143).

7.4 Properties

Properties have been at issue in much of the history of emergentist thinking. For example, consider the genesis of modern interest in emergence, P. W. Anderson's (1972), where he says:

> The behaviour of large and complex aggregates of elementary particles, it turns out, is not to be understood in terms of a simple extrapolation of the properties of a few particles. Instead, at each level of complexity, entirely new properties appear, and the understanding of the new behaviours requires research which I think is as fundamental in its nature as any other.
>
> (p. 393)

The sense of novelty expressed here is not the novelty of new values of a property. If we place an apple into a fruit bowl, the mass of the bowl plus apple is greater than the items separately. This new value of mass is an uninteresting form of novelty; instead, we want some sense of qualitatively novel features of a complex system emerging (see Reuger 2000a,b).[6] Mainwood (2006) has suggested that Anderson follows in the traditions of early twentieth-century British emergentists, such as C. D. Broad (1925). Broad's notion of emergence was also based on the idea that systemic properties of complex systems are different from the properties of the parts of those systems (and the properties of different systems made from those parts). Entities *A*, *B*, *C* in relation *R (A, B, C)* have properties that apply only to *R* and are not derivable in principle from a *general* summation rule. Famously, of course, his mechanism for novelty was specific laws of composition, and it is widely thought that empirically there is little evidence for such laws.

We believe that this focus on novelty is misplaced. Very often, emergence will involve novelty, but it is not novelty in and of itself that is most important about a property. In previous chapters we observed that it is the stability

[6] Anderson's notion of emergence is mostly weak or epistemic, so his brand of "novelty" is not in competition with the laws and properties of microphysics determining the properties of complex entities.

conditions and contextually emergent states and observables. These are both the only vocabulary for which specific modal facts can be stated, and are the ontological difference makers studied by scientists. In other words, scientists study what is modally relevant, and contextual emergence captures this tracking of the important properties: states, observables, and their stability conditions. As we have stressed throughout this book, it is these inherently relational properties that are the truth-makers for the modalities determining how systems can evolve.

The reductionist might express a concern at this point. The claim is that we can only ground a modality M at level L, scale S, or domain D by reifying a system/environment distinction at L, S, or D (LSD for short) because the modality is grounded by property compatibility/incompatibility relations that rely on that system/environment distinction.[7] One could imagine a parallel to the causal exclusion argument concerning properties—a property exclusion argument. The idea is that lower or smaller LSD property compatibility relations exclude or compete with higher or larger LSD property compatibility relations.

The contextual emergence response draws on previous chapters. Essentially, the precise modal facts grounded in the higher or larger LSD compatibility relations are not isomorphic with the lower or smaller LSD modality. A different set of modalities are grounded in the lower or smaller set of property incompatibility relations (and these may include the role of properties resulting from larger-scale LSD compatibility/incompatibility relations). The modalities may overlap in some respects, but the higher or larger LSD modalities are not a subset or approximation of the lower or smaller LSD modalities. The set of higher or larger LSD modalities contain modal facts which are not included in the set of modal facts below it. It only appears as if they cover the same set of modalities if one imports the system/environment distinction epistemically into the lower LSD from the higher LSD, but doing so amounts to a change of contextual topology for the lower LSD description. It is not the lower LSD simpliciter!

In terms of sets of degrees of freedom (DOFs), one can express the properties of a specific scale or domain as a set of dispositions. Of course, this notion of a disposition is implicitly modal—to say X has the disposition to do Y is to say that it is part of the possibility space that X inhabits that it may produce Y. Think of this as a truth-maker for the modal fact that Y is possible, and that X has the disposition to produce Y. Then the claim is that it is only with specific system/environment distinctions that we derive the right sets of dispositions to be the truth-makers for relevant modal facts. We

[7] Scientists may be more familiar with the concept of self-consistency/inconsistency. Compatibility/incompatibility relations are a generalization of this concept.

can think of dispositions expressed in terms of DOFs of a system; that is, a disposition is expressed as *a particular combination of DOFs*. Note that this is distinct from Jessica Wilson's (2010) use of DOFs to express identity, but it is compatible with it. It is the higher or larger LSD sets of DOFs that are compatible/incompatible in the correct way to express relevant modalities, not the different sets of DOFs at the lower or smaller LSD.

New larger-scale dispositions, then, can come from removing DOFs from the smaller scale, or by combining DOFs at the smaller scale to form brand new DOFs. This latter idea is not contained in Wilson (2010), but when microscale constituents come together sometimes their individual properties are combined in ways that produce new systemic properties (e.g., Bénard cells in convection or protein products in gene networks). Alternatively, one can think of them as situated in the relational network of the lower-level or smaller-scale properties. Unlike Humphreys' (2016) proposed fusion account of emergence, we do not require the lower-level, so-called basal, properties to be lost, but only be made inherently relational.[8] In terms of DOFs, then, the individual DOFs at the lower or smaller LSD can be removed by changing the context, or they can be combined through physical interactions to produce new, relational DOFs that access a specific region of possibility space not accessible to the uncombined interactions.

What matters is not simply whether DOFs are eliminated, but *how* they are eliminated, expanded, or otherwise transformed. The source of the constraint in a relevant context is what matters, so that, for instance, the level or scale at which a symmetry is possessed matters. Thus, the symmetry of a collective may be the reason why DOFs are eliminated. The reduced set of DOFs captures the collective disposition exercised by a particular freedom; nonetheless, this is not merely a sum of smaller-scale DOFs but requires the multiscale relations of the larger-scale context to reduce the DOFs.

This also means that the order in which DOFs are removed or transformed matters. This reflects the physically inequivalent ways in which transitions can occur. The production of brand new larger-scale DOFs reflects this inequivalence (e.g., the emergence of a Newtonian physics from QM in Chapter 4, Section 4.5). A new disposition/DOFs at the larger scale is a particular way of combining elements from the smaller-scale DOFs, plus a particular elimination of parts of that set by the larger-scale context. This is how stability conditions work to define a new context for properties and behaviors of systems in possibility space. Note that this picture of combining DOF also goes beyond Wilson's account in not regarding mere reduction (rather than elimination) of DOFs as uninteresting. For instance, we view the

[8] See the discussion in Chapters 4 and 5, as well as Lancaster and Pexton (2015).

reduction of DOFs by larger-scale symmetries to be just as important in the process to produce new larger-scale DOFs as pure elimination.[9]

In summary, then, if we pick a particular system/environment distinction S/E1 rather than S/E2, we can define or apply a language/conceptual framework to express modalities. But this is not the whole story, since the reason why we can express those modalities is because S/E1 is in the world, and the property compatibility/incompatibility relations that require S/E1 as a context can ground those modal facts. In contrast, we cannot ground those modal facts in property compatibility/incompatibility relations of S/E2. Only the compatibility relations of properties associated with S/E1 are the truth-makers for the modalities. If we deny that S/E1 is in the world, then we must deny those modal facts because their truth-maker is not real. Note this is not the usual reductionist position. Typically, reductionists inspired by the history of science, rather than by pure metaphysics, do not deny the modal facts of higher-level or higher-domain sciences. For example, they would not deny the regularities of chemistry; rather, they want to say that those regularities are ultimately just expressions of the regularities of the more "fundamental" LSD, whatever that is. But the whole point of contextually emergent states and observables expressing relevant modal facts is that those modal facts come out of the higher-level or larger-scale properties and their relations (e.g., in stability conditions).

This still leaves the reductionist the option to deny all modal facts except those of "fundamental" physics, but if this is the reductionist's position it leads to what we think is an unacceptable consequence. Why are the sciences effective for investigating the world? If scientific inquiry is not a generally reliable way of discovering modal facts, then how are the modal facts of the "fundamental" physical level even discoverable? After all, in such a scenario it is not that scientific inquiry discovers modal facts and then follows the rabbit down the hole to see that those modal facts are ultimately grounded in allegedly more "fundamental" principles. Such an "Alice in Wonderland" picture of reduction in scientific inquiry tells us a story about why such inquiry can lead us to the "fundamental" LSD for reduction. But if

9 Indeed, the difference between elimination and reduction is not as clear cut as Wilson's account suggests. She states that the difference is that in reduction the fixed DOFs still feature in the functional form of the laws, whereas in elimination the DOFs do not feature in the laws (2010). But this does not take account of generalized coordinate systems, where we write down coordinates expressing the dispositions of the system in their simplest terms. These are examples of good coordinates, or our contextual observables. When the laws of a reduced system are written in terms of generalized coordinates, the reduced DOFs do not appear. It is the generalized coordinates that represent the properties that matter for relevant modalities. Of course, the system does not know or care how you come to your description of its subsequent behavior, but it therefore seems hasty to build an ontological distinction on the notion of elimination when the difference with respect to reduction is not so clear-cut.

the reductionist denies modal facts of larger or higher LSD sciences, how can those larger or higher LSD modal fictions ever lead to the true modal facts?

7.5. Property incompatibility relations

Modalities can be expressed as property compatibility relations. Possession of a particular property immediately rules out possession of some other properties and rules in possession of still other properties. Being an entrant in the men's 100-meter final in the Olympics rules out being in the women's 100-meter final. This is true for noncategorical properties as well. Having a negative charge is incompatible with electrostatically being attractive to other negative charges, but compatible with being electrostatically attractive to positive charges. For these noncategorical properties, though, there are many properties which could be compatible but only in concrete contexts. The way systems explore modal space is through what is ruled in and out by their properties given the context in which the system is placed.[10]

One can think of possibility space as a Lego set, in which the first "physics" blocks constrain the final shape but do not uniquely determine it. Yet in the sciences, many types of physical impossibility are not expressed by laws in any strict sense. Instead, they are expressed by what James Woodward (2003) has called "invariances:" regularities in the world which apply (often strongly) in only a particular set of contexts but not others. This is because these "laws" are really tendencies which manifest only in contexts defining possibility space/subspaces. Or in other words, sometimes whether a property P is compatible with a property Q depends on the contextual feature C mediating between the two.

More on this subsequently, but for now note that although this picture is anti-Humean, it is not one in which laws "govern" or exist beyond the phenomena in question. Laws are simply patterns in property compatibility relations that can either be widespread or very specific, and either strong or relatively weak. The fact that they are strong or weak partly reflects how sensitive the property compatibility relations are to changes in contextual mediators. The fact that they are widespread or narrow simply reflects how commonly those properties that must be compatible/incompatible are produced. The laws of physics are more widespread because the relevant dispositions and their compatibility relations for the modalities of physics are simply widely found. Whereas the laws of biology—to the extent that

[10] It is worth keeping in mind that to even exist as a system requires some stability conditions that have already created access to the relevant region in possibility space for that system's existence (e.g., lasers; Chapter 4, Section 4.9.1).

there are any—apply only to systems that have manifested the biological properties whose compatibility anchors those modalities. Note also that biology appears nomically inhibited in comparison to physics simply because it wears its contextuality on its sleeve, whereas the contextual aspects of physics are usually subtle and overlooked, as our case studies have demonstrated.

A Humean response to this might be that there is radical openness to systems in what follows what. But Humeans tend to focus on what is compatible temporally rather than what is compatible with the properties of what has gone before. David Weissman (1978, p. 280) uses the example of spatial arrangements: "[Y]es, items of every sort may succeed one another in time, but, no, these same things are not universally connectible, each with every other, in space" (1978, p. 280).

We may also think of topological connectedness as displayed by the fractional quantum Hall effect discussed in Chapter 5. Here, not every transformation is possible. More generally, whatever properties follow from others must be compatible with the properties that have proceeded it and with the context into which the new properties are born (e.g., think of embryogenesis and development).

What is a context in the most general possible sense? A structured web of property relationships. Ontology contains both the properties and the structure of the relations. One "foliation" in the web of properties leads to one context; another foliation leads to another. Any foliation can produce properties entering into the web of relations in part of the ontology of the world. Contexts are defined by relations among property compatibility relations that involve property relations at multiple levels and scales. Hence, our ontology is one of properties, where contexts and systems are specific structurings of those properties which are in the world. This is why stability conditions are so crucial. They are lynchpin relations, if you will, defining the specifics of a foliation.

Properties are the ontology of contexts, while contexts are needed to produce properties. Modalities are grounded in properties, or the network of property relationships, and that network is necessary to produce other properties. This is not a vicious circle, but a virtuous circle.

An objection might be raised against the framework that we have set out. Insofar as compatible/incompatible are modal notions themselves, we have not grounded modality at all. We have no issue with this, because unlike the Humean our aim is not to give a reductive account of modality, nor is it to *ground or reify* modality in some extra-scientific metaphysical glue. For instance, contextual emergence is happy to embrace all of Aristotle's four types of causes and more. Instead, our goal is to explicate modality in a way that highlights its relational nature as revealed by the sciences. By thinking of modality as property compatibility, it is easier to see how context is important for modality. It is only in specific contexts that the properties whose compatibility expresses the relevant modality will be produced.

7.6 Nested possibility spaces

We have described and illustrated the relationship between modal spaces associated with particular LSDs: the possibility space of physics places the most general constraints on the possibility space of chemistry, while also enabling chemical possibility in the first place. In turn, the possibility space of chemistry constrains the possibility space of biology, while also enabling biological possibility, and so on. We can understand this in terms of the nested way in which property compatibility relations occur. Any property produced by combining properties of physics must be compatible with those properties. Nonetheless, the relationship between scales and domains is only a partially ordered hierarchy (Primas 1977).

Recall also that we did not describe this nesting of possibility spaces like a Russian doll, where chemical possibility just passively sits inside physical possibility. Rather, these possibility spaces are interleaved like a multiscale honeycomb, both at time *t* and over an extended time, such that to explore part of physical possibility the world must go through the "gatekeeper" properties associated with modalities of a larger scale or domain (i.e., the relevant stability conditions). For instance, consider the production of elements. The binding energy curve for combining nucleons means that in our universe the only physical way of producing elements heavier than iron is through stellar supernovae and kilonovae. Stars must be made to explore areas of chemical elemental possibility, which means going through astrophysical possibility to reach an area of the lower-level or smaller-scale possibility space. Another example of this would be the way in which some biological organisms are required to make chemical compounds that are not found in nature unless they have been synthesized biologically. To reach those areas of chemical possibility they must be compatible with biological possibility.

Or think of Dolly, the cloned sheep. To explore that area of biological possibility it was necessary to go through a particular zone of social possibility (which includes the invention of scientific inquiry and cloning techniques, ethical review panels, etc.). The property of being a sheep is compatible with the property of being a clone, though only in specific contexts outside the domain of sheep and clones allowing property bridges to be made. The properties that ground the modalities allowing sheep to be cloned include properties only found at a higher or larger LSD. If we sought to reduce those properties to only biological properties, we would not have the correct modal constraints necessary to bridge the separate regions of biological possibility space.

What is being illustrated is that if there are two length scales, i and j, going from property A at scale i to B at scale j might need the modal constraint C which is at scale j, *because that constraint is only grounded in properties of scale j*. If C is replaced with another constraint, C^*, that will not have the correct set of modal constraints to connect the relevant regions.

This is completely compatible with, and enables, the multiple realizability of properties. So, if the constraint we need associated with scale j is C and that is expressed in terms of the compatibility relations of a property (or properties) at scale j, P_C, then P_C is a *gatekeeper* or threshold property. If P_C is actualized, then the correct difference maker to explore the possibility space exists. It does not matter how P_C is produced, so long as it is. Since P_C is the property that grounds C, and C is the essential stability condition for the manifestation of some scale i property, we can produce P_C in many ways, and it will not make any difference. For instance, this is how the stability conditions for temperature work, enabling many kinds of molecules to exemplify thermodynamic states and observables. We encountered the same thing with spontaneous symmetry breaking in Chapter 5. Again, the microscale details do not matter for how the broken symmetry state arises. What matters is that the broken symmetry plays an important role in what possibilities are accessible in modal space.

7.7 Realization and multiple realizability

We have observed several examples of multiple realizability (MR) in our case studies ranging from physics to biology, and we have tried to show how contextual emergence enables and explains real world MR at multiple scales, in a way that tells against ontological reductionism, such as that posited by compositional and realization-based models of reality. At least in the context of the philosophy of biology and the philosophy of mind, many philosophers have expressed skepticism about MR and its antireductionist import. We will address those concerns in this section. But first, it is important to discuss what philosophers often mean by the terms "composition" (a synchronic determination relation pertaining to parts and wholes) and "realization" (a synchronic determination relation pertaining to the *properties* of parts and wholes). Contextual emergence, and the sciences, make clear that such metaphysical notions have little bearing on reality or scientific explanation.

More generally, as Ladyman points out in what follows, the metaphysician's synchronic account of composition, realization, and the like is not in keeping with the sciences:

> Composition is a real feature of the world and is in general *diachronic, dynamical, and domain specific*, since it depends on the relevant kinds of interaction among parts ... For example, roughly speaking, chemistry says that atoms compose molecules when there are (relatively enduring) chemical bonds between them, and economics says that agents compose a market when there are financial transactions between them. In both cases, the systems in question are considered over time, and composition is a result of interactions between the parts of the whole. *Hence, the answers that science gives to the special composition question make reference to causal processes, not to instantaneous facts about, for example, how close things are to each other ... Even compositional explanation is diachronic because it typically involves temporal limits and/or relative temporal scales*. For example, in statistical mechanics, macroscopic measurements must be carried out over time scales much greater than the mean time taken in atomic interactions. Hence, in science the part/whole relation is shorthand for a process. The scale relativity of ontology is with respect to both time and space because composition is diachronic and dynamical, and because at other length scales, the relevant entities do not exist. For example, there are no water molecules at the scale of quark interactions.
>
> (2017, Kindle Locations, 3491–3492, emphasis added)

This entire book is a testament to everything that Ladyman said. Therefore, whatever multiple realizability is, it has nothing to do with what philosophers call "composition" and "realization." And as we tried to make clear in Chapter 6, talk of brain-state types "realizing" mental-state types is simply a functionalist fiction.

Historically, multiple realization has been given as an argument against reductionism, and many philosophers have pushed back on such arguments. For example, Zerilli (2021, chapter 8) grants that the existence of MR is an empirical question. He grants much of what we have said herein, but is skeptical about "true multiple realization" and thus skeptical that it counts as an argument against reductionism or some sort of mind–brain identity theory. Part of the problem with this discussion is that while everyone agrees that MR means difference underlying sameness (i.e., truly causally distinct mechanisms underlying the same cognitive function or content of conscious experience), there is no set agreement on what counts as truly causally distinct or same function. That is, there is little consensus on the necessary and sufficient conditions for MR, what constitutes causally relevant differences in the mechanism, or what constitutes truly different ways to "bring about" the same function that defines the kind. In short, whether MR is true depends on how tightly or weakly one defines the "mechanism" and the "function" in question. No one would deny that lots of different animals and insects utilize different mechanisms such as electrical fields and magnetic fields to perform general functions such as navigation and communication (e.g., electroreception). The MR skeptic clearly has a different target.

What makes two causal mechanisms truly causally distinct? Polger and Shapiro (2016) assert that "multiple realization occurs if and only if two (or more) systems perform the same function in different ways" (p. 45). Thus, Zerilli argues that crossmodal plasticity and supramodal plasticity are not evidence of MR, because, for example, the auditory cortex came to resemble the visual cortex; that is, the auditory cortex came to take on the same columnar organization as the visual cortex (2021, p. 141), and therefore this is not a case of truly distinct causal mechanisms underlying the same function.

Our opening response is that if the brain or any complex biological system can, under specific contexts or conditions, convert itself or some subregion of itself into a distinct causal mechanism to mimic some damaged mechanism, that should count as MR *par excellence*! We would also reply that surely degeneracy counts as MR by definition. And as we saw in Chapter 6, Lisa Feldman Barrett acknowledges that degeneracy is a real feature of brains. This is an empirical claim, of course, and is subject to revision in light of further data. But for now we shall take the word of leading neuroscientists. Perhaps neural degeneracy is rare, but even so, it is proof in principle of MR. Even in cases of reuse, redeployment, or recontextualization, if a particular cognitive function can at different times have different workings subserve them, even if those workings do not morph into new mechanisms in the process, that should count as MR with respect to those cognitive functions.

Many of us would argue that convergent evolution provides more such cases of MR. The standard definition of convergent evolution is as follows: the causally independent evolution of similar functional features or evolutionary strategies in species of different periods or epochs in time or in different geographical regions in the same time period. Convergent evolution creates analogous structures that have similar form or function but were not present in the last common ancestor of those groups. The idea is that the same function, behavior, evolutionary strategy, or even roughly the same mechanism, comes into being in causally independent circumstances as the result of environmental and evolutionary pressures. Sometimes in such cases the same mechanism (or partial mechanism) arises independently to perform the same function, and sometimes the mechanism is different. For instance, flying as an evolutionary strategy appears again and again through convergent evolution, but the mechanism can be very different (e.g., pterosaurs, bats, birds, various insects). However, such cases do not rise to the level of true MR for the skeptic, because "flying" is too general a function.

There are other cases of convergent evolution where almost the same mechanism appears independently across time, to perform a very similar function. For example, the brain architecture of the dorsal ventricular ridge in reptiles developed independently of the mammalian neocortex, but they share a good deal of structure and function, in part due to evolutionarily conserved neural circuits (Yamashita and Nomura 2017, p. 302). As another example, birds, whose neural architecture also developed independently, do not have a mammalian neocortex as such, but the avian pallium contains circuits homologous to those of the mammalian neocortex (Wada et al. 2017, p. 285). However, it is very likely that the genetic and developmental mechanisms and pathways that form the avian columnar circuit formation are different (p. 285). That is, while the neocortex of mammals and the pallium of birds share similar mechanisms and functions, the neural and genetic mechanisms that explain the development and placement of those neural structures is different. There are other more straightforward cases of convergent evolution such as eyes like ours, photosynthesis, sex as a strategy, nervous systems, or nerve nets. But so long as the mechanism remains the same or the function is defined is too broad (e.g., dolphins and monkeys respectively herding prey), the MR skeptic is likely to be unimpressed.

There is an argument, however, that we believe shows that natural selection trumps the existence of specific mechanisms, and thus is a case of contextual emergence. What emerges are various universal mechanisms, behaviors, strategies, and cognitive functions as a result of the stability conditions defining contexts that constrain development in specific evolutionary environments.[11] Also, there are more general evolutionary principles such as organisms that are most adaptive and are most likely to survive and thrive—and the most adaptive will be the animals with the most general intelligence. Here is the argument in a nutshell.

First, natural selection and thermodynamics are, all things being equal, universal, including not only all of Earth but other planets. Hence, not knowing anything else about an environment or planet except the evolutionary environment of selection will allow us to predict traits that organisms are likely to possess. If the environment has tall trees, high elevations, deep dark cold waters, and a low oxygen level, natural selection will allow us to predict with a high degree of accuracy what sorts of traits and evolutionary strategies have emerged therein, regardless of knowledge of implementing mechanisms. Certainly, many different features of the environment—biological, physical, and otherwise—will place constraints on what is possible, but otherwise natural selection will be the determining factor.

[11] See our discussion in Chapter 6, Section 6.4.4.

Natural selection tells us that no matter what sort of mechanisms might be available in different planetary contexts or even on other planets, while the mechanisms can and do vary, particular traits will come into being via one set of mechanisms or another. And this has happened here on Earth in the form of convergent evolution. Natural selection transcends any specific mechanistic account of a particular cognitive function.

As Peter Godfrey-Smith puts it, "evolution sometimes builds a range of different structures that carry the same function" (2020, p. 91). As he points out, crustaceans and other animals do not have a cerebral cortex, visual centers, or pain-related mechanisms such as our own, and yet they have the same functions and probably subjective experience as well (p. 91). His favorite example of convergent evolution is the nervous system of the octopus, which is highly decentralized, with two-thirds of the neurons not in the brain but in the tentacles. Delegation of limb control is decentralized to some extent with each tentacle able to explore and multiply sense the environment independently. And there is no body map within their brain, thus orientation in space happens differently somehow, and they have a big optic lobe behind each eye with dedicated processing to that eye only (pp. 129–130).

There is also the example of creatures with very small brains (100,000 neurons or fewer), such as bees and spiders, who nonetheless, by neural mechanisms that must be quite different and perhaps less varied in many ways, manage to have sophisticated visual capacities and other sensory modalities, complex motor control, navigation, communication, problem-solving, and in some cases sociality and cooperation.

Second, specific biological mechanisms such as evolutionarily conserved neural circuits are going to be a function of various historically contingent constraints, and thus such mechanisms can differ greatly, especially if we consider radically different environments or different planets. There is no good evolutionary reason, for example, that DNA, RNA, or neurons need be the implementing mechanisms. These are contingent facts. For instance, here on Earth some bacteria-killing viruses called bacteriophages replace adenine with 2-aminoadenine. To quote Zhou et al. (2021), "life doesn't have to be GTAC, life can be more diverse." The genome of single-celled plankton (dinoflagellates) is organized into alternating unidirectional blocks, as opposed to being random. This changes the three-dimensional structure of the genome into rod-shaped chromosomes, and the structure is strangely dependent on transcriptional activity. Other oddities include few transcription factors in their genome, no alteration of gene expression in response to environmental changes, and, unlike other eukaryotes, instead of the exclusive use of histones to structure their DNA, they use viral proteins incorporated into their genome long ago. This may in part explain why their genome structure is so different. As Salazar et al. note: "It shows that nature can work in a completely different way than we thought. There are many possibilities for how life could have evolved" (2020). Many others make the same point: "The fundamental forces at work in evolution are independent of which molecules interact with which and how" (Kershenbaum 2020, p. 46).

Perhaps the most striking example of this is the comb jelly (*ctenophores*)—more specifically, the Pacific Sea Gooseberry (*Pleurobrachia bachei*). The ctenophore nervous system is quite different. Ctenophore nerve cells, unlike the nerve cells of other animals, do not use the usual chemical messengers, including serotonin, acetylcholine, dopamine, noradrenaline, adrenaline, octopamine, histamine, or glycine. The only neurotransmitter in common is glutamate. Other important differences are the absence of an innate immune system to fight off bacteria and viruses, the absence of the mechanisms that process regulatory micro RNAs used to control the expression of other genes, the absence of key regulator proteins of development such as hedgehog and JAK/STAT, and the absence of key genes involved in formation of muscles, all of which suggests a strong case of convergent evolution and multiple realizability (Moroz et al. 2014, p. 114).

Third, if an environment is complex, changing, and sufficiently challenging, natural selection will favor the most adaptive organisms—biologically, behaviorally, and cognitively. The most adaptive organisms will be the most plastic and robust organisms. Those whose nervous systems and embodiment are sufficiently plastic to enable the most universally adaptive strategies in the face of environmental changes, such as sociality, tool use, and language, will be the most fit. Therefore, it is not surprising that many disparate animals on Earth, such, as octopuses, bees, crows, dogs, dolphins, and primates, evolve, to one degree or another, universal strategies, skills, and modes of learning, such as motor control and perception, motility, predictive coding perhaps, memory of various sorts, mental time travel, sociality and cooperation, communication and language, culture, and tool use. The selective advantage of general intelligence, and the selective advantage of all the aforementioned skills and cognitive functions, is well documented across all life on Earth. Natural selection dictates that when stability conditions allow, these same strategies and cognitive functions will emerge over and over across disparate environments (including on different planets) with disparate mechanisms as they already have on Earth.

Take the strategy of sleep, for example. While experts still debate the various possible functions of sleep, it is ubiquitous across species. There is evidence now that sleep even predates animals with brains and central nervous systems; witness Hydra (Kanaya et al. 2020, p. 1). Amongst those animals who sleep and do have nervous systems, there is evidence of MR. Sleep-like states are conserved across most species, but using different mechanisms (Hayashi and Lui 2017, p. 343). For instance, hypocretin has a crucial role in maintaining waking states in mammals. Invertebrates have no homologue, while in fruit flies neuropeptide pigment dispersing factor is critical in regulating circadian rhythms and wakefulness (p. 360). Different neural circuits, neurochemicals, hormones, and genes are engaged in sleep regulation in different animals. There is simply no universal mechanism or molecular pathway that encodes sleepiness or wakefulness.

All of this provides us with a big clue as to the nature of the human brain and why humans are quantitatively if not qualitatively more intelligent than other species on Earth. We are the most adaptive creatures with the most general intelligence in part because our brains, nervous systems, and bodies maximize plasticity and robustness. Thus we are very responsive and adaptive to internal changes and environmental changes—our brains and bodies are "engines" of plasticity and robustness. Which means, in fact, that our cognitive systems and ways of being are very open to environmental changes; and what is more, as social neuroscience predicts, there will be all sorts of positive feedback relationships between our cultural advances such as tool use, sociality and language, and our brain development and activity. In this biopsychosocial process of potentially, ever-increasing plasticity, various boot-strapping processes involving specific cognitive functions, biological mechanisms, and specific technologies converge, and become integrated, repurposed, and iterated into ever more complex patterns of development. Which is to say, we human beings are not just brains or brains and bodies, but, as argued in Chapter 6, are temporally extended multiscale biopsychosocial networks. Key "design" principles, such as plasticity and robustness, are themselves MR and not fully explicable in any mechanistic fashion.

One might wonder why neomechanists resist the reality of MR. Perhaps part of the answer lies in the fact that the best argument for MR harkens back to a time when functionalism (generally computational) in the philosophy of mind and cognitive science was ascendant. During that period, functionalists often felt justified in ignoring neuromechanistic details and the brain more generally (the stuff that goes in those boxes in the cognitivist "boxological" computational diagram), because implementing mechanisms were viewed as secondary. Jerry Fodor and others famously used functionalist-based MR as an argument against type–type identity theory. Regardless of one's evaluation of such arguments in the past, it should be clear that contextual emergence does not invoke functionalism of any sort. Instead, the focus is on neuroscience and evolution and the stability conditions defining the contexts central for the phenomena.

Given the case that we have presented, the belief that cognitive functions, such as specific types of memory, will always involve the same mechanisms, is unsupported. As well, the belief that such cognitive functions arising on other planets where there is life would involve the same mechanisms is unsupported. The idea that there is some law or identity relation that universally links cognitive functions or, even worse, specific mental contents and specific mechanisms or brain states, is not scientifically grounded. This is the sort of magical metaphysical thinking that allows some to take Putnam's brains-in-vats and Boltzmann brains seriously.

Let us return to MR and neural networks discussed in Chapter 6. Debates about MR rarely have multiscale networks in mind.[12] As we argued, the brain's networks featuring "small-worldness," "rich-clubbyness" with hubs, and "scale-freeness," are universal topological/organizational features. Indeed, the neural architecture of many different insects and animals exhibit such topological features, and such features appear early in the development of individual organisms. The reason for this is clear: Those topological features maximize plasticity and robustness with respect to both internal and environmental perturbations. Thus they maximize adaptability.

Certainly, mechanistic explanations can advert to such organizational features in principle, if by "organizational features" one means some representation or model (however abstract) of the spatiotemporal, causal, and dynamical relationships between various components. We argued in Chapter 6 that network explanations are not simply maps or representations (however abstract) of such componential relationships. This is true not only because the nodes and edges in network explanations need not refer to components and their relations directly, but also because the behavior of the components is often determined by or *constrained by* the *global* organizational feature of networks (e.g., small-worldness). Again, the stability conditions provided by the network architectures define the context for component behavior.[13] Network properties can both embody and alter the global state of the system (or some subset of it), meaning that the behavior of the various components can be changed by tweaking the network properties (and the more dynamical order parameters involved in defining the context for the network).

Why are such global networks so prevalent, so often multiply realized in complex biological and cognitive systems, even though processes at smaller scales often happen at very different time scales as the processes they support at larger scales? There is often a rapid turnover of entities and states at the smaller scales, creating genuine-world multiple realizability that belies any simplistic account of composition and realization.[14] Those global topological features define contexts that, once in place, constrain the behavior of the ever-changing constituents to maintain the relevant efficacious

[12] Batterman (2018) is an exception.

[13] The stability conditions defining the contexts in which these architectures arise are also crucial.

[14] Recall that individual fluid parcels will be shifted among different Bénard cells, meaning that even in these relatively simple cases the cells exhibit MR.

topological features. The multiscale and global codetermination and causal dependency involved in such systems looks nothing like the functionalist's MR wherein the same cognitive functions are "realized" by distinct biological mechanisms. Nor, as we stressed in Chapter 6, does genuine-world MR in networks look like localization and decomposition. The autonomy of functional and topological networks, and the autonomy of other global organizing principles in the brain, have nothing to do with realization relations or brute identity relations. It has everything to do with contextually emergent causally open networks.

For any particular synchronic frame or still-shot of a biological system at a time *t* with some duration *d*, the determining features include *diachronic* multiscale interactions (context sensitivity) and *global stability conditions* outside the time-slice in question that cannot even be assigned a scale or "level."[15] That is, when it comes to such complex biological systems one should take the word *process* very seriously and understand that such systems are spatially, temporally, functionally, and, in a thin sense, teleologically extended. This is not to deny that there are a variety of both global-to-local and local-to-global determination relations involved in such systems. This alone should be enough to dispense with talk of realization and the like, and to appreciate contextual emergence.

Thus, once we see that global topological network properties and other organizing principles do not "supervene" on and are not "realized" by structural components as such, there is little reason to think that topological explanations are reductionist or mechanistic in the sense of localization and decomposition. As Love notes, the mutual interaction between physical and genetic causes does not "'bottom-out' in lower-level activities of molecular entities. The interlevel nature of the causal dynamics between genetic and physical factors runs counter to this expectation and is not amenable to an interpretation in terms of nested mechanisms realizing another mechanism." (2018, p. 341). This is consistent with contextual emergence and our arguments for how the function of mechanisms are shaped by their contexts. He goes on to point out that "the reciprocal interaction between genetic and physical causes does not require stable, compositional organization, which is a key criterion for mechanisms" (p. 341). Again, this is consistent with our arguments in Chapter 6:

> The productive continuity of a sequence of genetic and physical difference-makers can be maintained despite changes in the number and types of elements in a mechanism. Although compositional differences can alter relationships of physical causation (fluid flow or tension), these relationships do not require the specificity of genetic interaction predominant in most mechanistic explanations

[15] Think of the KMS conditions—stability conditions independent of any scales (Chapter 4).

> from molecular biology ... Standard mechanistic strategies of representation and explanation appear inadequate to capture these mechanisms.
>
> (p. 341)[16]

We have seen this pattern countless times, where the components at the smaller scales do not determine the behavior and properties at the larger scales—the contextual emergence pattern. In this case, global stability conditions in the form of constraints and organizing principles define a multiscale context that determines the behavior of the parts.

With contextual emergence, global stability conditions and context sensitivity are *fundamentally* at play. Contextual emergence makes sense of C. D. Broad's claim that, "an emergent quality is roughly a quality which belongs to a complex as a whole and not to its parts" (Broad 1925, p. 23). For him, the global or systemic properties *P* of a system *S* are reducible only when the parts *in isolation* are sufficient to explain the existence of *P*. That is, there is reducibility when *P* can be derived or predicted in principle from the parts of *S* in isolation or when embedded in simpler systems (Stephan 1992, p. 55).

Contextual emergence emphasizes the ontological and explanatory fundamentality of multiscale stability conditions and the contexts that they define. Often these conditions operate globally over interconnected, interdependent, and interacting entities and their relations at multiple scales (e.g., topological constraints and organizational constraints in complex biological systems). In Chapter 6 we argued that multiscale contextual constraints and *global* organizing principles such as network topology, plasticity, degeneracy, robustness, autonomy, and universality cannot possibly be explained in terms of localization and decomposition.[17] These principles are multiply realized in complex biological systems, repeatedly instantiated over and over in many different biological systems and kinds, and remain invariant over vast changes in structural dynamics.

Thus, such organizing principles are themselves MR. We would expect that, as with natural selection, if there is life on other planets in the universe, such global organizing principles will be at work regardless of the physical or biological nature of the organisms in question. The autonomy and invariance of these organizational principles is best explained by their cocausal and codeterminative nature as stability conditions defining contexts into which biological phenomena come to expression and function.

We do not doubt that there are many universal limits on plasticity and other organizing principles of cognitive systems, but this is an empirical question, not something that can be determined by metaphysical assumptions. As noted, we also do not doubt the existence of universally conserved

[16] See also Love (2012, p. 120) and Love and Hüttemann (2011).
[17] See Silberstein (2021) for more discussion.

mechanisms across evolutionary time scales, at least relative to particular evolutionary histories such as here on Earth. What this indicates is that some mechanisms are more invariant than others. Such mechanisms do what they do not for any nativist or internalist reasons, but because of their being shaped for maximal invariance by contingent evolutionary processes—which is just more context. Part of how this works is that such conserved mechanisms carry with them the internally relevant context that allows them to function with relative autonomy. It is contextual emergence that allows complex multiscale networks to maximize robustness and autonomy and thus adaptability.

It is logically possible that after intense analysis of all the different animals on Earth and actual alien species, that we will discover some real mechanistic invariance across all of them. Nevertheless, even that invariance would be context-dependent for its appearance and functioning. Such a discovery would not obviate the need for global stability conditions.

Contextual emergence implies that focus on which LDS neural mechanisms subserve particular cognitive functions simply misses the point. And as we have seen, many neuroscientists and cognitive scientists agree that the bottom-up story is inadequate. For instance, in network neuroscience the explanation for various cognitive functions and mental contents is through multiscale networks, and the determination in such networks is also multiscale. The conclusion is that there are no strong arguments based in neural mechanisms against MR properly understood. On the contrary, given the multiscale contextual emergence pattern, MR is to be expected on every scale, and this is exactly what we have observed.

7.8 Laws

Historically, one of the roles that laws have played is as a source of order in the universe. The philosophical tradition has analyzed laws of nature mainly along two incompatible lines. One is nomological realism, where the laws have their own metaphysical existence and account for the patterns and regularities of nature. This is a form of metaphysical glue holding the universe together and providing for its order. The competing view is the regularity tradition, where laws are taken to be descriptions of the regularities and patterns observed in nature.[18] But there is no presumed metaphysical glue underlying

[18] While it is common to refer to regularity accounts as "Humean," this misrepresents Hume's view. He believed that we come to associate patterns such as cause and effect through our experience of observing events occurring together (e.g., that tree branches move when the wind blows). Yet Hume was a firm believer in laws as representing necessities. What he questioned was our empirical access, hence our ability to know these necessities.

these regularities. There are many, many varieties of both positions as well as the previously discussed dispositional account of laws.

The philosophical tradition has largely framed a false forced choice between nomological realism and regularity accounts, however. A contextual emergence account of laws of nature as extrinsic dispositions involving stability conditions and contexts is a rejection of this false forced choice. Both of those traditions share the assumption that there are absolutely fundamental entities at bottom whose behavior determines the behavior of everything else—they both accept hierarchical and foundationalists models of reality critiqued throughout this book.[19] In a contextual emergence world, all necessity is natural and involves multiscale contexts.

Although these two traditions involve differing relationships to the various theories of causation on offer, a very important class of laws in scientific theories has been dynamical laws, usually conceived as governing or specifying the history of state transitions that a system makes given some initial starting conditions (and, perhaps, specific boundary conditions too). Such dynamical laws often play important roles in philosophical analyses of laws of nature as well as metaphysical accounts of reality. The focus on dynamical laws as fundamental has fueled much of the belief that such laws plus the initial conditions provide necessary and sufficient conditions for the behavior of constituents regardless of context (i.e., the laws in the "most fundamental" domain of reality ultimately control all behaviors in all domains) and boundary conditions can simply be ignored.

One of the implications of contextual emergence is that such law fundamentalism is misguided. As we have observed, even if there are basic, universal dynamical laws, they can only be fundamental in the sense of contributing some of the necessary conditions making actual-world properties and behaviors possible. This implies that dynamical laws can only function as such in relevant contexts (e.g., where stability conditions guarantee the existence of relevant states and observables). Hence, dynamical laws do not govern in any sense other than that in specific contexts where the compatibility relations enforce self-consistency of the behavior of the target system. Even in elementary particle physics, dynamical laws are not autonomous or context-free.

The main role that laws of nature play is that the most basic laws determine the boundaries of possibility space (i.e., defining what is outside the realm of possibility for a universe with this set of laws). Physics-based dynamical and conservation laws, then, are "fundamental" and "universal" in the sense that the relevant space of possibilities for the behaviors of constituents in the universe are delineated by these laws, though these

[19] We say more about fundamentalism later.

laws do not determine the actualization of the delineated possibilities. Those actualizations are only fully determined by concrete contexts into which the laws come to expression. For instance, gravity determines the space of possible motions for an apple falling from a tree, but the concrete context where you reach out your hand and catch the apple actualizes a particular possibility among all those allowed. So, laws come to expression in concrete contexts as part of the network of relations. The most basic universal laws of physics, then, are part of every context; hence, barring extraordinary contexts, these laws cannot be violated as we have stressed throughout this book. Nevertheless, it is also the case that these laws are not determiners or governors of those networks of relations nor the behaviors of systems.

The point should be familiar by now: "fundamental" physical laws, both dynamical and adynamical, only establish necessary conditions for the possible behaviors of systems; contingent contexts at multiple scales establish the rest of the jointly necessary and sufficient conditions for actual behaviors. For perfectly understandable pragmatic reasons, in the sciences we typically have a target system whose behavior we want to explain for which we might invoke a dynamical law or a causal mechanism, and we regard the rest as boundary conditions or background conditions. We have seen, however, that there will be many such possible and useful carvings even involving the same "chunk" of reality. Nonetheless, we could even go further and suggest that there is no ultimate metaphysical distinction between dynamical/causal laws and adynamical laws, because both simply represent different constraints. Recall that it is stability conditions that are the gatekeepers opening or closing doors to specific regions of possibility space. Dynamical and causal processes can only *bring things about* subject to such gatekeeping.

As we discussed in Chapter 1 and illustrated in several examples throughout, many of these gatekeepers are adynamical or acausal (e.g., spatial dimension). Discussions about the relative roles of dynamical versus adynamical and causal versus noncausal features have been going on in physics for quite some time.[20] Let us return to an example mentioned in Chapter 1: the relativity principle and the light postulate in special relativity (SR). In Einstein's work on SR, notice that he never mentions exactly how clocks slow, or why meter sticks shrink—the kinds of constructions or reductions one might expect if all physical effects have their grounding in laws at the smallest length and time scales. Nevertheless, the principles of SR were so compelling that physicists eventually accepted the light postulate and the relativity principle as fundamental, without a causal–mechanical explanation of time dilation and length contraction. In Einstein's own lingo, SR provides a "principle" as opposed to "constructive" explanation. Eventually,

[20] See Silberstein, Stuckey, and McDevitt (2018) for discussion.

most physicists accepted that even the physics at the smallest microscales must be subject to the relativity principle and the light postulate. In other words, the relativity principle and the light postulate of SR are examples of stability conditions that constrain the possibilities for behaviors of systems, such as clocks and rods, across multiple scales.

Moreover, these stability conditions are adynamical and acausal. There is no corresponding differential equation for the relativity principle or light postulate; rather, any differential equation corresponding to a dynamical law is treated by physicists as subject to the constraints of these stability conditions when any system approaches a velocity that is an appreciable fraction of the speed of light. The point is that dynamical laws do not make "the world go round," as is so often assumed in discussions of laws.[21]

Contextual emergence implies a constraint-based account of all laws. The idea of laws as constraints is to pursue a path between nomological realist and regularity accounts of laws. In scientific practice, such as Hamiltonian formulations of physical theories, laws typically take the form of differential equations. Such equations express relations among observables and forces describing the evolution of a system state. Think of Newton's second law, $F = ma$, where F represents a mathematical model for the force acting on a particle, m is the particle's mass, and a its acceleration. The modeling of forces and system states is usually carried out in terms of energy and constraints in a relevant state space. There are exceptions to this, of course, such as Lagrangian or least-action formulations of dynamical laws, that begin with a future boundary condition rather than time evolving the system in question.[22] But no matter how these laws are formulated, their function is one of constraint. In the example of Newton's law of gravity, the law constrains or restricts the force of gravity to have a particular form. Similarly for Coulomb's law: the force between charges is constrained to be of a specific form.[23]

Laws of nature, according to contextual emergence, constrain the action of forces along with other conditions constraining system behaviors (i.e., through the various stability conditions acting as modal gatekeepers). As we have observed, these conditions can be physical, dimensional, dynamical, topological, and so forth. Physicists pay attention to the conditions defining the particulars of a physical context as part of what it means to understand

[21] There are some subtleties in how one pragmatically distinguishes between dynamical and adynamical or causal and acausal explanations that we do not have space to discuss here. See Silberstein, Stuckey, and McDevitt (2018, and 2021) for details.

[22] See Silberstein, Stuckey, and McDevitt (2018) for details

[23] Of course, for something more "fundamental" one would want to appeal to how Einstein's field equation in general relativity (where Newton's law is a limiting case) expresses the constraints on the ways in which the bending and warping of space can take place. Or appeal to the constraints on the form the electroweak force (where Coulomb's law is supposed to be a limiting case, though this relationship is much more subtle than the gravity case).

system behaviors. Nevertheless, as we have seen throughout this book, many of the important details, such as stability conditions defining the contexts, are often missing from analyses of system behavior, largely because they are taken for granted.

It bears repeating that what we have observed throughout this book is that the most "fundamental" or microscale laws are those that define the boundary for the space of physical possibilities for the universe as a system and its subsystems. Nonetheless, these laws do not dictate or fully determine the behavior of any systems, but rather constrain those behaviors to be somewhere within physical possibility. Further constraints from the concrete contexts defined by stability conditions restrict or give access to both the kinds of systems that can exist and their behaviors. Laws as constraints, then, are not mere regularities. Yet no metaphysical or other kinds of glue are needed to explain why systems such as the solar system behave as they do. Constraints, forces, and energy in their contingent configurations are primarily responsible for the behavior of systems where the most basic laws plus any laws that emerge over time contribute some of the constraints for system behavior. Additional multiscale contexts are always needed to explain the properties and behaviors of geological or biological systems. Hence, while there are essential pragmatic reasons for distinguishing between initial conditions, background conditions, and system to be explained, given contextual emergence these distinctions do not "carve nature at the joints."

One further implication of contextual emergence is that many laws are emergent. Laws will come into being over time as contexts arise. They then may form necessary conditions or constraints on what is to follow (e.g., playing a role in redefining the possibility space). Given the emergence of the right set of conditions, a new law may be born within that previously given context. The earlier laws and constraints would be no more fundamental in any absolute sense than any new laws, though the former might contribute necessary conditions for the latter. And, in turn, these newly emergent laws could contribute necessary conditions for the birth of still more subsequently emergent laws.

Notice that contextual emergence does not endorse a view such as Smolin's (2013) where the earlier laws and constraints, values of physical constants, and so on, evolve over time. What changes over time is the growth of new contexts and their interlocking/interleaving with each other in complex, partially overlapping, and multifaceted ways. Emerging laws and emerging multiscale contexts go hand in hand—there is no dualism here.

Moreover, the contextual emergence picture of laws is different from Cartwright's (1983, 1999) view that laws are always local rather than universal. For her, there are no "laws that are unconditional and unrestricted in scope" (1999, p. 59). Cartwright has in mind "universal" in the sense of universally and unalterably governing all phenomena rather than as defining the general space of physical possibilities across all energy and length scales. While we agree that no laws ever operate unconditionally—context-freely—we see no need to deny that some laws may be involved in phenomena across all length and energy scales (and in that sense are "universal" or most general). Hence, instead of a world with a variety of patches of causal capacities (1999), a contextually emergent world has domains with nontrivial overlaps, where laws and concrete contexts lead to the behaviors we see across physics, chemistry, geology, biology, and all the other domains that provide for their own organizational coherence, not unlike self-organizing behavior in complex systems. The world is not "dappled"; it is not a bunch of autonomous domains. Instead, it is interleaved and interweaved from multiple scales.

7.9 Causation, laws, and context

For contextual emergence there is no hard and fast distinction between causal relations and nomological relations. These are different forms of modal constraints. As we have observed throughout this book, contextual emergence also implies that causal processes and causal mechanisms are contextual (the biological examples in Chapter 6 illustrate this particularly well). Causal fundamentalism goes the way of law fundamentalism: "Fundamental" causes are those that contribute some possibly necessary conditions to a contingent set of necessary and sufficient conditions for the properties and behavior of a system, and may be present across a wide variety of contexts (e.g., gravity). Large-scale and global stability conditions play important roles defining the contexts structuring the exact form of causal action that any "fundamental" causes or mechanistic entities might contribute.

The arrow of determination and explanation is not exclusively bottom-up but multiscale and multidirectional, since any causal process is bounded by relational constraints which can be top-down, bottom-up, or side-to-side (as it were). There are no discrete causally closed or absolutely autonomous scales or domains of reality. Rather, there is a relation of mutual integration, interdependence, and reciprocal conditioning. This is why the best "causal" explanations often involve multiscale contexts.

Contexts are essential for causal relations. Without modal constraints provided by contexts, causal action is not possible. Put another way, like laws, causes must come to expression in concrete contexts. There are no context-free causes. Given that contextual features come from above and below per a particular LSD of causal connections, this means that there is *no causation without top-down constraint*, though there is more going on than just top-down influence. Hence we have used expressions such as "large-scale stability conditions," global constraints," and so forth, because the emphasis should be on multiscale interactions as stability conditions and contexts operate in all directions.[24]

All causal relations, whether thought of as counterfactuals, conserved quantity interactions, mechanisms, or difference-makers require well-defined states and observables to be the subject of causal relations. Those well-defined robust states require stability conditions for their production and maintenance. Indeed, one can think of *causation as a particular structure in the network of property compatibility relations*. A cause is a property which is placed in such a modally constrained context that the only property compatible with the cause property is the effect property.[25] This means that *causation is not a strict statement of necessitation simpliciter, but, rather, a matter of contextual constraint contributing to a contingent sufficient condition for an effect* (think Mackie's INUS conditions, 1974, p. 62). An effect does not follow its cause if other factors intervene. The effect is a property that is compatible with the causal property only in specific contexts but not others.

Therefore, while all types of explanations do differ in important ways in many cases, given contextual emergence, causal explanations, nomological explanations, and explanation in terms of adynamical or acausal constraints get a relatively unified account in the sense of all being or involving different kinds of multiscale stability conditions. This is not to say that every explanation or determination relation should be thought of as essentially causal, dynamical, and adynamical/acausal to the same degree and in the same way. Rather, we can think of all these explanations as arising from contextual accessibility of possibilities, so these different forms of explanation are not as essentially distinct from each other as the literature often treats them.

In contextual emergence, then, causation is a network of property compatibility relations, and property compatibility relations are contextual in the sense that they depend on relational structure with background properties not directly part of the causal chain. That is, those specific causal interactions must be ruled into the possibility space which, in large part, is determined by concrete stability conditions defining the context. Therefore, to say *A* causes *B* we need not only *A* and *B* but the stability conditions that make *A* and *B* viable states/observables and the modal constraints that screen off intervening features that would prevent *A* from causally connecting to *B*. That is, the properties associated with *A* are compatible with the properties associated

[24] Cartwright has expressed a similar sentiment (2015 DEP summer school and private communication).

[25] Even the property of the absence of the effect is incompatible.

with *B* only in specific contexts (e.g., Bénard cells and their influence on individual fluid elements).

Let us remember again that the world is not dappled *à la* Cartwright, nor does it contain brute bridge-laws *à la* C. D. Broad. Instead, it is a highly unified world while also being a pluralistic world in many respects. According to contextual emergence, following William James (more on this shortly), it is a mistake to treat monism and pluralism as always mutually exclusive; they go together.[26] That is, the contextual nature of reality is responsible for both its unity and plurality. It is also a mistake to think that a world is either hierarchically arranged in some *absolute* sense or it is dappled (another false forced choice).

What contextual emergence entails is that there will be multiple ways to carve up the world in terms of mereology, levels, systems, hierarchies, dynamics, domains, and so forth, depending on varying explanatory and pragmatic interests and perspectives.[27] In other words, given contextual emergence one would expect integrative pluralism (Mitchell 2002) and many other varieties of explanatory and causal emergence to be the case. We have discussed several detailed examples of this in the previous chapters. Note that it does not follow from any of this that scientific explanations are merely mind- or theory-dependent. Contextual emergence is a mind-independent fact about nature, but the world does not tell us how to represent it or dissect it for analysis.[28]

Nevertheless, note again that contextual emergence also makes it reasonable to look for theories, models, and tools that can unify phenomena horizontally across different disciplines and scales (e.g., many of the formal tools used in physics to model universality are now being used in biology and cognitive science). While the relevant stability conditions and their contextual constraints are often very different in some cases, we have provided a universal model of scientific explanation that subsumes various sorts of causal explanations. Note that nothing in our account of causation implies that causal explanations need involve manipulability as either a necessary or sufficient condition. Also note that our account of causation easily subsumes all the types of causal explanations that we have observed herein, including structural or mechanistic causal explanations as well as more formal causal analyses such as statistical and other mathematical models as given by network models and causal modeling (e.g., Pearl 2009 and others). Again, we

[26] There has been a longstanding struggle with unity and plurality at least since Parmenides and Heraclitus that still plagues us today. See Gunton (1993) for an insightful discussion.

[27] See Huttemann and Love (2011) for illustrative examples of this in biology.

[28] Some of our carvings may come closer to "nature's joints" than others, but those joints are dependent on stability conditions.

think that the metaphysical game of trying to delineate the essence of laws or causes is not very scientifically useful.

7.10 Determinism as contextually emergent

Contextual emergence highlights the role of contingent conditions. This has implications for how we think of determinism as a global thesis about the character of the universe (Bishop and beim Graben 2016). As with dispositions, forces, laws, and causation, contextual emergence implies that determinism or the absence thereof is also a contextual feature of reality. The conditions making dynamical determinism possible (e.g., Earman 1986; Bishop 2005a) are not universal or context-free. It is only by investigating contexts that we can determine whether or not concrete systems have the property of determinism.

Since the seventeenth century it has been customary to use physical determinism—the form of determinism characterizing theories in physics—as our best guide to metaphysical determinism. Physical determinism is supposed to be a property that laws of nature can have, yet if laws of nature are contextual then the property of determinism will also be contextual.

Consider Laplacean determinism, modeled after classical particle mechanics (CPM), whose key property is known as *unique evolution* (Bishop 2006a, 30–32):

> (UE) A given state is always followed (and preceded) by the same history of state transitions.

The basic idea is that given the laws of CPM, a frictionless pendulum, say, will repeat the same behavior if it is started with the same initial conditions. More precisely, the mathematical model in the same initial state (or any state in the series of state transitions) will undergo the same history of transitions from state to state. Let M denote the collection of all models sharing the same set L of physical laws, and suppose that P is the set of relevant physical properties for specifying the time evolution of a system characterized by L:

> A model $m \in M$ exhibits unique evolution iff every model $m' \in M$ isomorphic to m with respect to P undergoes the same evolution as m.

There are several conditions forming the context for UE to hold in CPM. For instance, dropping absolute space from Newtonian spacetime yields a spacetime structure that admits UE-violating symmetries consistent with the laws of motion. If the state descriptions in CPM have a many-to-one correspondence to physical states, then gauge transformations connect all the descriptions that correspond to the same physical state. It is then possible to have solutions to the equations of motion that admit arbitrary functions of time leading to violations of UE. And one can investigate the conditions leading to "space invaders" (Earman 1986, 34) and gravitational ejections of particles from the universe (Xia 1992).

The bottom line for these failures of UE in the dynamical laws of CPM is the fact that the existence and uniqueness theorems for the equations of motion only guarantee the existence and uniqueness of solutions for finite times. Therefore, the equations of motion only have unique solutions for some interval of time (Arnold 1988).[29] This interval might be short, and, as time goes on, the time interval for which such solutions exist might get shorter. This situation can be diagnosed by observing that the Lipschitz continuity condition—providing an upper bound for the spread of the distance between to initial points x and y in state space for some vector field F—is a stability condition for determinism (Bishop and beim Graben 2016).

There are other cases and theories to explore (e.g., Bishop 2006a), but the point is that UE depends on specific kinds of conditions holding. Hence, even for CPM—supposedly the archetypal deterministic theory—one must investigate on a case-by-case basis whether the conditions guaranteeing UE exist.[30] These conditions are given by stability conditions defining the relevant context; thus, determinism—and indeterminism—are contextual affairs in physics and other scientific domains. More generally, it is possible to show that stochasticity can contextually emerge from deterministic descriptions and that determinism can contextually emerge from stochastic descriptions (Bishop and beim Graben 2016).

Hence, if our best scientific accounts indicate that physical determinism is contextual, we have no evidence for any global metaphysical determinism from the sciences.[31] That determinism and indeterminism are contextual rather than absolute features of the physical world, though surprising at first glance, is a consequence of a contextually emergent world. If properties emerge from the network of relations and conditions in a specific context, it is expected that determinism/indeterminism would be a contextually emergent property of the network. Furthermore, since modalities contextually emerge, we should expect metaphysical determinism and indeterminism to be contextually emergent properties as well. Hence, we find no

[29] Something students are never told in their undergraduate physics classes!

[30] Newton did not believe that the mechanics he created and described in his famous *Principia Mathematica* was deterministic. It seems that his intuitions were correct.

[31] We also point out that arguing from unique solutions in some formal equations to the claim that a particular physical system, or the whole universe, is absolutely and metaphysically deterministic is deeply problematic. For one thing, there is no guarantee that the equations have the right relationship to their target physical systems (Bishop 2011). Moreover, in a world without metaphysical glue, reading that glue into such equations to secure a thesis about metaphysical determinism is a question-begging inference.

basis for arguing that our world is necessarily metaphysically deterministic (or indeterministic). At most, we can say that many of the systems and contexts of our experience are deterministic (or indeterministic, e.g., quantum contexts).

It is worth pointing out that just because a concrete network or context is indeterministic this does not imply that the system is without determination. The determination of the behaviors of systems and their constituents depends on the particulars of their contexts, the contingent networks of conditions and relations.

7.11 Fundamental ontology?

What we have said so far makes clear that contextual emergence implies an ontology of properties and relations, where properties themselves often are relationally determined. Ontologically, a context is a structured web of property relationships. Contexts, then, are neither mysterious nor brute; rather, they are relations among properties entering into compatibility relations that involve properties and relations that are not just in one domain, scale, or at one "level" of reality. Contextual emergence crucially involves interlevel relations among scales and nontrivial overlaps among different domains. The basic pattern is that properties are key elements of the ontology of contexts, while contexts are necessary for producing properties. The ontology implied by contextual emergence contains no metaphysical glue or other forms of reductive grounding for the properties, relations, and possibilities that are features of the actual world.

Clearly, an implication of contextual emergence is that in addition to there being no metaphysical grounding, there is no single, fundamental physical ontology in the sense of a basic set of building blocks from which everything else is constructed and fully determined in a "bottom-up" fashion. In most philosophical accounts, this "fundamental" ontology is taken to be the most elementary particles and forces at rock bottom or whatever some theory of quantum gravity claims are the ultimate building blocks. Contextual emergence rejects the foundationalist and hierarchical conception of reality, as we described in Chapter 1 and have illustrated throughout the book.

Elementary particles and forces, for example, provide some of the necessary conditions for molecular structure, while the concrete chemical context provides the rest of the necessary and sufficient conditions for the molecules that chemists explore and work with in everyday laboratory contexts. Similarly, molecular chemistry provides some of the necessary conditions for the behavior of cells, while the concrete cellular context provides the rest of the necessary and sufficient conditions for the existence and behavior of cells. Likewise, cells provide some of the necessary conditions for organs and their function, while the concrete context of the organs in an organism and their environments provide the rest of the necessary and sufficient conditions for the existence and function of organs. And so forth. This is the contextual emergence pattern.

It is equally important to keep in mind that entities, processes, and laws at any scale or level will depend contextually on stability conditions at larger scales or "higher levels," and that includes the microphysical. For instance, elementary particles only have the properties they do because of the spatial and temporal dimensionality of the world. Fermions (quarks and electrons) only have mass because of their interaction with the Higgs boson field. Without that interaction, quarks and electrons could not contribute to the existence of atoms and molecules. According to the Unruh effect, a consequence of quantum field theory (QFT), the vacuum for an accelerating observer will appear to have a temperature. In other words, if observers are in an accelerated reference frame they will detect particles, whereas in a non-accelerated frame they will not.[32]

Even particles themselves do not represent a "fundamental," autonomous entity. According to QM and QFT, matter can be a particle, field, or wave, depending on multiscale context, theory, and observation. This implies that *in concrete contexts and under specific conditions,* matter sometimes behaves like a particle, field, or wave, as discussed in Chapter 4. Furthermore, the properties of particles themselves are not intrinsic but contextual. As Colosi and Rovelli (2009) note, particles are not entities with intrinsic properties as defined by n-particle Fock space. Rather, they are defined relationally/contextually by their experimental context. So, given two different detectors (immersed in a cloud chamber or not), we can get two different "particles"—that is, two different classical trajectories defined by relational properties—even though we have the same source.

In fact, the typical notion of a particle is associated with the global particle state of n-particle Fock space and, according to Colosi and Rovelli, "the notion of global particle state is ambiguous, ill-defined, or completely impossible to define" (Colosi and Rovelli 2009, p. 14). In a laboratory setting, what we mean by "particle" is a collection of detector hits forming a space-time trajectory in the presence of colliding beams and a detector. This does not entail the existence of an object with intrinsic properties, such as mass and charge, moving through the detector to cause the hits.

That particles are best modeled by local particle states rather than n-particle Fock states computed over infinite regions, squares with the fact

[32] See Silberstein, Stuckey, and McDevitt (2018) for details.

that particle detectors are finite in size and experiments are finite in time. The advantage to this approach is that one can unambiguously define the notion of particles in curved spacetime as excitations in a local (flat) region. Moreover, this contextual conception of particles is more compatible with the quantum notion of complementary observables, in that every detector has its own Hamiltonian and therefore its own particle basis (unlike the unique basis of Fock space). According to Colosi and Rovelli: "In other words, we are in a genuine quantum-mechanical situation in which distinct particle numbers are complementary observables. Different bases that diagonalize different HR [Hamiltonian] operators have equal footing. Whether a particle exists or not depends on what I decide to measure" (Colosi and Rovelli 2009, p. 14). In "the wild," nature provides the necessary contexts for creating various types of particles.

Contextual emergence thus rejects the very idea of a "fundamental" single-scale ontology out of which everything else is composed of, constituted by, or realized by some dynamically or causally closed "fundamental" entities and properties with metaphysical autonomy. For contextual emergence, there is no absolute terminus of dependence relations whether ontological or explanatory, and that includes nomological dependence, mereological dependence, causal dependence, explanatory dependence, conceptual dependence, and the like.

We believe that the search for a fundamental, autonomous, context-free ontology is misguided. Is contextual emergence then a rejection of physicalism? It is certainly a rejection of ontological reductionism and a rejection of modal metaphysics in its entirety. It also goes without saying that the multiscale nature of contextual emergence is a rejection of causal/dynamical closure of the physical and microphysical. Nothing in everyday experience, actual scientific practice, or multiscale explanations suggests closure. Nor does science suggest a world of well-ordered hierarchical levels of organization with simple reductive compositional, constitution, and realization relations. There is thus no worry here about causal inheritance or causal over-determination.

As we have made clear throughout, *ultimate* explanation in terms of composition and realization are not the norm in physics, nor in any other sciences. For instance, as Humphreys notes, molecules are not best conceived as being *composed of atoms* when there is covalent bonding because of the spatial overlap of the wavefunctions of the shared electrons: "Quantum-mechanical calculations show that having the electron density of the outermost electrons spread out over the entire molecule, rather than being concentrated in the regions around individual atoms, decreases the overall energy and enhances the stability of the molecule" (2016, p. 84).

Aside from the wrongheaded and antiscientific hierarchical picture of reality often assumed by analytic metaphysics, we have argued throughout this book how multiple stability conditions and their multiscale contexts and constraints work together to explain the emergence of a phenomenon. Consider some simple examples of causal cooperation to produce specific effects. If a hunk of iron is dropped from a height, gravity is sufficient to bring about the effect of falling to the ground at a constant acceleration. Suppose that right after the hunk is dropped, a magnetic field of appropriate strength and orientation is switched on such that the iron hunk comes to rest suspended above the floor. Typically, in the absence of the magnetic field, gravity would be sufficient for the motion of the iron hunk. Likewise, in the absence of gravity, the magnetic field typically would be sufficient for the motion of the iron hunk. Nonetheless, the causal sufficiency of neither gravity nor the magnetic field has disappeared. Instead, the causal sufficiency of the two forces combine to produce the levitation.

Now suppose you stick your hand out and grab the iron hunk. The gravity and the magnetic field are no longer sufficient for the motion of the iron hunk. Gravity, the magnetic field, and your intentions and hand motion combine to be sufficient for the motion of the iron hunk. The sufficiency of neither gravity nor the magnetic field have been violated or overridden. And there is no causal exclusion. Rather, the sufficiency of all the causes *cooperate* to produce a different behavior of the iron hunk distinct from the behavior due to gravity or the magnetic field alone because the context has changed.[33]

Such examples are legion. Consider biological constraints, such as chirality associated with DNA, developmental processes of gene regulatory networks, and natural selection. In isolation, these constraints are each sufficient to determine several kinds of outcomes. Jointly, these factors work together to largely determine the development of DNA base-pair sequences. And as we have known for decades, genes express themselves differently in isolation than in the presence of other genes in a genome and different physical, biological, and even cultural environments: the causal effects of genes are entirely context-dependent![34] The autonomic nervous system, in the absence of intentions and desires, is sufficient to determine the behavior of arms and legs. Yet in a voting context, for instance, such processes combine with your values and desires to raise your hand enthusiastically in favor of your preferred candidate. What such examples illustrate is that the general causal structure of our world is one of causal cooperation rather than causal competition. Exclusion is the rarest of causal phenomena in the universe!

[33] This implies that causal closure has a typicality structure. More on this shortly.
[34] Recall examples from Chapter 6.

More generally, one of us has argued at length elsewhere that the supposed causal closure of physics (CoP) is not well-supported by either experience or physics (Bishop 2006b, 2010, 2012a). The crux of the argument is that CoP is an independent metaphysical assumption or idealization made by some scientists and philosophers. Physics does not imply its own closure; rather, CoP is a metaphysical doctrine, an add on. Physics tells us what happens when specific forces are accounted for, but tells us nothing about what happens when other unaccounted-for influences are present (e.g., human minds changing the conditions of an experiment). All CoP really says is that *when nonphysical or non-microphysical influences are absent (those influences outside the domain of the theory in question), physical events will proceed typically*. To state CoP's character explicitly:

> **CoP** In the absence of nonphysical or nonmicrophysical influences, the physical event $e(t_1)$ and elementary physical laws will produce the physical event $e(t_2)$ in the usual fashion, provided that the system in question is appropriately isolated during its evolution from t_1 to t_2.

That is, CoP is a *typicality condition*, not a scientifically established ontological fact. Typicality is all that scientific practice licenses, even in high-energy physics. It is one thing for a physicist to say that in some idealized model there are four forces determining the behavior of all matter; it is quite another to infer that the actual world is like that without exception based on that idealized model. Such an inductive leap requires substantial argument.[35]

Over the years, many philosophers of mind and metaphysicians have argued that CoP is well-supported by science and that the burden of proof is on those who would argue against it (e.g., Loewer 2001, pp. 152–153; Papineau 2002). As CoP is supposed to be an empirical claim, we can start by asking: what would constitute evidence of a violation of CoP? Barry Loewer gives us a clue: some deviation or gaps in the laws. We will say more on why this move is suspicious, but just consider the vagueness of the criterion. No one ever specifies what would count as a genuine violation or gap in the laws. Typically, any proposed violation or gap is dismissed as "the effect of an unknown physical cause." If one is already convinced of the truth of CoP, then that is the stock response one would have to give.

Perhaps the answer is that some clear-cut failure of intertheoretic, mereological, or mechanistic reduction would be disconfirming evidence against CoP. If so, this entire book constitutes such evidence.

Leaving aside the closure of microphysics or closure within the physical, one might wonder why the everyday fact of mental causation does not constitute evidence against CoP. Of course, for the defender of CoP, mental

[35] See Bishop (2010), Horst (2007), and Silberstein (2006).

causation is just evidence that the identity theory or functionalism must be true. And yet the utter failure to reduce intentionality, much of cognition, and conscious experience to neural processes is actually disconfirming evidence against CoP.

What of the classic chestnut argument for CoP that autonomous or efficient mental causation would violate inviolate conservation laws? Nothing in contextual emergence says that mental causation properly understood must violate conservation laws. Nor does contextual emergence entail that conservation laws, which, while often highly invariant and general, must obtain in every possible context. For various reasons, even physicists have sometimes seriously considered exceptions to conservation laws. For example, at present there is a class of solutions to the quantum-mechanical measurement problem, such as the Ghirardi–Rimini–Weber (GRW) theory of QM (Ghirardi, Rimini, and Weber 1986), that require spontaneous collapse of the wavefunction and therefore violates the conservation of energy. Conservation of energy is upheld during unitary evolution of the wavefunction, and yet fails during collapse events. One version of GRW uses nonlinear terms in the fundamental equation of motion to produce stochastic inhomogeneities in the flow of time and, thus, collapse. Energy conservation is violated when those inhomogeneities are present. The point is that conservation laws only support CoP as a typicality condition.

Another classic argument for CoP is from the four "fundamental" physical forces. The claim is that there are only four causally/dynamically efficacious forces in the universe. Thus, all efficient or ultimate causation and causal patterns in the world are the result of the interplay of said forces. Not only do we not know if there are only four basic physical forces (e.g., witness dark matter and dark energy), but the claim that basic physical forces are causally closed and sufficient to explain all other efficient causation in the world is question-begging. The argument presupposes that "fundamental" forces supply all the necessary and sufficient conditions for the existence and behavior of things, period—and this is precisely what the contextual emergence pattern in nature shows is false.

Furthermore, nothing in physics rules out the addition of new "fundamental" forces. For instance, Newton's laws of motion are general conditions constraining all motions, regardless of how they are produced, for relatively slow speeds and small masses. These laws are incomplete with respect to *how* and *why* things move. The laws of motion themselves do not place any limit on what kinds of forces can operate on bodies. The only thing required by classical mechanics is that motion produced by such extra forces must not violate Newton's laws. That is a low bar, to be sure. All this needs to be appreciated because when we are debating the possibility of mental "forces" we need to be specific about exactly which notion of force we have in mind.

Nothing in contextual emergence suggests that mental causation is force-like in the sense of classical Newtonian gravity or force-bearing particles in the standard model. There is nothing occult going on here. When physicalists attack the idea of mental causation, they often conjure up the spookiest conception of what a "mental force" might be. If only tacitly, they characterize the mind as a source of, and a sink for, energy that can be converted from material to immaterial forms (and vice versa), just as gravitational potential energy can be converted to thermal energy or mass–energy to kinetic energy. But the mind does not have to be like this to be a genuine *causal* influence in the world.

Perhaps the most compelling argument in support of a broader notion of closure, the closure of the physical conceived as including physiology and neurobiology—call it CoP′—is the argument from physiology. The argument goes like this. If there were *in-principle* explanatory and causal gaps in biological, physiological, and neurochemical processes that might be exploited or filled by irreducible mental causes, then after hundreds of years of investigation we would have discovered these violations of CoP′, and no such violations have ever been discovered (e.g., Papineau 2001; Melnyk 2003, pp. 160–161). At some point, the absence of evidence becomes evidence of absence, or so the argument goes.

However, the argument from physiology also begs the question. The interpretation involved in the kinds of theoretical and empirical research marshaled in support of the induction presuppose that only physical causal factors are effective. Physics, chemistry, biology, physiology, and neuroscience—the kinds of sciences relevant to this induction—exclude the mental as a matter of methodology. For instance, physiological research focuses exclusively on the interplay of physical factors in the behavior of organisms, and has traditionally encompassed not only general physiology but also endocrinology and neurobiology. The only kinds of hypotheses, methods, data, and analyses admitted in physiology and neurobiology—as in physics, chemistry, and biology—strictly presuppose physical causes, and, in particular, *mind-independent* causes, which is the proper focus of physiology and neurobiology. In this light, the question-begging induction in the argument from physiology is more like: "We aren't looking for it, didn't find it, so it's not there!"[36]

Again, the defender of CoP′ never stipulates what counts as evidence against it. We think that it is easy to come up with counter-examples against the argument from physiology, even if, puzzlingly, one is not persuaded by everyday mental causation. Consider Valerie Gray Hardcastle and Matthew Stewart (2002). They argue that depression, somatization, and the

[36] For more details, see Bishop (2010).

like, provide evidence that explanations for mood disorders and human behavior require irreducibly social and mental conditions and processes. Add to this the growing literature on placebo effects and conversion disorders (Silberstein 2001, 2011) and we have plenty of *prima facie* evidence against CoP.

There are several interesting cases from psychology and psychiatry where irreducible mental causation is by far the best explanation: the placebo effect (Harrington 1997), pain control under hypnosis (Wegner 2002), biofeedback (Thompson and Varela, 2001), and social psychological experiments involving perception, cognition, social context, authority, and group effects. Consider conversion disorders (CD)—conditions which start as a mental or emotional stress or crisis but transform into a variety of physical manifestations (e.g., your leg may become paralyzed after falling from a ladder even though you suffered no physical injury). There is virtually no analysis of the philosophical implications of CD in the entire philosophical literature. This is surprising, as CD brings together almost every major concern of philosophy of mind, philosophy of psychology, and philosophy of psychiatry, and has clear implications for whether physiology is causally sufficient for human behavior:

> [CD] played a great role in medical history, from Hippocrates in the 3rd century BC to the French neurologists Charcot and Janet, and finally Sigmund Freud. It was part of the scientific discovery of hypnosis and placebo effects, entangled with trauma and sexuality ... Of all neurological disorders, only this one changes the patient's consciousness with no visible damage to the nervous system. It is of course conversion hysteria, the mysterious malady that can change from day to day, making physically healthy people feel ill, leading to temporary blindness, tunnel vision, paralysis, false pregnancy, imagined pain, or numbness to touch. From 1 to 3% of neurological patients still receive this diagnosis.
>
> (Baars, SCR, September, No. 2, 2003, p. 1)

This apparent lack of current interest among philosophers is surprising for several reasons. First, in the last ten years there has been a resurgence of research into CD in cognitive neuroscience. Mostly utilizing brain imaging, the research has provided some tenuous clues into the possible neural correlates or mechanisms involved.[37] Gleaning the pathophysiology of this disorder is a challenge given the broad range of symptoms and the obvious difficulties of large-scale studies.

Second, the lack of a clear consensus model hampers the development of conceptual analysis and therapies—an invitation for philosophical contributions.

[37] Remember, as we argued in Chapter 6, that finding neural correlates does not yield reduction of the phenomenon in question.

Third, and most importantly, as Sean Spence notes about CD, "... it immediately sets up certain dualisms, dichotomies which may prove to be both inaccurate and unhelpful: is it a disorder of body (or brain) or mind, one that is organic or 'psychogenic', or conscious or unconscious" (2009, p. 268). Of course, CD also raises questions about whether the agent's will is being thwarted in some way, given the failure of the body to follow some command, or whether in fact the agent's will is being carried out because, on some level, they desire the deficit. This strikes us as another false forced choice fallacy. All this makes CD almost unique in subsuming philosophical concerns pertaining to mind–body and related sciences.

CD is often defined as a motor or sensory symptom that has no detectable organic basis or demonstrable physical cause, sometimes leading people to ask how a physical symptom can be produced without an efficient physical cause. As we shall see, this characterization borders on the misleading. Indeed, the term "conversion disorder" is taken from Sigmund Freud's belief that the "hysterical" unconscious mind *converts* some psychic trauma into a physical malady for some hidden psychological motive unknown to the conscious mind. There has always been a question as to whether CD is truly psychogenic or an organic brain illness, as Jean-Martin Charcot suggested *contra* Freud. This is just another instance of the long history of tension in psychology and psychiatry between mechanistic biological accounts of explanation/understanding and hermeneutical accounts (Bishop 2007; Richardson, Guignon, and Fowers 1999).

In contrast to the reductionism presupposed by CoP′, explaining CD fully requires ineliminable and irreducible reference to multiple interacting scales as follows: 1) codetermining brain "mechanisms," or rather their failure, 2) attention, 3) beliefs, desires, emotions, 4) social context, and the 5) complex nonlinear interconnections among all these. CD shows that ignoring any of these features and their nonlinear connections yields impoverished explanations. CD is only something that could happen to a conscious being with beliefs and values embedded in a social environment that provides societal norms, expectations, and values. When it comes to much of human behavior generally, and treating aberrant and abnormal behavior specifically, it therefore will be impossible to ignore or eliminate historical aspects and relationships of the patient as well as the psychosocial etiology of the condition in question.

Let us begin with a summary of clinical and cognitive neuroscience data regarding CD. Well-known conversion symptoms include "hysterical" blindness, paralysis, "anesthesia," dysphagia, gait disturbance, "seizure," localized weakness, aphonia, difficulty in swallowing, urinary retention, deafness, double vision, and (everybody's favorite) hallucinations. To be more clinical, all the following is summarized in Spence (2009, chapter 7) and includes key criteria from DSM IV (300.11) for CD[38]:

1. Acute onset typically follows some emotional or life crisis, and there is no detectable organic explanation.
2. Biologically implausible symptoms not consistent with any known organic cause for the same symptoms (e.g., tremors come and go but are not consistent with any known organic disease). For example, unlike Parkinson's disease, in which tremors have multiple oscillator systems that oscillate at different frequencies, CD tremors depend on single oscillators such that they all oscillate at the same frequency and disappear when the patient is instructed to tap out rhythms with their contralateral limbs—all of which suggests that CD symptoms behave more like voluntary behavior in some ways.
3. Symptoms not reflective of neurological reality but rather manifest according to patient's knowledge or cultural conception of relevant neurology. For instance, if the symptom is an abnormality of gate, the form of locomotion in question is not consistent with the implied deficit. That is, conversion symptoms typically do not conform to known anatomical pathways and physiological mechanisms, but, rather, follow the patient's conceptualization of a condition.
4. Patients deny authorship or agency of symptoms, and deny any desire to manifest them (e.g., really believe they are paralyzed).
5. Symptoms appear, manifest variably, or disappear altogether, depending on social context and specific individuals present, illustrating the irreducibly social and interpersonal character of CD.
6. When not attending to the deficit in question because of distraction, or when attention is directed elsewhere, say via sedation, normal function shows itself in some fashion, making it clear that the patient has the potential at any given time for normal behavior and function, as well as showing that attention is central to the maintenance of the abnormal effect.
7. CD is often correlated with other diagnoses such as depression, anxiety, or personality disorders.

The fact that CD symptoms behave more like voluntary behaviors, and require attention, suggests where researchers might look for neural correlates or mechanisms. That is, one might think that these abnormal behaviors are "controlled" by "higher" brain centers (prefrontal executive systems, supervisory attentional systems, etc.) that are involved with voluntary or

[38] Of course, there is a DSM V that retains most of the essential points, but we cite IV to remain consistent with Spence.

volitional behavior. This is the assumption that was made in what few imaging studies exist of CD. The results are summarized here later. In the 1960s, electrophysiological studies found a central corticofugal inhibition of afferent stimuli responsible for hysterical sensory loss. These evoked potentials showed abnormalities that disappeared after resolution of the symptoms. In the 1990s, more evidence of central inhibition came from a single-photon emission computed tomography study in a woman with sensorimotor hemi-syndrome, which demonstrated hypoperfusion in the contralateral parietal region and increased perfusion in the frontal region. This pattern was supported by a positron emission tomography study of a patient with hemi-paresis, where once again increased activity was found in the frontal region with activation of the anterior cingulate and orbitofrontal cortex. The authors suggested that frontal regions were inhibiting the motor and premotor areas when the patient tried to move the affected limb.

Patrik Vuilleumier and colleagues (2001) performed a relatively large sample single-photon emission computed tomography study on seven patients with sensorimotor conversion disorder, when they were symptomatic, and when some of them were asymptomatic. They found reduced blood flow in the subcortical structures (contralateral thalamus, putamen, and caudate) when a vibration stimulus (which normally activates both sensory and motor cortices) was applied to the affected limb. The pattern normalized in patients who subsequently recovered. This study also suggests central inhibitor mechanisms, involving not only cortical areas but also corticosubcortical circuits. In summary:

1. Where sensory deficits exist, such as "hysterical blindness," there is correlated hypoactivation of the relevant sensory cortices, and evidence of abnormal function in the "higher" frontal regions.
2. When the dysfunction is related to motor disturbance, there is evidence of excessive activation of inhibitory prefrontal (orbitofrontal) cortices, or a suppression of activity in prefrontal (dorsolateral) cortices involved in the generation of action.
3. In short, the consensus is that "higher" brain centers somehow inhibit or suppress "lower" brain centers.
4. Attempting to use brain imaging to distinguish between "feigners" (voluntary) and those with CD (relatively involuntary), one study found that patients with CD exhibited reduced prefrontal activation in the left dorsolateral and ventrolateral prefrontal regions, whereas feigners had reduced activity in the right prefrontal regions.

Since CD symptoms are highly correlated with the aforementioned neural correlates and the abating of such symptoms is highly correlated with the disappearance of the said abnormal neural correlates, there is no reason to say that CD abnormalities involve no mechanistic physical cause. Nevertheless, CD provides us with a case where we have an efficient causal mechanism or at least neural correlates, yet there is no mechanistic, organic, or medical explanation *in terms of the neural correlates in question*. Take the following summary of the meta-analysis of neuroimaging data as it applies to MCD (conversion disorder related to motor control). The authors conducted a systematic literature review on CD and carried out a meta-analysis of functional neuroimaging studies on MCD. The goal was to discern differences between patients and healthy controls, between affected versus unaffected sides, and an overall analysis to identify neuronal areas related to MCD:

> Patients with MCD differ from healthy controls in the amygdala, superior temporal lobe, retrosplenial area, primary motor cortex, insula, red nucleus, thalamus, anterior as well as dorsolateral prefrontal and frontal cortex. When comparing affected versus unaffected sides, temporal cortex, dorsal anterior cingulate cortex, supramarginal gyrus, dorsal temporal lobe, anterior insula, primary somatosensory cortex, superior frontal gyrus and anterior prefrontal as well as frontal cortex show significant differences. Our results show that emotional, motor planning, and inhibitory processes are involved in MCD. Instead of single miss-functioning of a specific neuroanatomical area, a complete network of areas seems to influence the presentation of MCD symptoms. Patients with MCD seem to primarily differentiate from healthy controls in the frontal and prefrontal cortices, ACC, and amygdala relevant for motor-planning and selection, intentional behaviour, volition, and autonomic responses. This effect, as well as the results from all included experiments, is similar to the emotional unawareness theory of Perez and colleagues, which states that the large-scale brain network mediates emotional and cognitive mechanisms and is modulated by experience-dependent neuroplasticity.
>
> (Boeckle et al. 2016, p. 1)

In fact, we have no reason to believe that the relevant neurological changes in the brain are caused in some bottom-up neurological fashion. There is no evidence of genetic or neural malady as such behind CD symptoms. Indeed, there is every reason to believe that CD symptoms are typically triggered by psychosocial crises and causes, while their abatement is due to changes in the psychosocial context.

For pragmatic reasons, psychiatric and neurosciences and medicine are often concerned with explanation as intervention and manipulation, and therefore it often makes sense to look for some localized mechanism that one can treat as an independent variable for the purposes of intervention. For instance, this is the *modus operandi* of psychopharmacology, which is much less efficacious than the public appreciates. But one should not assume that such interventions therefore get at *the cause* of the phenomena in question. In the case of CD, manipulating social contexts, interpersonal settings, beliefs, and emotions have profound and often immediate effects. For example, CD symptoms of hysterical paralysis have been relieved with a placebo (Rommelfanger, 2012).

Certainly, where CD is concerned, we ought to be multiscale-explanatory and causal pluralists. It seems clear that explaining CD involves emotions, desires, beliefs, social contexts, neural correlates, attention, and more—it is a malady of body, mind, and brain in a social context. CD is both organic and psychogenic. It involves both conscious attention and subconscious stressors that one should not be afraid to invoke for fear of a slippery slope to Freudianism.[39] In the case of CD, the patient has no idea that there is a voluntary and attentional component to maintaining their symptoms. In short, CD is best explained from a biopsychosocial and 4E cognition perspective.

As stated previously, there have been some attempts to uncover the causal mechanisms (or the lack thereof) of a CD involving (one-sided) paralysis in one arm. The study involved seven patients with the same symptomology. All the patients had recent onset of their symptoms, and most recovered within a few months (normally recovering from local nerve damage is very rare). As revealed by MRI scans, no physical basis could be found for either their symptoms or their recovery. All the patients in question were facing stressful life events when their paralytic symptoms occurred. None of the patients had major psychiatric or neurological disorders. Each subject served as his or her own control, in that brain scans were used to compare the left and right (healthy and unhealthy) sides of the same person, including their left and right hemispheres. In each subject, brain images were taken with vibrations applied to both arms. Given that hands and arms are connected to the opposite sides of the brain, the tested side was in the "contralateral" hemisphere. The patients were scanned during bouts of hysterical paralysis as well as when they were no longer paralyzed or plagued by symptoms.

What were the results? *No differences were found between the two hemispheres of the seven subjects at the level of the cortex*. However, unusually low bloodflow was found in the thalamus (the gateway to the cortex) and the basal ganglia. Voluntary commands to move the affected limb were blocked from activating the pathways that control the paralyzed arm either via blockage in the thalamus or the basal ganglia which involves automatic muscle control. The CD subjects could feel (in the cortex, so to speak) that they were trying very hard to move their arm; nevertheless, all their efforts were blocked in the output pathway to the muscles. Control of the paralyzed arm was blocked not in the cortex itself, but in the cortical gateway (the thalamus),

[39] Upon reflection, is there really any denying that in some sense our behavior or mental states are sometimes a function of some subconscious machinations?

and in the basal ganglia which partly control unconscious details of movement (Vuilleumier et al. 2001, pp. 1077–1090). What more positive evidence could you want for mental causation than this?

The defender of CoP′ will surely remind us here that, nonetheless, the brain scans did reveal unusually low blood-flow in the thalamus and the basal ganglia and, therefore, even in such cases we have a purely physiological causal mechanism at work. The CoP′ proponent will say that had the brain scans revealed absolutely no physiological causal mechanisms at work anywhere in the brain or body, then that would start to look like evidence for irreducible mental causation.

In reply, notice that the CoP′ defender's response presupposes a radical form of emergence, where there is no connection whatsoever between the emergent phenomenon and some physical–causal mechanism. In other words, the defenders of CoP′ help themselves to an absurdly strong criterion for positive evidence of mental causation. Even substance dualists, such as Descartes, assumed that mental causal forces must bring about some physiological change in the brain to alter physiological and mental states. The mental will do this by "initiating" physiological causal mechanisms or "initiating" physical causal chains such as low blood-flow in the thalamus. In short, there is no reason to assume that whenever some efficient causal mechanism, such as low blood-flow, can be pinpointed that it *excludes* mental states from being a part of the causal chain or process.[40] The fact that mental causation always involves physiological causal mechanisms ought to render it more respectable not epiphenomenal.

Equally important is the fact that the causal mechanisms at work in the case of CD, such as low blood-flow in the thalamus, are always extremely subtle (hard to detect and at relatively small scales). In CD there is never any organic or neurological explanation for their appearance or their effects. For instance, the partial paralysis patients did not have low blood pressure nor any other physical disease or condition that explained the low blood-flow. And this would remain true even if, in principle, you could trace some physical causal chain back that led to the low blood-flow. The point here is that the CoP′ defender really does not have any alternative physiological (i.e., nonmental) explanation for the CD effects, even *if* they have what *appears* to be a sufficient, closed, and complete physiological causal chain. How did that causal chain get there? How did it start? The best explanation leads back to the mental stress of the CD patient, and in the conditions in the wider social environment that brought about that *emotional–intentional* stress. Finally,

[40] That is, unless one assumes causal exclusion. But with CoP being only a typicality condition, and the actual evidence for ontological reductionism weak, any basis for causal exclusion is also weak (e.g., Bishop 2012a).

there is no common organic cause that explains both the mental stress and the low blood-flow.

The CoP′ defender will surely ask: *how* do the *phenomenal–intentional features* of a person "initiate" a physiological chain or causal mechanism? When they ask *how* the mental can do this, they are surely asking: *in virtue of what causal mechanism* can the mental bring about low blood-flow in the thalamus or some physiological causal chain that does bring it about? This question assumes that the mental acts on the physiological in the same way that particles act on one another in Newtonian mechanics; that is, they are assuming that to influence the physical, the mental must operate with efficient causal–mechanical mechanisms. There is no reason to make this assumption unless you are convinced that everything is physical or causal–mechanical. If agents and agency are emergent in the sense of contextual emergence consistent with the rest of the physical, chemical, biological, neural, and cognitive phenomena in the universe, mental-to-physical causation can be some other kind of cause than efficient causation, and therefore the word "initiate" is probably misleading here. In other words, the states, broadly construed, of embodied, embedded, extended, and enactive contextually emergent agents at time t_1 cause, broadly construed, the states of such agents at time t_2. For instance, your anxiety over what the results may show can alter your blood pressure and pulse checked at your physical examination.

Perhaps, then, such causation is more like Aristotle's formal or final causation, which is not exclusively mental. Such a broadening of the notion of causation or influence makes much better sense of the social and behavioral sciences than does restricting ourselves to just efficient causation (Bishop 2007 and 2022; Richardson, Fowers, and Guignon 1999; Taylor 1985a, 1985b). Maybe this seems mysterious, but it is always going to be mysterious if one merely assumes that only efficient mechanical–causal mechanisms can act in the world (i.e., assumes the hidden premise that only physical causes are effective); this is just to assume CoP′! CoP′ as a typicality condition is consistent with considering such alternatives and trying to figure out how to characterize and explore them rigorously.

What CD and the like indicate is that *somehow* an individual's beliefs, desires, emotions, fears, values, conception of what reality is like—in their concrete social context—get "expressed" or "implemented" neurophysiologically and not (always) the other way around. We do not claim to currently have answers to all the empirical questions about mental causation. But at least we can propose a framework that can recognize the reality of mental causation and begin to help account for it (e.g., Allefeld, Atmanspacher, and Wacherman 2009; Harbecke and Atmanspacher 2012). Viewing the individual and their complete environment (self-interpretive, diachronic social, physical, and evolutionary environment) as *one* physical, social, functional, and teleological whole, goes a long way towards making final and formal (mental) causation seem more natural and plausible compared to just relying on efficient material causation alone. It is not *brains* that possess phenomenal–intentional experiences and states, such as CD; it is embodied, embedded, extended, and enactive persons.

So, yes, contextual emergence is a rejection of physicalism. However, it is not a rejection of monism in the broadest possible sense. Indeed, this is part of the point we are trying to make in our discussion of CD. One of us has argued at length that contextual emergence is best interpreted as a kind of neutral monism (Silberstein 2019; Silberstein and Stuckey 2020; Silberstein, Stuckey, and McDevitt 2020). We do not have space to discuss neutral monism at great length, nor the hard problem of consciousness to which it most relates. Suffice it to say that while the hard problem or "generation problem" is often a driving force in discussions of emergence, that has explicitly and intentionally not been the case in this book, for two reasons. First, we do not think that the weirdest and most contentious case should drive the discussion of emergence in the sciences. Second, we simply do not believe that experience or subjectivity *emerges* from some *essentially* nonsubjective material substance.[41] This is *not* a plea for panpsychism, mind you, which we regard as even more odious than pure physicalism. It is a suggestion for neutral monism *à la* William James and Bertrand Russell. We will discuss this briefly, because people often ask us how contextual emergence bears on the mind–body problem or hard problem.

7.12 Neutral monism

The basic idea of neutral monism is that the mental and physical are nondual (not essentially different) and are neutral in virtue of being neither essentially mental nor physical, and in virtue of the fact that they are grounded in something neutral. This ubiquitous "something" does not "pervade" spacetime nor "generate" spacetime, because it is co-extensional with spacetime. According to neutral monism, our conscious experience of an external world is not some virtual model or construction of the world trapped in the mind and generated by the brain, while the actual external physical world lies outside us forever as I-know-not-what noumena. In neutral monism there is

[41] Early in his career on the job market, Bishop had a standard answer to interview questions about why he was interested in reduction and emergence as a philosopher of science. His answer was that if the conceptions of reduction and emergence in the philosophy of mind literature do not work in physics and chemistry, they are not worth much in philosophy of mind and action. This book represents an extended argument for an ontological version of emergence that works across the sciences, so has plausibility for helping us with issues in the philosophy of mind and action.

only the spatiotemporally extended world of experience, of which we are a part.

James says that neutral monism is "a radically pluralistic monism of pure experience" (Taylor 1996, p. xiv). His poorly chosen expression "pure experience," however, is not meant to invoke sense-data theory or idealism. Rather, it is his expression for what Russell calls "events." All that exists is pure experience, and pure experience is all that exists, according to James. This is what makes James' view monistic. But it is also "radically pluralistic" in that "pure experience" is infinitely varied in its nature. It is, as James says, simply "made of that, of just what appears, of space, of intensity, of flatness, of browness, of heaviness, or what not." It is "the instant field of the present ... plain, unqualified actuality or existence, a simple that" (1996, p. xv). Events, such as everyday experiences of flat tires, concerts, and conversations, which are neither essentially physical nor mental, are fundamental and exhaustive.

As James puts it: "Subjects" knowing, "things" known, are "roles" played, not "ontological" facts (James 1905, p. 110). From the perspective of neutral monism, the claim that the world is carved at the joints as physical/mental, inner/outer, of subject/object, is not a datum, but rather a false inductive projection. As he puts it: "Subjectivity and objectivity are affairs not of what an experience is aboriginally made of, but of its classification" (1905, p. 1208). So-called "inner" experience is not inherently or essentially mental, and the so-called "external" world is not inherently non-mental or physical. "Pure experience," as James calls it, in itself "is no more inner than outer ... It becomes inner by belonging to an inner, it becomes outer by belonging to an outer, world" (1988, p. 217).

For James, the explanation for why we experience the world as divided essentially into inner-subjective-mental and outer-external-physical, is going to be about various constraints. Consider Kant, for instance. Given his unity of apperception, time is an *a priori* condition for experience with no subjectivity, no time or space. Kant, here, is providing a transcendental analysis in mentalistic terms. This means that the dynamical character of thought/experience and the world are two sides of the same coin. Kant's transcendental arguments in *The Critique of Pure Reason* are supposed to show that we must conceive of the world in a specific way, structured internally according to categories such as time, space, and causation. Those arguments are fraught with many interpretative perils and controversies, to be sure, but the basic idea is that experience is possible for you only if some experiences are conceptualized as being of enduring objects, enduring through time and space. Likewise, to experience a world of enduring objects there must be some sense of an enduring self. You cannot have one without the other (2013, chapter 24). This is an example of a constraint-based explanation for the very possibility of, and most basic features of, subjectivity. You can think of Kant as giving his version of the stability conditions required for subjectivity.

Kant says that time cannot be perceived in itself, yet one cannot perceive events without spatial arrangements and temporal coordination. Time and space are necessary *a priori* conditions of the possibility of experience. As he puts it, space and time are pure forms of intuition. Space is the form of outer sense, time the form of the inner sense. This is to say that in defining the notion of an object or thing one must include the concept of time and space, and vice versa. The concepts of space, time, and thing cannot be defined in isolation without bringing in the other concepts—they are only codefinable. Likewise, the idea of an enduring subject or perceiver can only be codefined with respect to an enduring world or object that is being perceived, and vice versa.

One can view James' account of experience as Kant-like minus the internalist, mentalist cognitivism. Here is James' ultimate statement about Kant's transcendental idealism and its relation to radical empiricism (James 1904, p. 535):

> If neo-Kantianism has expelled earlier forms of dualism, we shall have expelled all forms if we are able to expel neo-Kantian in its turn. For the thinkers I call neo-Kantian, the word "consciousness" today does no more than signalize the fact that experience is indefeasibly dualistic in structure. It means that not subject, not object, but object-plus-subject is the minimum that can actually be. The subject–object distinction meanwhile is entirely different from that between mind and matter, from that between body and soul. Souls were detachable, had separate destinies; things could happen to them. To consciousness as such nothing can happen, for, time itself, it is only a witness of happenings in time, in which it plays no part. It is, in a word, but the logical correlative of "content" in an experience of which the peculiarity is that fact comes to light in it, that awareness of content takes place. Consciousness as such is entirely impersonal—"self" and its activities belong to the content. To say that I am self-conscious, or conscious of putting forth volition, means only that certain contents, for which "self" and "effort of will" are the names, are not without witness as they occur. Thus, for these belated drinkers at the Kantian spring, we should have to admit consciousness as an "epistemological" necessity, even if we had no direct evidence of its being there.

James makes it explicit that he thinks that Kant is on the right track, but he wants to be rid of the inherent dualisms in Kant such as that between phenomena/noumena or consciousness/content. That is what radical empiricism (neutral monism) is all about. As James puts it: "... that the only things that shall be debatable among philosophers shall be things definable in terms drawn from experience" (1909, p. xii). He is rejecting any reified notion of consciousness or self as existing apart from the content of experience. He maintains that the reification of self arises only when some "bit" is abstracted from the flow of experience and reflectively considered in a different context from that flow, a context of relations external to the "bit" separated out, but internal to the flow of experience as a whole. For James, "the dualism of knower and known is an external dualism of experienced relations not an inner dualism of substance" (Taylor 1996, p. xvi).

Kant and the others are right that we do not experience things in time and space; rather, we experience them temporally and spatially. But given neutral monism they are wrong to say that this is an imposition of individual minds and their categories upon some unknowable noumenal world (the thing-in-itself). The mistake, in one form or another, which we find in both the analytic and continental tradition, is *representationalism* (e.g., Bishop 2007; Taylor 1995, 2002). Once we take neutral monism on board, we can immediately see that there is no need for (and no sense for) representationalism with a capital R, to explain the experience of enduring subjects and objects in space or time. The point is that subject and object coexist as a subject in a world in space and time, so the agent is not trapped behind "a veil of perception" but is directly part of the world, and the external world is not some noumenal container onto which the subject projects a virtual reality. Given neutral monism (transcendental), phenomenology cannot be, and should not be, divorced from natural science, and experience cannot be separated or bracketed from the natural world, as there is no primary/secondary distinction to be upheld.

Following Kant's unity of apperception, which the reader will notice is a principle, constraint-based explanation itself, we can go a step further. It is only when the subject/object division exists that one gets a world in space and time. So subject/object and world in space/time are both necessary and sufficient for one another.

James' answer to an objection as to the impossibility of thoughts and things being nondual will help us see that he agrees with Kant about the relationship between time, space, objecthood, and subject (James 1904, p. 489):

> If it be the self-same piece of pure experience, taken twice over, that serves now as thought and now as thing—so the objection runs—how comes it that its attributes should differ so fundamentally in the two takings. As thing, the experience is extended; as though, it occupies no space or place. As thing it is red, hard, heavy; but who ever heard of a red, hard or heavy thought? Yet even now you said that an experience is made of just what appears, and what appears is just such adjectives. How can the one experience in its thing-function be made of them, consist of them, carry them as its own attributes, while in its thought-function it disowns them and attributes them elsewhere ... To begin with, are thought and thing as heterogenous as is commonly said? Their relations to time

> are identical ... Sensations and apperceptive ideas fuse here so intimately that you can no more tell where one begins and the other ends.

James' neo-Kantian point is that what we call the "external" flow of time, so-called "things" changing, and the "internal" flow of time, so-called "thoughts" and "feelings" changing, is one and the same experience. We do not need any Kantian categories to explain this experience of change. As James puts it: "The 'I think' which Kant said must be able to accompany all my objects, is the 'I breathe' which actually does accompany them ... That entity [consciousness] is a fictitious, while thoughts in the concrete are fully real. But thoughts in the concrete are made of the same stuff as things are" (1904, p. 543). As he says, "the parts of experience hold together from next to next by relations that are themselves parts of experience" (1904, p. 534). No internal cognitive filters are required. As he says: "According to radical empiricism, experience as a whole wears the form of a process in time" (1904, p. 540). Of course, James is not denying that there are real and important differences between thoughts and things. You can lift the latter but not the former.

At least for a time, Russell held a similar view to that of James. What we call spacetime is nothing but the events therein, and those events are neither inherently mental nor inherently physical. Russell calls such "neutral" 4D finite events "unstructured occurrences," such as "hearing a tyre burst, or smelling a rotten egg, or feeling the coldness of a frog" (1927, p. 287). Contrast this with panpsychism (which Russell also held at one point in time), in which what is fundamental is both inherently physical and mental. Russell goes on to say: "Matter and motion ... are logical constructions using events as their material, and events are therefore something quite different from matter in motion" (1927, p. 292).

Note that for Russell, events are not substances, things, bare particulars, or particles—they are fundamental reality. As he puts it: "It is a mere linguistic convenience to regard a group of events as states of a 'thing', or 'substance', or 'piece of matter', or a 'precept'" (1927, p. 284). Going further, he says that "electrons and protons ... are not the stuff of the physical world" (1927, p. 386). And again, "bits of matter are not among the bricks out of which the world is built. The bricks are events, bits of matter are portions of the structure to which we find it convenient to give separate attention" (1956, p. 329). As he puts it: "The whole duality of mind and matter ... is a mistake; there is only one kind of stuff out of which the world is made, and this stuff is called mental in one arrangement, physical in the other" (1913, p. 15). Compare this to the words of James: "Things and thoughts are not fundamentally heterogeneous; they are made of one and the same stuff, stuff which cannot be defined as such but only experienced; and which one can call, if one wishes, the stuff of experience in general ... 'Subjects' knowing, 'things' known, are 'roles' played, not 'ontological' facts" (1905, p. 63).

The point is that when Galileo, Robert Boyle, John Locke, and others insisted on the primary/secondary distinction or any other kind of dualism of inner mental experience and external physical world, they doomed us to the mind–body problem and the hard problem. Such dualisms are not empirical data that we experience directly; they are cognitive illusions, inductive projections of theorizing minds. The way to undo this mistake is not to merely move the location of the mysterious dualism—qualia and subjects—from brains (strong emergence) to fundamental physics (panpsychism), but to reject completely and thoroughly the primary/secondary property distinction with neutral monism.

Contextual emergence does not entail neutral monism or vice versa, but they are consistent with each other. First, contextual emergence as a thoroughly relational view eschews a hierarchical and foundationalist conception of reality and implies that matter has no intrinsic material essence (e.g., through the Higgs mechanism, mass is relational). Jamesian and Russellian "events" are every bit as basic or "fundamental" in a contextually emergent world as anything else. Second, neutral monism also rejects a hierarchical and foundationalist conception of reality in favor of a "radical pluralism." The relationality of contextual emergence offers a plausible ontology for neutral monism. Third, the combination of contextual emergence and neutral monism completely deflates the hard problem and the generation problem. Fourth, neutral monism rejects the primary/secondary distinction as being basic, leading to a contextual view of the properties in the world. That is, both "physical" and "mental" properties are nondual and equally contextual. Fifth, contextual emergence and neutral monism both deflate CoP in all its forms. Sixth, contextual emergence and neutral monism reject both reductionism and dualism in favor of a pluralistic unity. Seventh, both contextual emergence and neutral monism allow us to be sophisticated commonsense realists, while giving up the odious and ever problematic idea that there is a noumenal reality hiding behind appearances. We can give up the search for some "beables" that will magically make sense of everything.

Taking all this in, we are just left with all the scientific questions of how and why we experience things as we do—the contents of experience, basic and otherwise. As we alluded to in Chapter 6, this will no doubt involve contextual emergence and integrative pluralism. The cognitive neuroscience of subjective experience will be very important here, but it will be only one of many sciences involved. And there will be no need to continue the doomed quest for neural correlates of consciousness conceived as minimally sufficient conditions, a first step to a purely *neuro-mechanistic explanation* of subjective experience that we have yet to achieve. Note that the most promising neural correlates (in the ordinary scientific sense of various causally and dynamically relevant explanatory observables) are large-scale or global properties such as the different kinds of networks discussed in Chapter 6.

The bottom line is that we do not think that subjective experience is some sort of qualia-like entity or state that jumped into being fully formed at some point in the history of the universe from an essentially non-mental material substance as the result of strong or radical emergence. Experience evolved and became more varied and complex over time as a function of the multi-scale contexts that have developed over time reflected in biological evolution and cultural diversity. We obviously do not know at exactly what point in the unfolding chain of being experience of some sort appeared. Given neutral monism, we speculate along with others that even the most minimal experience, however nonhuman-like, requires some sort of physical or biological embodiment. As cognitive neuroscientist Marc Wittman points out: "Temporal experience, self-consciousness, and the perception of bodily states and feelings are tightly bound to each other; they cannot be experienced separately" (2016, p. 135). Also: "Research on consciousness inevitably shows that our concepts of self, time, and body are interrelated. Presence means becoming aware of a physical and psychic self that is temporally extended. To be self-conscious is to recognize oneself as something that persists through time and is embodied" (p. 104).

This is very much in keeping with James' own neutral monism:

> The individualized self, which I believe to be the only thing properly called self, is a part of the content of the world experienced. The world experienced (otherwise called the "field of consciousness") comes at all times with our body as its centre, centre of vision, centre of action, centre of interest. Where the body is "here," where the body acts is "now;" what the body touches is "this;" all other things are "there," and "then" and "that."
>
> (1912, p. 170)

The point that James and Wittmann are making, in part, is that being embodied agents/subjects in an environment is a concrete instantiation of the phenomenology of breaking the world into subsystems consisting of subject and object. As philosopher Jennifer Windt notes, embodiment or spatiotemporal localization of one sort or another is a necessary constraint—part of the explanation for how and why the subject/object partition discussed by Kant and James exists:

> The experience of being in a world only makes sense if we assume that someone, a phenomenal self, is experienced as being located in this world. Phenomenal selfhood requires the experience of a self-world boundary; phenomenal selves are always situated selves, located at some point in the experienced world and distinguished from other persons or objects in it. They are always firmly located

> in the present and bear a particular relation to remembered past and anticipated future selves ... Spatiotemporal situatedness thus is a plausible candidate not only for identifying the phenomenal core of dreaming but also for understanding conscious experience and phenomenal selfhood as such.
>
> (Windt 2015, p. 522)

Contextual constraint-based accounts of experience are found widely in philosophy from Hinduism, Buddhism, and Augustine, to phenomenology and philosophical hermeneutics.[42] One major advantage to neutral monism is that we can finally jettison the odious assumption made by many in the field of neuroscience and cognitive science that experience, or subjectivity, is irrelevant to explaining the cognitive abilities of organisms, and vice versa.

We do not discuss the emergence of life in this book. We do not know what the earliest most minimal example of life was evolutionarily. We do not know if life in one form or another emerged more than once on Earth. And we do not know which of the many theories of life's emergence is true. We do note that every account of the emergence of life involves a spatiotemporally extended multiscale process that potentially involves changes in Earth's orbit, atmospheric conditions, and environmental conditions, or likely all three in concert. In short, the best account for the emergence of life will exhibit the contextual emergence pattern. The emergence of DNA, RNA, proteins, and whatever else counts as life could have happened in more than one way. But no matter how it happened, it involved a long multiscale evolutionary process, wherein more and more stable complex systems got iterated into even greater systems of complexity given relevant stability conditions defining contexts rich in constraints.

By now we are all aware that for millions of years the atmosphere, meteorological patterns, the physical environment, geological patterns, and life have all evolved and changed together as a complex nonlinear system. We are also all aware that animals and other organisms and their environments shape one another profoundly over time, on evolutionary, developmental, and other time scales. Thus, whether we are talking about the emergence of life, mind, or culture, the essential story is going to involve the contextual emergence pattern, not strong or radical emergence, not modal magic, and not any sort of emergence that defies scientific explanation. Given contextual emergence, the existence of life and mind, and their ever-growing complexity, are not improbable or surprising. Indeed, it was ontological reductionism and physicalism that made these "hard" problems.

[42] We note that several people have been pursuing contextual constraint-based accounts combined with science recently (e.g., Godfrey-Smith 2020; Rovelli 2021).

7.13 Objections and frequently asked questions

In the remainder of this chapter we will address the two most frequent concerns we get when presenting contextual emergence to an audience. One is an objection to our ontological claim that the relational and contextual nature of reality at all scales is inherent and ineliminable, even for microphysics—the idea here being that while multiscale contexts might be explanatorily essential in science, in principle every stability condition and context can themselves be reduced to microphysics at the smallest scale in a way that is context free. The other asks for clarification as to how contextual emergence is different from Humphreys' and others' "transformational emergence," and likewise how distinct is contextual emergence from OSR.

7.13.1 *Ontology or epistemology?*

The first concern is essentially that while the picture we have painted of the scientific enterprise may be spot on, it suggests only epistemic emergence and nonreductive physicalism. That is, the objector grants that science often needs the multiscale stability conditions and their contexts, and the higher-level theories or concepts to explain things, but this is all compatible with ontological reduction in principle. The failure to produce reductive explanations across the board is a failure in practice, not in principle, and thus contextual emergence is just epistemic emergence or nonreductive physicalism, which allows for lots of "in practice" failures of reductionism.

The key to making this objection stick is to argue either that in every case, what we call the contextual features (e.g., stability conditions) could themselves be reduced in principle in some fashion to physics at the smallest scale, or at least treated as nothing more than initial conditions, boundary conditions, or background conditions. The underlying idea here is that everything, however concrete or abstract, will be reducible in principle to entities or laws in the "fundamental" physics of the smallest scale. The claim here is that everything (e.g., crystals, human bodies, cookies, ecosystems, monetary currencies, etc.) are nothing but microphysical states of affairs whose behavior is fully determined by properties and the various laws governing them at the smallest scale. In one version of this objection, all larger-scale or higher-level contextual factors will somehow, at least in principle, wash out at the end as nothing but a function of ignorance or epistemic emergence (let us call this strictly vertical reductionism).

In another version of the objection it is acknowledged that while larger-scale or higher-level contextual factors are here to stay at all scales, that does not rule out some "fundamental" theory, such as a theory of quantum gravity, that while inherently contextual could still, functionally speaking, be to all higher-level or larger-scale phenomena what the rules and cells in Conway's Game of Life are to all the "real patterns" in the game such as Gliders[43] (call it vertical plus horizontal reductionism).

In both proposed versions of reductionism, unlike the Game of Life with its simple cells and rules, the relationship between "actual patterns" in the "actual" world and the "fundamental" theory (whatever it is) will be much more complex. In vertical plus horizontal reductionism, for instance, the "fundamental" theory can admit of all kinds of stability conditions defining contexts and their constraints, so long as it is all within the realm of "fundamental" physics. Indeed, perhaps the "fundamental" physical theory will explain or ground the need for stability conditions and contexts at other scales.

Thus, implicit in both versions of the objection is the assumption that the most basic physical entities and laws (whatever they turn out to be) are unaffected by, or autonomous from, irreducible multiscale stability conditions and contexts at larger scales. At the end of the day, for the ontological reductionist to win the day, it had better turn out that while contextuality is true in QM regarding the properties of QM systems, that classical measuring devices themselves are nothing but quantum systems (i.e., at least in principle their behavior requires no interactions with classical objects or stability conditions outside the realm of QM). It had better turn out that multiscale contexts are fully dischargeable or fully explicable from within the "fundamental" physical theory alone.

We have argued at length in this book that this implicit assumption is not well supported in any scientific endeavor or context at any scale. Despite the many cases mustered in our defense, and the inductive strength of our book-length argument, might we nonetheless be wrong? Yes, of course everyone engaged in these debates could be wrong; we are no exception. However, our reply is that across the board, based on science itself, ontological reductionism is not the way to bet. The best reasons for believing in ontological reductionism are evidence of an interlinked bottom-up chain of successful intertheoretic reductions, mechanistic or mereological reductions from physics on up that fully discharge multiscale contexts, or powerful evidence for CoP. We believe that this entire book stands as an argument that none of those conditions obtain. Indeed, we would say that at this point in history the burden of proof lies with the ontological reductionist and the defender of CoP.

Let us, however, be more specific in our reply to this objection. The Game of Life analogy presupposes, at the end of the day, that there are only three determining factors:

[43] See our discussion in Chapter 1

1. The fundamental dynamical laws.
2. The fundamental entities.
3. The fundamental initial conditions.

Throughout this book we have argued at length that this conception of reality is likely false, even in "fundamental" physics.[44] Instead, reality as portrayed by science suggests a world of irreducible multiscale contexts. First, we argued that at all scales, stability conditions and the contexts that they define are not reducible to dynamical laws, initial conditions, or boundary conditions in any fashion. Second, we argued that dynamical laws themselves function as contextual constraints rather than as rules "governing" temporal evolution. Third, dynamical laws themselves require stability conditions. Indeed, this is true of all constraints. For instance, while laws provide some constraints for how relevant states and observables behave, they provide no conditions guaranteeing the existence of the said states and observables. Consider time-translational symmetry. It does not exist independently of and prior to physical interactions. But it does not follow from this that such symmetries are merely epistemic. Symmetries and conservation laws also have stability conditions that must obtain (e.g., no symmetry breakings without contexts). Fourth, we have presented several case studies demonstrating that stability conditions and their related contexts and constraints cannot all be reduced to nothing but initial conditions, boundary conditions, or background conditions.

Fifth, pursuing the Game of Life analogy to the end would leave us with no explanation for the initial condition of our universe known as the Big Bang, nor would it leave us with any explanation for the "fundamental" laws and constants. Of course, there is nothing wrong with having fundamental explainers in a scientific theory, but such a picture would leave *many* things unexplained in this case. If, on the other hand, one takes the contextual emergence pattern seriously, there is no hard and fast distinction between dynamical laws, initial conditions, boundary conditions, background conditions, and other kinds of constraints. This leaves open the possibility for a multiscale and global contextually emergent explanation of the Big Bang, constants, and so forth.[45] Given contextual emergence, there is also no reason why the multiscale relationality in question might not be spatiotemporal. Thus, the arrow of explanation need not be exclusively from "past"

[44] The analogy is problematic even for the Game of Life itself. The basic rules of the Game of Life must come to expression in the context of software that must be implemented in hardware. Similarly for the initial conditions. The rules and conditions cannot write software and build hardware. Bonus points for those readers who spotted this problem with the Game of Life analogy early on.

[45] See Silberstein, Stuckey, and McDevitt (2018) for more details.

(the Big Bang) to "future." Perhaps such explanations are best viewed not on analogy with the Game of Life, but more like a spatiotemporal crossword puzzle whose content is dictated in part by global spatiotemporal constraints. For example, the universe as a whole started with some stability conditions, such as its dimensionality, crucial to defining the universe as a context. This context and its conditions, in turn, played some role in what stability conditions existed in the early universe. Think, for example, of the gatekeeper function of the number of spatial dimensions for the number, types, and properties of particles that are possible.

Sixth, one could attempt to place all the essential adynamical stability conditions into the reduction base, just as panpsychism attempts to put experience into the reduction base of "fundamental" physics. Perhaps some ontological reductionists can accept putting experience into the reduction base because no other option seems plausible to them. Nonetheless, to make this same move with so many nonmental features of reality would simply be cheating and render ontological reductionism trivial. Not to mention the fact that simply stuffing all the conditions into the reduction base does not explain the transitions to new states and observables, the unfolding of access to new regions of possibility space, and other changes that scientists spend their careers working on.

So, while we understand the temptation to argue that anything treated as a multiscale context could itself be reduced in principle, it just is not so. Even if we consider reducing macroscopic objects and living things to microphysics, this "just reduce the context" argument will not fly. Any attempt to perform the reduction of the contextual feature in any particular case is going to require marshalling *other* multiscale contextual features relevant for the new case—it is contextual emergence all the way down, as it were.

Return to the example of QM contextuality and everything we learned in Chapter 4 about measurement and environmental decoherence. QM is telling us that some form of multiscale contexts cannot washout. Yes, you can treat the classical measuring device as a QM system. Indeed, if you could manage to screen off the measuring device it would behave like a QM system. But whether the measuring device behaves quantum-mechanically or classically is a function of multiscale contexts—the presence or absence of multiscale stability conditions.

Modulo logical or physical constraints on computability, perhaps, in principle, God or the Mathematical Archangel could start with nothing but some "fundamental" physical theory and derive or predict the behavior of everything else. Of course, we cannot completely rule out that possibility. Our point, however, is that not only is this not the way to bet, given the state of the sciences, but there is in principle another way the world *could* be: contextual emergence. Contextual emergence is not only consistent with the sciences but is supported by them, as we have observed in the previous chapters. This book has explored the contextual emergence pattern bottom-to-top, from the contextual nature of QM to the recognition in neuroscience that global brain networks need wider multiscale networks involving embodied and cultural contexts.

That is, if one believes in reading their metaphysics from the best sciences, then one ought to believe in contextual emergence. Of course, those prone to a reductionist world-view can continue to hold out hope that some future science will make ontological reductionism look more promising, but as someone once said, sometimes absence of evidence is evidence of absence.[46] Note, again, that contextual emergence provides a unified and coherent picture of the sciences and reality, just not unification via standard forms of reductionism. Bottom line, we promised the reader a possible and plausible ontological alternative to epistemic emergence or radical emergence, and we have delivered.

The contextual emergence pattern has epistemic implications for explanation, to be sure, as we have shown throughout this book. The pattern of explanation reflecting contextual emergence is found across the sciences—a pattern involving interleaving scales and contexts in explanations that can be described somewhat more formally as relative onticity (Atmanspacher 2016; Atmanspacher and Kronz 1999). There is no "fundamental" ontology that provides the explanatory—reductive—key to explanations in the sciences. As explained in the previous chapters, this is not because there are in-practice difficulties with working out such explanations. It is because there is an ineliminable role of contexts in scientific explanations even if scientific practice is to *presuppose* the stability conditions defining these contexts more often than making them explicit.

7.13.2 *Contextual emergence compared to transformational emergence and ontological structural realism*

This brings us to the question regarding clarification of where contextual emergence stands with respect to transformational emergence (TE) and OSR. Of course, one can turn this into an objection as well. Perhaps the challenge

[46] As we saw in Chapter 6, as regards the fate of mechanistic explanation in biology and neuroscience, there are those who grant the essentiality of multiscale contextual emergence and constraint-based explanation. They agree that localization and decomposition often fail in such cases because of multiscale contexts and constraints, but argue that this is still consistent with such explanations being mechanistic. We have no problem with this; they can use the word "mechanistic" any way they want. However, there are those who acknowledge multiscale contextuality but still insist that we can provide reductive explanations in neuroscience and biology via 'localization' and 'decomposition' in *some* sense of those terms. As we laid out in Chapter 6, we disagree. For a lengthy argument, see Silberstein (2021).

here is to question the uniqueness of contextual emergence. Let us begin with how contextual emergence compares with TE (e.g., Guay and Sartenaer 2016; Humphreys 2016; Santos 2015). Various advocates of TE have related unique accounts, and we do not have space to engage all the details. For the sake of brevity we will focus on Humphreys' (2016) cornerstone work on TE, and Gil Santos' (2020) more recent version that is somewhat closer to contextual emergence. We should say that TE and contextual emergence, despite their differences, certainly share a similar spirit.

Beginning with Humphreys' (2016) account, we can summarize the differences as follows. Contextual emergence stresses the following:

1. That the pattern of contextual emergence is fundamental to the nature of reality and is therefore the deepest explainer for how and why ontological emergence is possible.
2. That there are no fundamental physical entities from which everything else emerges: no entities possessing intrinsic properties, primitive thisness, and complete ontological autonomy from constraints and contexts at larger scales.
3. The irreducibly multiscale nature of emergence including global stability conditions (e.g., KMS conditions).
4. That stability conditions and the like can include both dynamical/adynamical and causal/acausal factors, all of which are best viewed as forms of modal constraint.

None of these four essential features of contextual emergence are compatible with, or at least entailed by, TE as generally presented (Humphreys 2016).

Humphreys defends TE as diachronic/dynamical and ontological, though he notes that such emergence can also have synchronic aspects.[47] And while diachronic, Humphreys also emphasizes that the interaction in question need not be causal mechanical or involve direct, contiguous interactions. Indeed, he stresses emergence via novel dynamical laws. As he says: "Those fundamental entities, as a result of a dynamic process, interact with one another and as a consequence of that interaction the entities diachronically change" (Humphreys 2016, p. 135).

TE entails the falsity of what Humphreys calls "generative atomism,"

> the position that everything in a given domain results from composing properties possessed by immutable and individually distinguishable atoms, and that molecular entities are constructed from those atoms. It therefore requires that compositionality holds for the properties, and that immutability for the atoms

[47] His (1997) "fusion" relation is a subset of TE, but for our discussion we will focus mostly on his more general version.

> holds within all contexts created by systems that are generated by the relevant rules or laws.
>
> (p. 142)

By "immutability" Humphreys means that the "composite" system is made up of, and determined by, unchanging and numerically distinct individuals with "primitive thisness" and an individual essence. "Compositionality" means the same for the properties of the individual parts (e.g., there is no "fusion" in his sense of the word). Add in the realization picture of properties and one has generative atomism (p. 135).

Contextual emergence certainly rejects generative atomism. We should note, however, that while Humphreys defines immutability in terms of "essential properties," we have emphasized the failure of what we called "fundamentalism" and the "hierarchy thesis." That is, in contrast to TE, contextual emergence rejects the idea of a "fundamental" physical reduction base with complete metaphysical autonomy and essences distinct from various entities, processes, or laws across multiscale contexts. In metaphysical parlance, this is a rejection of autonomous physical entities with fixed intrinsic properties. Contextual emergence stresses the essentially multiscale nature of determination relations, unlike TE. And contextual emergence certainly rejects the very idea of essential properties, unlike TE. We would say that contextual emergence or multiscale contexts defined by stability conditions is the underlying explanation for why generative atomism fails, whereas Humphreys would not agree, as we shall see.

To appreciate these differences more, we need to dig a little deeper. Humphreys takes it that "contextualism is ill-suited to the task of understanding emergence because it directs our attention away from important diachronic features" (Humphreys 2016, p. 131). He understands contextualism and holism as imposing synchronic structures on entities so that novelty is a consequence of synchronic structure. Hence, we cannot understand emergence in its dynamical aspects.

While some accounts of contextualism and holism do emphasize synchronic determination relations, throughout this book we have emphasized that stability conditions defining multiscale contexts are often dynamical and involve multiscale interactions. It is true that there are dynamical and adynamical stability conditions, but these are all just species of modal constraints—modal constraints that lead to genuine transitions to new states and observables as a consequence. It is important to appreciate that in contextual emergence even the adynamical and acausal constraints, such as conservation laws or the graphical property of "small-worldness," do not constrain in virtue of being synchronic, and all such constraints affect the diachronic behavior and interactions of the components. Contextual emergence can provide a basis for a dynamic contextualism, in contrast to Humphreys' worries. But unlike TE, contextual emergence holds that multiscale stability conditions and their contexts are doing the work here, not new transformational dynamical laws popping into existence that change the essence of entities.

For Humphreys, it is "the effect of interacting with at least one other thing that is important, not being part of the whole. Conversely, it is possible for something to be changed merely as a result of being part of a whole without interacting with the other parts of that whole" (pp. 116–117). As our case study of Rayleigh–Bénard convection illustrates, multiscale interactions are key to emergence, and being a part of whole is neither necessary nor sufficient for emergence.

Humphreys contrasts relations with interactions, saying that for TE "it is not relations but dynamic interactions and their outcomes that matter" (p. 77). Here is how he states the essence of TE:

> Here is the basic idea. Transformational emergence occurs when an individual a that is considered to be a fundamental element of a domain D transforms into a different kind of individual a*, often but not always as a result of interactions with other elements of D, and thereby becomes a member of a different domain D*. Members of D* are of a different type from members of D. They possess at least one novel property and are subject to different laws that apply to members of D* but not to members of D. The fact that a fundamental entity can be transformed makes transformational emergence essentially different from generative atomism, in which the fundamental entities are immutable.
>
> (p. 60)

The key to TE is the dynamical interaction of two or more elements, such as particles. Those dynamical interactions might be with other elements in the whole, or they might be with something external to the whole. Those dynamical interactions, because of the failure of immutability, create *essentially new entities* which obey *different dynamical laws*—this is what makes such interactions more than mere "relations." As Humphreys puts it, "transformational emergence allows that the laws operating at the initial state of the system did not fix the laws at all times" (p. 68).

For TE, the dynamical laws operating at initial states of the system do not fix the laws at all times, such that the new dynamical transformational laws can override the original dynamical laws, thus bringing truly novel and autonomous entities into existence, such as mental processes from brain processes (2016, chapters 1 and 2). The focus in TE is thus on new dynamical "laws" coming into being through novel interactions—new dynamical laws that in turn now govern the new system in question.

Clearly, for Humphreys, relations are assumed to be static while interactions are dynamical. However, in this book we have presented several examples of contextual emergence where relations are dynamical. Furthermore, there is no necessary requirement for new dynamical laws popping into being or the transformation of essences. New laws may emerge in larger-scale contexts with the laws of microscale physics serving as some of the necessary conditions for this emergence, but there is no overriding of these microscale laws—they are also contextual and part of the new context! And laws do not govern, as we have observed; they constrain. To put it another way, Humphreys' TE very much keeps the picture of reality, wherein there is an essential and objective distinction between initial conditions, boundary conditions, background conditions, dynamical laws, and system to be time evolved.

What is it about the world that makes TE and the failure of generative atomism possible? Why do some interactions engender TE, and not others? Of course, Humphreys will invoke his transformational laws (of which fusion is just one possible instance) as the answer to that question. However, in reading Humphreys, sometimes it seems to us as if TE happens via new dynamical interactions *and then* new transformational laws spring into being for the new domain D*, and sometimes it seems as if the new transformational laws are supposed to do the explaining (i.e., no real ontological emergence happens until the new transformational laws come into being). This order of operations problem naturally arises if one maintains a "governing" account of dynamical laws as opposed to viewing them as extrinsic dispositions resulting from multiscale stability conditions and their contexts.

However, either way it goes for Humphreys, we still want to know the following: Why and how does TE happen? What is it about the world that makes it possible? Is this just a brute fact? What is the nature and origin of these new transformational laws? Are these laws like C. D. Broad's brute transordinal laws (1925), or the novel, brute causal relations of Timothy O'Connor's strong causal emergence (2005), or the dispositions-based laws in Cartwright's "nomological machines" (1997)? In other words, is TE just another form of the strong/radical kind of ontological emergence, where we are forced to accept brute transformational laws or causes? Recall that contextual emergence invokes no brute laws or causes, and no modal glue, metaphysical or otherwise, to explain ontological emergence.

Contextual emergence completely rejects the "foundationalist" or "hierarchical" picture of reality. Moreover, it rejects the very idea of ontological autonomy at any scale, and the idea that determination and explanation must involve some extra-added new metaphysical dynamical laws or causal relations to bring about transformation. Contextual emergence highlights the inherently interactive, multiscale interdependence of phenomena at all scales. Stability conditions act as gatekeepers for systems to access modal possibilities. The nature of reality is a contingent multiscale web of inextricably interconnected and interdependent *extrinsic* dispositions, most of which are in constant flux. In short, contextual emergence is the reason whatever TE happens is possible, not some extra-added dynamical "governing" laws transforming the essence of entities, say, like the rules governing finite automata but with the added power to transform essences.

Maybe in the end, what really separates contextual emergence from TE is that Humphreys really believes that it is the *transformational laws themselves, even in a vacuum* so to speak (a metaphysical vacuum, not the vacuum of space), that is somehow the source of ontological emergence. Let us try to make this clear with an example. In Humphreys' classic 1997 paper "How Properties Emerge," he notes that quantum entanglement is an ideal candidate for TE. TE would emphasize that the ontological emergence in this case—the entangled state—results as a brute fact because of a specific kind of interaction, say, between electron 1 and electron 2. Humphreys originally conceived of quantum entanglement as an instance of transformation via a "fusion" law, but that is not necessary to make our point. Indeed, TE, especially fusion, seems like it could have been inspired by the textbook Hamiltonian representation of quantum entangled states, wherein the states of the individual particles do not fix the state of the two-particle system in the relevant product space. TE invites us to imagine a possible world in which quantum entanglement could occur wherein there is nothing in that world but the two electrons interacting and a particular transformational law.

However, recall our discussion of quantum entanglement in Chapter 4 (Section 4.7). Given contextual emergence, there could not be a world with just two electrons in it, because electrons cannot exist outside of a rich context defined by the stability conditions for electron states and observables. As we emphasized, QM entangled states arise only if the system is screened off from the environment in the right way (i.e., there is no environmental decoherence). This screening-off requires multiscale stability conditions. The existence of the environment and screening-off conditions are essential here—these are the model space of possibilities and the gatekeepers allowing access to entangled states, respectively. In other words, there are no such things as electrons defined as entities with essential properties that are then transformed into something essentially different by a transformational law. There can be no electrons and no entanglement without relevant multiscale stability conditions and their context. In fact, as we indicated in Chapter 4, entangled states can be considered as equally "fundamental," with the separable electrons viewed as emergent.

Hence there are significant differences between Humphreys' account of TE and contextual emergence, and any cases of the former turn out to be examples of the latter on our view. Let us turn to Santos (2020) account of TE. He, of course, focuses on the dynamical and diachronic nature of such emergence, but explicates features that move closer to contextual emergence. First, he emphasizes the relational or contextual nature of ontology, denying the existence of ontologically autonomous entities in favor of reality as a web of interdependent relations. For Santos, TE emphasizes a dynamic relational view, which assumes the occurrence of structured networks of transformative and interdependent relations.

Second, Santos acknowledges the essentially multiscale and interlevel nature of contexts. Indeed, he explicitly acknowledges contextual emergence in this regard:

> This notion of structural conditioning converges with the ontic sense of the notion of contextuality as elaborated by Silberstein (2017). In such a view, contextuality is conceived as "a particular confluence of circumstances that produce a combination of constraints and stability conditions," which will in turn "limit the modal space available to the system (reduce its degrees of freedom) and open up new possibility spaces closed off outside of that context (add new degrees of freedom)" (2017, 167). In this sense, I side with Silberstein when it comes to denying the existence of discrete causally closed and absolutely autonomous levels or scales of reality (2020, 125). In a fundamentally dynamic relational view, there is rather a relation of mutual integration, interdependence, and reciprocal conditioning between the different levels (Silberstein 2017, 167), and that is the reason why "the best explanations often involve multiscale contexts."
>
> (187–188)[48]

Lastly, this leaves us with the comparison of contextual emergence with OSR, or rather, particular brands of OSR. One major metaphysical motivation that accounts of OSR have that contextual emergence does not, are concerns about scientific realism. From our perspective, generally speaking, this is not an interesting topic. While many scientists are realists about their theories, and we follow the sciences for implications about the actual world, contextual emergence is not putting forward an account of scientific realism. We have made explicit our ontological commitments and implications throughout this chapter.

Motivationally, contextual emergence does share with OSR a desire to explain novelty, order and stability, and a desire to paint a unified picture of how all of reality, across various scales and levels of organization, hangs together. More specifically, what we share with OSR is the idea that reality is not composed of, constituted by, or realized by "fundamental" physical

[48] In further email communication, Santos has acknowledged that his "relational–transformational account is pretty much what you call contextual emergence." What major differences, if any, between contextual emergence and Santos' TE remain to be worked out, but we certainly see his account having significant overlap with this book.

entities with intrinsic properties and primitive thisness. Things or entities in OSR are nothing but webs of relations and interactions, or codifications of modal structures (Ladyman and Ross 2007). The major difference between contextual emergence and OSR is what precisely replaces an ontology of self-existent and autonomous entities or "things."

Contextual emergence invokes a multitude of multiscale contexts that have equal claim to "fundamentality," each one essential for grounding relevant modal constraints, as we have observed throughout this book. In this respect we share something with the OSR of Ladyman and Ross (2007), in that just like their "rainforest realism" there is a multiplicity of modal truths at all scales in contextual emergence. *However, in contextual emergence we do not have primitive modal relations or structures* hiding behind the multiscale net of interdependent relations—no metaphysical or modal glue of any kind. Contextual emergence, as we have discussed, has modalities grounded in or expressed by properties and structured relations among properties involving multiscale contexts defined by stability conditions.

To derive a sense of what the OSR of Ladyman and Ross (2007) appears to be committed to that contextual emergence is not, consider the following:

> There is still a distinction between structure and non-structure. Merely listing relations among locators does not state anything with modal force ... Physical structures exists, but what is it? ... What makes the structure physical and not mathematical? That is a question that we refuse to answer. In our view, there is nothing more to be said about this that doesn't amount to empty words and venture beyond what the PNC [the principle of naturalistic closure] allows. The "world-structure" just is and exists independently of us and we represent it mathematico-physically via our theories ... *But OSR as we develop it is in principle friendly to a naturalized version of Platonism.*
>
> (p. 158, emphasis added)

> One distinct, and very interesting, possibility is that we become truly used to thinking of the stuff of the physical universe as being patterns rather than little things, *the traditional gulf between Platonistic realism about mathematics and naturalistic realism about physics will shrink or even vanish.* The new wave of structuralism in the philosophy of mathematics, which has a number of supporting arguments in common with OSR, adds substance to this speculation."
>
> (p. 237, emphasis added)

By way of contrast, contextual emergence is not friendly to Platonism, naturalized or otherwise. This brand of OSR is just another (admittedly subtler) form of glue or grounding that contextual emergence rejects in principle. The idea behind the OSR of Ladyman and Ross is that while our successful theories might get the details about "things" and "entities" wrong, they do get the "modal structure" or nomological bones of reality right, such as group-theoretic structures. For contextual emergence there is no such dualism or division to even discuss.

As OSR, in part, is driven by worries about scientific realism and how to distinguish "real" nomological relations from mere regularities, it cannot merely assert dispositionality—dynamical relations and multiscale contexts—as fundamental as contextual emergence does. OSR reifies some underlying structure that can survive undetermination, theory change, interpretational differences, and so forth. Ladyman and Ross are very clear that what defines their view is a principled distinction between the modal structure of the world versus occurrent structure (2007, p. 188). For contextual emergence, the occurrent structure of the world with its essential dispositionality and contexts *just is* the modal structure of the world.

Of course, Ladyman and Ross do not go so far as to say that nature is literally "made of math" or the universe is an automaton of some sort. But recall that they do say that "information is a fundamental concept, for understanding the objective modality of the world" (2007, p. 189). And: "Hence, information may in some sense be all there is at the scale where we describe most generally the modal structure of the world" (2007, 186). Contextual emergence rejects all such claims.

Rovelli, whom you may recall also defends a thoroughly relational and contextual model of reality, has this to say about OSR:

> The difference from contemporary structural realism, for instance, seems clear: I can imagine Nāgārjuna adding a short chapter to a contemporary edition of his book entitled "All Structures are Empty." They exist only when you are thinking about organizing something else. In his terms: "They are neither precedent to objects; nor not precedent to objects; neither are they both things; nor, ultimately, neither one nor the other thing."
>
> (2021, p. 153)

Rovelli is here making the same point that we are about OSR. There are no "intrinsic" or "primitive" modal structures; there is just the net of relations and interactions themselves that dynamically refine the modal space of possibilities. Nothing at all is required to "ground" this net or explain the stable patterns within it. Even OSR has some substantial metaphysical commitments!

The basic argument of Ladyman and Ross (2007) is as follows:

P1. Physics (especially QM and relativity) is telling us that OSR is true.

P2. Physics (especially QM and relativity) really is fundamental, at least in the sense that it is the key basis upon which we should draw the most universal metaphysical inferences about the nature of reality.

C1. Therefore, OSR is a universal and fundamental fact.

> C2. Therefore, while the entities-centric special sciences appear to be a counterexample to OSR, such entities must really be nothing but irreducible "real patterns" in the more fundamental physical modal structures.

In contrast, as we have argued, for contextual emergence physics has no priority in this regard. Contextual emergence rejects the the primacy of physics constraint endorsed by OSR. The contextual and dispositional nature of reality is equally exhibited at every scale, in every domain, and across every scientific discipline.

And because the OSR of Ladyman and Ross is primarily motivated by what they believe are the fundamental mathematically rendered structures of physics, such as group theory and related symmetries, no-go theorems in QM, the relativity principle, and so on, they are forced into conceiving of "fundamental physics" and the "special sciences" as being ontologically different in important ways, as suggested by C2. For instance, they insist that talk of causation and causal structure is exclusive to the special sciences, while talk of modal structure (i.e., nomological structure) is exclusive to physics or at least fundamental physics (2007).

By way of contrast, as we make clear throughout this this book, modal talk in contextual emergence includes pluralistic nomological and causal relations of all sorts, and at all scales and levels of organization. In contextual emergence there is no essential dualism of laws versus causation, and *no essential ontological difference* between the domain of physics and the domains of the special sciences. We have argued that physics is more "fundamental" in the sense of being more universal in defining and refining possibility space, but that sort of fundamentality does not warrant the OSR of Ladyman and Ross.

Steven French summarizes his view of OSR as follows:

> From this perspective, then, laws and symmetry principles are simply features of the underlying structure (or, better, law statements, etc., are descriptions of significant features of that structure). The properties of purported objects have the ontological features they do because of the law-like (and other) relations they enter into and the purported objects themselves, qua substantive metaphysical entities (or at the very least qua the kinds of entities that might be said to have an "individual profile"), are reconceptualized as nothing more than nodes or metaphorical "crossing points" in this network of relations (and hence can be eliminated). The "governing" metaphor that is used to articulate the relationship between the laws and these purported objects and their properties is then replaced by the relation of metaphysical dependence. The purported objects are dependent on the structures (and here the role of symmetries in presenting that dependence is fundamental) and the properties are dependent on the laws themselves... Furthermore, these laws and symmetries encode the relevant

> possibilities and insofar as they are stable or robust under certain changes to initial conditions can be regarded as necessary, in a certain sense. In this way the structure of the world can be said to be modally informed.
>
> (2014, p. 302)

According to French, the fundamental symmetries and laws (those of general relativity, QFT, and some future theory of quantum gravity) are "inherently modal," and the relationship between the laws and symmetries is such that "there is a kind of 'reciprocal interweaving and bonding' between laws and symmetries and that it is this that yields those entities formerly known as objects" (2014, p. 264). We assume this means that dynamical laws and the relevant symmetries are cofundamental. That is, symmetries do not merely yield or constrain laws, and symmetries are not just byproducts of laws. French's OSR is here primarily concerned with metaphysical unity. There is no object/property dualism, and no object/laws dualism either.

Again, in contextual emergence what is fundamental are not inherent modalities grounded in symmetries and the like, but multiscale contexts and properties. Thus, contextual emergence is not eliminativist about entities/objects or their properties; indeed, extrinsic dispositions are fundamental for us.[49] We share with OSR the rejection of intrinsic properties and primitive thisness. Rather than primitive physical structures with inherent modality, contextual emergence has multiscale relations and contexts which can change on a case-by-case basis, depending on the stability conditions and relevant contexts in play.

We think that the structure-talk in OSR makes sense only if one imbues it with inherent modality, which contextual emergence rejects. We share with OSR the desire to eliminate object/property dualism and object/laws dualism. We do this not by reducing, grounding, or eliminating any of the latter, but by making entities, properties, and multiscale modality inherently contextual and stability conditions-based (e.g., as with the constraint account of laws given previously).

In the brand of OSR as described by French, the eliminativism about entities in OSR comes from the idea that they are just modes of fundamental physical symmetries, not from the elimination of primitive thisness and intrinsic properties. The elimination of the latter should not vex any philosopher of a special science, as this reduction in metaphysical baggage does not threaten the autonomy of any special science or the reality of its entities. In fact, it should be clear that contextual emergence guarantees the in-principle, at least partial, autonomy of every scientific discipline and branch, yet without sacrificing an ordered, coherent picture of the world. Entities and laws are

49 Ironically, in some sense this makes contextual emergence *more* realist than OSR!

inherently contextually multiscale; they are extrinsic dispositions, and the multiscale contexts cannot be eliminated. The "elementary" particles of the standard model (whatever they may be) and the group theoretic structures that describe them, are not *absolutely* more fundamental than Bose–Einstein condensates, genes, or volcanoes. There is no absolute compositional or modal hierarchy of the world. In the world of contextual emergence, there are elements that are neither axiom nor theorem in any absolute sense.

7.14 Conclusion

We hope the reader feels that the foregoing questions and concerns have been rightfully discharged. And we hope the reader will join us in embracing contextual emergence as an account of both the sciences and the ontology of the world that is unified and coherent while also accounting for novelty. As captured by "Sonnet for The Perch," the world has a multicontextual order to it that bespeaks emergence on all scales.

Contextual emergence offers us a vision of the sciences and the world that opens up new possibilities for moving forward on many fronts, and discharges ancient occult conundrums grounded in the hyperrationalism of modal metaphysics. Yet contextual emergence does not remove the wonder and delight of the sciences and the world that stokes the curiosity and joy of the firefly-loving philosopher and scientist.

References

Ahaaronov, Y., Cohen, E., and Tollaksen, J. (2018). Completely top–down hierarchical structure in quantum mechanics. *PNAS*, 115(46):11730–11735.

Ahmadzai, M., Seguella, L., and Gulbransen, B. (2021). Circuit-specific enteric glia regulate intestinal motor neurocircuits. *Proceedings of the National Academy of Sciences*, 118(40):e2025938118 DOI:10.1073/pnas.2025938118.

Alexander, S. (1920). *Space, Time, and Deity, The Gifford Lectures at Glasgow 1916–1918*, Vols. 1 and 2. Macmillan, London.

Alexandrov, D. V., and Alexandrova, I. V. (2020). From nucleation and coarsening to coalescence in metastable liquids. *Philos. Trans. Royal Soc. A*, 378:20190247.

Allefeld, C., Atmanspacher, H., and Wackermann, J. (2009). Mental states as macrostates emerging from EEG dynamics. *Chaos: An Interdisciplinary Journal of Nonlinear Science*, 19:015102.

Allen, J.-M. A., Maroney, O. J. E., and Gogioso, S. (2017). A stronger theorem against Macro-realism. https://arxiv.org/abs/1610.00022v3 [quant-ph]. Quantum 1:13.

Amann, A. (1993). The Gestalt problem in quantum chemistry: Generation of molecular shape by the environment. *Synthese*, 97:125–156.

Anderson, M. L. (2010). Neural reuse: a fundamental organizational principle of the brain. *Behavioral and Brain Sciences*, 33:245–313.

Anderson, M. L. (2014). *After Phrenology: Neural Reuse and the Interactive Brain*. MIT Press, Cambridge, MA.

Anderson, M. L. (2016). Précis of after phrenology: neural reuse and the interactive brain. *Behavioral and Brain Sciences*, 39:1–22.

Anderson, M. L. (2020). Neural reuse: a fundamental organizational principle of the brain. *Behavioral and Brain Sciences*, 33(4):245–313.

Anderson, P. W. (1972). More is different. *Science*, 177(4047):393–396.

Anderson, P. W. (1987). The Resonating Valence Bond State in La2CuO4 and Superconductivity *Science*, 235:1196–1198.

Anderson, P. W. (1997). *Basic Notions of Condensed Matter Physics*. Westview, Boulder.

Anderson, P. W. (2000). Brainwashed by Feynman? *Physics Today*, 53(2):11–14.

Anderson, P. W. (2011). *More and Different: Notes From a Thoughtful Curmudgeon*. World Scientific, Hackensack.

Appelquist, T. and Carazzone, J. (1975). Infrared singularities and massive fields. *Physical Review D*, 11(10):2856–2861.

Applebaum, A. (2020). *Twilight of Democracy*. Anchor Books, New York.

Araki, H., Kastler., D., Takesaki., M., and Haag, R. (1977). Extension of KMS states and chemical potential. *Commun. Math. Phys.*, 53:97–134.

Armitage, N. P., Mele, E. J., and Vishwanath, A. (2018). Weyl and Dirac semimetals in three-dimensional solids. *Reviews of Modern Physics*, 90:015001.

Arnold, V. I. (1988). *Geometrical Methods in the Theory of Ordinary Differential Equations*, 2nd edn. Springer, New York.

Atasoy, S., Roseman, L., Kaelen, M., et al. (2017). Connectome-harmonic decomposition of human brain activity reveals dynamical repertoire re-organization under LSD. *Scientific Reports*.

Atmanspacher, H. (2012). Identifying mental states from neural states under mental constraints. *Interface Focus*, 2:74–81.

Atmanspacher, H. (2016). Relevance criteria for reproducibility: The contextual emergence of granularity. In Atmanspacher, A. and Maasen, S. (eds.), *Reproducibility—Principles, Problems, Practices, Prospects*, New York: Wiley, pp. 525–538.

Atmanspacher, H. and beim Graben, P. (2007). Contextual emergence of mental states from neur dynamics. *Chaos and Complexity Letters*, 2(2/3), 151–168.

Atmanspacher, H. and Kronz, F. (1999). Relative onticity. In Atmanspacher, A. Amann, A. and Müller-Herold, U. (eds.), *Quanta, Mind and Matter*. Dordrecht: Kluwer, pp. 273–294.

Avnir, D., Biham, O., Lidar, D. and Malcai, O. (1998). Is the geometry of nature fractal? *Science*, 279:39–40.

Aztekin, C. (2021). Appendage regeneration is context dependent at the cellular level. *Royal Society Open Biology*, 11:210126.

Baars, B. (2003). Hysterical Conversion, Consciousness and the Brain.

Ball, P. (2021). How big can the quantum world be? Physicists probe the limits. *Quanta Magazine* (August 18) https://www.quantamagazine.org/how-big-can-the-quantum-world-be-physicists-probe-the-limits–20210818/.

Bail, C. (2021). *Breaking the Social Media Prism: How to Make Our Platforms Less Polarizing*. Princeton University Press.

Bain, J. (2013). Emergence in effective field theories. *European Journal for Philosophy of Science*, 3(3):257–273.

Balasubramanian, K. (1991). Combinatorial methods in ESR spectroscopy. *J. Magn. Res.*, 91:45–56.

Balasubramanian, K. (1992). Combinatiorics of NMR and ESR spectral simulations. *J. Chem. Inf. Model.*, 32:296–298.

Bardeen, J., Cooper, L. N., and Schrieffer, J. R. (1957). Theory of superconductivity. *Physical Review*, 108(5):1175.

Barnes, E. (2012). Emergence and fundamentality. *Mind*, 121(484):873–901.

Barsbai, T., Lukas, D., and Pondorfer, A. (2021). Local convergence of behavior across species. *Science*, 371:292–295.

Bassett, D. S. and Sporns, O. (2017). Network neuroscience. *Nature Neuroscience*, 20(3):353–364.

Bassett, D. S., Zurn, P., and Gold, J. I. (2018). On the nature and use of models in network neuroscience. *Nature Reviews Neuroscience*, 19:566–578.

Batchelor G. (1967). *An Introduction to Fluid Dynamics*. Cambridge University Press, Cambridge.

Bateson, P. and Gluckman, P. (2011). Plasticity, robustness, development and evolution. *International Journal of Epidemiology*, 41(1):219–23.

Batterman, R. (2002). *The Devil in the Details. Asymptotic Reasoning in Explanation, Reduction, and Emergence*. Oxford University Press, Oxford.

Batterman, R. (2015). Philosophical issues in condensed matter physics and complex systems. In Falkenburg, B., and Morrison, M. (eds.), *Why More is Different*, Springer, Berlin, pp. 115–135.

Batterman, R. (2018). Autonomy of theories: an explanatory problem. *Noûs*, 52:858–873.

Batterman, R. and Green, S. (2020). Steel and bone: Mesoscale modeling and middle-out strategies in physics and biology. *Synthese*, 199(1-2):1159–1184.

Battiston, F., Amico, E., Barrat, A., Bianconi, G., Ferraz de Arruda, G., Franceschiello B., Lacopini, L., Kefi, S., Latora, V., Moreno, Y., Murray, M., Peixoto, T., Vaccarino F., and Petri, G. (2021). The physics of higher-order interactions in complex systems. *Nature Physics*, 17:1093–1098.

Baumgartner, M. and Gebharter, A. (2015). Constitutive relevance, mutual manipulability, and fat-handedness. *The British Journal for the Philosophy of Science*, 67(3):731–756.

Baumgartner, M. and Casini, L. (2017). An abductive theory of constitution. *Philosophy of Science*, 84(2):214–233.

Bechtel, W. (2010). Dynamic mechanistic explanation: computational modeling of circadian rhythms as an exemplar for cognitive science. *Studies in History and Philosophy of Science*, 41(3):321–333

Bechtel, W. (2011). Mechanism and biological explanation. *Philosophy of Science*, 78(4):533–558.

Bechtel, W. (2017). Explicating top-down causation using networks and dynamics. *Philosophy of Science*, 84(2):253–274.

Bechtel, W. (2017). Systems biology, negotiating between holism and reductionism. In Green. S. (ed.), *Philosophy of Systems Biology*. Springer, pp. 25–36.

Bechtel, W. (2017). Top-down causation in biology and neuroscience: control hierarchies. In Paolini, M. P. and Orilia, F. (eds.), *Philosophical and Scientific Perspectives on Downward Causation*. Routledge.

Bechtel, W. (2018). The importance of constraints and control in biological mechanisms: insights from cancer research. *Philosophy of Science*, 85(4):573–593.

Bechtel, W. (2019). Analysing network models to make discoveries about biological mechanisms. *The British Journal for the Philosophy of Science*, 70(2):459–484. The University of Chicago Press, Chicago, IL.

Bechtel, W. (2020). Hierarchy and levels: analysing networks to study mechanisms in molecular biology. *Philosophical Transactions of the Royal Society B*, 375:20190320.

Bechtel, W. and Abrahamsen, A. (2005). Explanation: a mechanist alternative. *Studies in History and Philosophy of Biological and Biomedical Sciences*, 36(2):421–441.

Bechtel, W. and Abrahamsen, A. (2011). Complex biological mechanisms: cyclic, oscillatory, and autonomous. In Hooker, C. (ed.), *Philosophy of Complex Systems*, Vol. 10. Elsevier, Amsterdam, pp. 257–285.

Bechtel, W. and Richardson, R. C. (2010). *Discovering Complexity: Decomposition and Localization as Strategies in Scientific Research*, 2nd edn. MIT Press, Cambridge, MA.

Becker, A. (2018). *What is Real? The Unfinished Quest for the Meaning of Quantum Physics*. Basic Books, New York.

Bedau, M. (2008). Is weak emergence just in the mind? *Minds and Machines*, 18:443–459.

Bedau, M. (2010). Two varieties of causal emergentism. In, Corradini, A., and O'Connor, T. (eds.), *Emergence in Science and Philosophy*. 2010(14):46–63. Routledge, London.

Bedau, M. and Humphreys, P. (2008). *Emergence: Contemporary Readings in Philosophy and Science*. MIT Press, Cambridge, MA.

Bennett, G. W., Brown, N. H., Bunce, G., Danby, G. T., et al. (2006). Final report of the e821 muon anomalous magnetic moment measurement at BNL. *Physical Review D, Particles Fields*, 73(7):072003.

Bernstein, J. (2021). Bad news: selling the story of disinformation. *Harper's Magazine*, https://harpers.org/archive/2021/09/bad-news-selling-the-story-of-disinformation /, September 4, 2021.

Berry, M. (1994). Asymptotics, singularities and the reduction of theories. In Prawitz. B. Skynns, and D. Westerståhl (eds.), *Logic, Methodology and Philosophy of Science IX D*. Elsevier, Amsterdam, pp. 597–607.

Berry, M. (2002) Singular limits. Reference Frame. *Physics Today (May 2002)*, pp. 10–11.

Bertolaso, M. (2016). *Philosophy of Cancer*. Springer, Netherlands.

Bertolero, M. A. and Bassett, D. S. (2020). On the nature of explanations offered by network science: a perspective from and for practicing neuroscientists. *Topics in Cognitive Science*, 12(4):1–22.

Bickhard, M. H. (2020). Why human inquiry is different than natural science inquiry. In Slife, B., Yanchar, S., and Richardson, F. C. (eds.), *Routledge International Handbook of Theoretical and Philosophical Psychology*, Routledge, Abingdon, UK, pp. 125–150.

Binney, J. J., Dowrick, N. J., Fisher, A. J., et al. (1992). *The Theory of Critical Phenomena: An Introduction to the Renormalization Group*. Oxford Science Publications, Oxford.

Birbriar, A. (2020). *Tumor Microenvironments in Organs: From the Brain to the Skin*. Springer.

Bishop, R. C. (2004). Nonequilibrium statistical mechanics Brussels–Austin style. *Studies in History and Philosophy of Modern Physics*, 35:1–30.

Bishop, R. C. (2005a). Anvil or onion? determinism as a layered concept. *Erkenntnis*, 63:55–71.

Bishop, R. C. (2005b). Patching physics and chemistry together. *Philosophy of Science*, 72:710–722.

Bishop, R. C. (2006a). Determinism and indeterminism. *Encyclopedia of Philosophy*, 2nd edn. Thomson Gale, Farmington Mills, MI, pp. 29–35.

Bishop, R. C. (2006b). The hidden premise in the causal argument for physicalism. *Analysis*, 66:44–52.

Bishop, R. C. (2007). *The Philosophy of the Social Sciences*. Continuum, London.

Bishop, R. C. (2008a). What could be worse than the butterfly effect? *The Canadian Journal of Philosophy*, 38:519–548.

Bishop, R. C. (2008b). Downward causation in fluid convection. *Synthese*, 160: 229–248.

Bishop, R. C. (2010). The via negativa: not the way to physicalism. *Mind and Matter*, 8:203–215.

Bishop, R. C. (2011). Metaphysical and epistemological issues in complex systems. In Hooker, C. (ed.), *Philosophy of Complex Systems*, Vol. 10. Elsevier, Amsterdam, pp. 119–150.

Bishop, R. C. (2012a). Excluding the causal exclusion argument against nonreductive physicalism. *Journal of Consciousness Studies*, 19:57–74.

Bishop, R. C. (2012b). Fluid convection, constraint and causation. *Interface Focus*, 2: 4–12.

Bishop, R. C. (2019). *The Physics of Emergence*. IOP Concise Physics Series. Morgan & Claypool Publishers: San Rafael, CA.

Bishop, R. C. (2022). Why human inquiry is different than natural science inquiry. In Slife, B. D., Yanchar, S. C., and Richardson, F. C., *Routledge International Handbook of Theoretical and Philosophical Psychology: Critiques, Problems, and Alternatives to Psychological Ideas*. Routledge, New York, pp. 412–431.

Bishop, R. C. and Atmanspacher, H. (2006). Contextual emergence in the description of properties. *Foundations of Physics*, 36:1753–1777.

Bishop, R. C. and beim Graben, P. (2016). Contextual emergence of deterministic and stochastic descriptions. In Atmanspacher, H., and Müller-Herold (eds.), *From Chemistry to Consciousness: The Legacy of Hans Primas*. Springer International Publishing, Cham, Switzerland, pp. 95–110.

Bishop, R. C. and Silberstein, M. (2019). Emergence and complexity. In Gibb, S., Hendry, R. F., and Lancaster, T. (eds.), *Routledge Handbook of Emergence*. Routledge, Abingdon, UK, pp. 145–156.

Bissell, M. J. and Hines, W. C. (2011). Why don't we get more cancer? A proposed role of the microenvironment in restraining cancer progression. *Nature Medicine*, 17:320–329.

Bliss, R. and Priest, G. (2018). The geography of fundamentality: an overview. In *Reality and Its Structure: Essays in Fundamentality*. Oxford University Press, pp. 15–30.

Boeckle, M., Leigl, G., Jank, R., and Pieh, C. (2016). Neural correlates of conversion disorder: overview and metananalysis of neuroimaging studies on motor conversion disorder. *BMC Psychiatry*, 16:195.

Boehme, M. et al. (2021). Microbiota from young mice counteracts selective age-associated behavioral deficits. *Nature Aging*, 1:666–676.

Bohm, D. and Hiley, B. (1993). *The Undivided Universe: An Ontological Interpretation of Quantum Theory*. Routledge, London.

Boi, L. (2017). The interlacing of upward and downward causation in complex living systems: on interactions, self-organization, emergence and wholeness. In Paolini,

M. P., and Orilla, F. (eds.), *Philosophical and Scientific Perspectives on Downward Causation*. Routledge, pp. 1–23.

Bokulich, A. (2008). *Reexamining the Quantum–Classical Relation*. Cambridge University Press, Cambridge.

Bong, K., Utreras-Alarcón, A., Ghafari, F., Liang, Y., Tischler, N., Cavalcanti, E. Pryde, G., and Wiseman, H. (2020). A strong no-go theorem on the Wigner's friend paradox. *Nat. Phys.*, 16, 1199–1205.

Boone, K. N. and Richardson, F. C. (2010). War neurosis: a cultural historical and theoretical inquiry. *Journal of Theoretical and Philosophical Psychology*, 30:109–121.

Brand, M. (1980). Simultaneous causation. In van Inwagen, P. (ed.), *Time and Cause*. Springer, Dordrecht, pp. 137–153.

Braunstein, S. L. and van Loock, P. (2005). Quantum information with continuous variables. *Rev. Mod. Phys.*, 77:513–577.

Brigandt, I. (2013). Systems biology and the integration of mechanistic explanation and mathematical explanation. *Studies in History and Philosophy of Biological and Biomedical Sciences*, 44(4):477–492.

Brigandt, I., Green, S., and O'Malley, M. (2018). Systems biology and mechanistic explanation. In Glennan, S., McKay, P., and Illari, B., *The Routledge Handbook of Mechanisms and Mechanical Philosophy*, Routledge, New York, pp. 362–374.

Britten, R. J. and Davidson, E. H. (1969). Gene regulation for higher cells: a theory. *Science*, 165:349–358.

Broad, C. D. (1925). *The Mind and Its Place in Nature*. Psychological Press, Routledge and Kegan Paul.

Buchholz, D. (1982). The physical state space of quantum electrodynamics. *Commun. Math. Phys.*, 85:49–71.

Budroni, C., Cabello, A., Gühne, O., Kleinmann, M., and Larsson, J. (2021). Quantum Contextality. https://arxiv.org/pdf/2102.13036.pdf.

Burnston, D. C. (2016a). Computational neuroscience and localized neural function. *Synthese*, 193:3741–3762.

Burnston, D. C. (2016b). A contextualist approach to functional localization in the brain. *Biology & Philosophy*, 31:527–550.

Burnston, D. C. (2021). Getting over atomism: functional decomposition in complex neural systems. *British Journal for the Philosophy of Science*, 72:743–772.

Busse, F. (1978). Non-linear properties of thermal convection. *Reports on Progress in Physics*, 41:1929–1967.

Butterfield, J. (2011a). Emergence, reduction and supervenience: A varied landscape. *Found. Phys.*, 41:920–959.

Butterfield, J. (2011b). Less is different: Emergence and reduction reconciled. *Found. Phys.*, 41:1065–1135.

Cacioppo, S. and Cacioppo, J. T. (2020). *Introduction to Social Neuroscience*. Princeton University Press.

Calzavarini, F. and Viola, M. (2020). Studies in the brain and mind. *Neural Mechanisms: New Challenges in the Philosophy of Neuroscience*. Springer.

Cao, T. Y. (1999). Why are we philosophers interested in quantum field theory? In Cao, T. Y. (ed.), *Conceptual Foundations of Quantum Field Theory*. Cambridge University Press, Cambridge, pp. 28–33.

Cao, T. Y. and Schweber, S. S. (1993). The conceptual foundations and the philosophical aspects of renormalization theory. *Synthese*, 97:33–108.

Carbone, M. and Bragg, A. D. (2020). Is vortex stretching the main cause of the turbulent energy cascade? *J. Fluid Mech.*, 883:R2.

Cartwright, N. (1983). *How the Laws of Physics Lie*. Oxford University Press, Oxford.

Cartwright, N. (1997). Where do laws of nature come from? *Dialectica*, 51(1):65–78.

Cartwright, N. (1999). *The Dappled World: A Study of the Boundaries of Science*. Cambridge University Press, Cambridge.

Casini, L. (2016). How to model mechanistic hierarchies. *Philosophy of Science*, 85(5): 946–958.

Castellani, E. (2002). Reductionism, emergence, and effective field theories. *Studies in History and Philosophy of Science Part B: Studies in History and Philosophy of Modern Physics*, 33(2), 251–267.

Caston, V. (1997). Epiphenomenalisms, ancient and modern. *Philosophical Review*, 106(3):309–363.

Caston, V. (2001). *Dicaearchus' Philosophy of Mind*. Routledge.

Caudy, W. and Greensite, J. (2008). Ambiguity of spontaneously broken gauge symmetry. *Physical Review D*, 78(2):025018.

Caves, C. M., Thorne, K. S., Drever, R. W. P., Sandberg, V. D., and Zimmermann, M. (1980). On the measurement of a weak classical force coupled to a quantum-mechanical oscillator. I. Issues of principle. *Rev. Mod. Phys.*, 52(2):341–392.

Chalmers, D. (1996). The conscious mind: in search of a fundamental theory. *The Journal of Mind and Behavior*, Oxford University Press, New York.

Chalmers, D. (2006). Strong and weak emergence. In, Clayton, P., and Davies, P. C. W. (eds.), *The Re-Emergence of Emergence*, Oxford University Press, Oxford, pp. 244–256.

Chandrasekhar, S. (1957). *An Introduction to the Study of Stellar Structure*. Dover Publications, New York.

Chemero, A. P. (2009). *Radical Embodied Cognitive Science*. MIT Press, Cambridge, MA.

Chemero, A. and Silberstein, M. (2008). After the philosophy of mind: replacing scholasticism with science. *Philosophy of Science*. 75(1):1–27.

Chen, X., Gu, Z.-C., and Wen X.-G. (2010). Local unitary transformation, long-range quantum entanglement, wave function renormalization, and topological order. *Physical Review B*, 82:155138.

Chung, L. W. K., Baseman, B., Assikis, V., and Zhau, H. E. (2005). Molecular insights into prostate cancer progression: the missing link of tumor microenvironment. *The Journal of Urology*, 173:10–20.

Colombo, M. and Weinberger, N. (2018). Discovering brain mechanisms using network analysis and causal modeling. *Minds and Machines*, 28(2):265–286.

Colosi, D. and Rovelli, C. (2009). What is a Particle? arXiv preprint arXiv:0409054.

Compagner, A. (1989). Thermodynamics as the continuum limit of statistical mechanics. *Am. J. Phys.* 57(2):106–117).

Cotter, J. P., Brand, C., Knobloch, C., Lilach, Y., Cheshnovsky, O., Arndt, M. (2017). In search of multipath interference using large molecules. *Sci. Adv.*, 3:e1602478.

Crall, J. D., Gravish, N., Mountcastle, A. N., Kocher, S. D., Oppenheimer, R. L., Pierce, N. E., and Combes, S. A. (2018). Spatial fidelity of workers predicts collective response to disturbance in a social insect. *Nature Communications*, 9(1):1201.

Cramer, John G. (1986) The Transactional Interpretation of Quantum Mechanics. *Rev. Mod. Phys.*, 58:647–687.

Craver, C. F. (2007). *Explaining the Brain: Mechanisms and the Mosaic Unity of Neuroscience*. Oxford University Press, New York.

Craver, C. F. (2014). The ontic account of scientific explanation. In *Explanation in the Special Sciences*. Springer, Netherlands, pp. 27–52.

Craver, C. F. (2016). The explanatory power of network models. *Philosophy of Science*, 83(5):698–709.

Craver, C. F. and Bechtel, W. (2007). Top-down causation without top-down causes. *Biology and Philosophy*, 22:547–563.

Cross, M. and Hohenberg, P. (1993). Pattern formation outside of equilibrium. *Reviews of Modern Physics*, 65:851–1112.

Damicelli, F., Hilgetag, C. C., Hütt, M. T., and Messé, A. (2019). Topological reinforcement as a principle of modularity emergence in brain networks. *Network Neuroscience*, 3(2):589–605.

Davidson, E. H. (1968). *Gene Activity in Early Development*. Academic Press, Cambridge, MA.

Davidson, E. H. (1994). Molecular biology of embryonic development: how far have we come in the last ten years? *BioEssays*, 16:603–615.

Dawkins, R. (1976). *The Selfish Gene*. Oxford University Press, Oxford.

Dingle, R. (1973). *Asymptotic Expansions: Their Derivation and Interpretation*. Academic Press, New York.

Dirac, P. A. M. (1929). Quantum mechanics of many-electron systems. *Proc. R. Soc. Lond. A*, 123:714–733.

Dirac, P. (1949). Forms of relativistic dynamics. *Reviews of Modern Physics*, 21:392–399.

Dorato, M. (2005). *The Software of the Universe: An Introduction to the History and Philosophy of Laws of Nature*. Ashgate, Routledge.

Dorato, M. (2016). Bohr's Relational Holism and the Classical-Quantum Interaction. arXiv preprint arXiv:1608.00205v1.

Dral, P. O. (2020). Quantum chemistry in the age of machine learning. *J. Phys. Chem. Lett.*, 11(6):2336–2347.

Dubin, D. A. and Hennings, M. A. (1990). *Quantum Mechanics, Algebras, and Distributions*. Volume 238 of Pitman research notes in mathematics series. Longman Scientific & Technical, Harlow, UK.

Duhem, P. (1954). *The Aim and Structure of Physical Theory*. Princeton University Press.

Earman, J. (1986). *A Primer on Determinism*. Reidel, Dordrecht, Holland.

Earman, J. (2004). Laws, symmetry, and symmetry breaking: invariance, conservation principles, and objectivity. *Philosophy of Science*, 71(5),1227–1241.

Ehrenfest, P. (1917). In that way does it become manifest in the fundamental laws of physics that space has three dimensions? *Proceedings of the Royal Amsterdam Academy of Arts and Sciences* 20:200–209.

Einstein, A., Podolsky, B., and Rosen, N. (1935). Can quantum-mechanical description of physical reality be considered complete? *Phys. Rev.*, 47:777–780.

Ellis, G. F. R. (2012). Top-down causation and emergence: some comments on mechanisms. *Interface Focus*, 2:126–140.

Ellis, G. F. R. (2016). *How Can Physics Underlie the Mind: Top-Down Causation in the Human Context.* Berlin: Springer.

Emch, G. G. and Liu, C. (2002). *The Logic of Thermo-Statistical Systems*. Springer, Berlin.

Euler, L. (1735). Solutio problematis and geometriam situs pertinentis. In *Commentarii Academiae Scentiarum Petropolitanae*, pp. 128–140.

Favela, L. H. (2019). Integrated information theory: a complexity science approach to consciousness. *Journal of Consciousness Studies*, 26:21–47.

Favela, L. H. (2020). The dynamical renaissance in neuroscience, *Synthese*, 199 (1–2):2103–2127.

Fazekas, P. and Kertesz, G. (2018). Are higher mechanistic levels causally autonomous? *The 26th Biennial Meeting of the Philosophy of Science Association in Seattle, WA on November 1-4, 2018.*

Feehly, C. (2021). Brains might sync as people interact — and that could upend consciousness research. *Discover* (July 26, 2021). https://www.discovermagazine.com/mind/brains-might-sync-as-people-interact-and-that-could-upend-consciousness

Feigenbaum, M. J. (1978). Quantitative universality for a class of nonlinear transformations. *J. Stat. Phys.,* 19:25–52.

Feldman Barrett, L. (2020). *Seven And A Half Lessons About The Brain.* Houghton Mifflin Harcourt.

Feldt Muldoon, S. and Bassett, D. S. (2016). Network and multilayer network approaches to understanding human brain dynamics. In Smeenk, C. (ed.), *Philosophy of Science*, 83(5):710–720.

Feng, D. and Jin, G. (2005). *Introduction to Condensed Matter Physics.* Vol. 1. World Scientific.

Fletcher, S. C. (2020). Similarity structure and diachronic emergence. *Synthese*, 1–28.

Fox Keller, E. (2010). *The Mirage of a Space Between Nature and Nurture*. Duke University Press.

Francis, R. C. (2011). *The Ultimate Mystery of Inheritance: Epi-Genetics*. W. W. Norton & Company.

Franklin, A. (2018). On the renormalization group explanation of universality. *Philosophy of Science*, 85:225–248.

Franklin, A. (2019). Universality reduced. *Philosophy of Science*, 86:1295–1306.

Franklin, A. and Knox, E. (2018). Emergence without limits: the case of phonons. *Studies in History and Philosophy of Modern Physics Part B*, 64:68–78.

French, S. (2014). *The Structure of the World: Metaphysics and Representation*. Oxford University Press, Oxford.

Friederich, S. (2013). Gauge symmetry breaking in gauge theories—in search of clarification. *European Journal for Philosophy of Science*, 3:157–182.

Frierson Faust, L. (2017). *Fireflies, Glow-worms, and Lightning Bugs*. University of Georgia Press.

Ganeri, J. (2011). Emergentisms, ancient and modern. *Mind*, 120:671–703.

Garg, M. and Kern, K. (2019). Attosecond coherent manipulation of electrons in tunneling microscopy. *Science*, 367:411–415.

Georgi, H. M. (1989). Effective quantum field theories. In Davies, P. (ed.), *The New Physics*. Cambridge University Press, Cambridge, pp. 446–457.

Gerasimova-Chechkina, E., Toner, B. C., Batchelder, K. A., White, B., Freynd, G., Antipev, Arneodo, I. A. and Khalil, A. (2021). Loss of mammographic tissue homeostasis in invasive lobular and ductal breast carcinomas vs. benign lesions. *Frontiers in Physiology*, 12:660883.

Gervais, J.-L. and Zwanziger, D. (1980). Derivation from first principles of the infrared structure of quantum electrodynamics, *Phys. Lett.*, 94B:389–393.

Ghirardi, G., Rimini, A., and Weber, T. (1986). Unified dynamics for microscopic and macroscopic systems. *Physical Review D*, 34:470–491.

Gilbert, S. and David, E. (2009). Ecological developmental biology: integrating epigenetics, medicine and evolution. In *Integrative and Comparative Biology*. Sinauer Associates, Inc, Sunderland, MA, pp. 231–232.

Gillett, C. (2010). Weak emergence and context-sensitive reduction. In Corradini, A., and O'Connor, T. (eds.), *Emergence in Science and Philosophy*. Routledge, London, New York, pp. 25–46.

Gillett, C. (2016). *Reduction and Emergence in Science and Philosophy*. Cambridge University Press, Cambridge.

Godfrey-Smith, P. (2020). *Metazoa: Animal Life and the Birth of the Mind*. Farrar, Straus and Giroux, New York.

Goldford, J. E., Lu, N., Bajić, D., Estrela, S., Tikhonov, M., Sanchez-Gorostiaga, A., Segrè, D., Mehta, P., and Sanchez, A. (2018). Emergent simplicity in microbial community assembly. *Science*, 361:469–474.

Green, S. and Betterman R. (2021). Making sense of top-down causation: Universality and functional equivalence in physics and biology. In Voosholz, Jan and Gabriel, Markus (eds.), *Top-Down Causation and Emergence*. Springer, Cham, Switzerland, 39–63.

Green, S., Serban, M., Scholl, R., Jones, N., Brigandt, I., and Bechtel, W. (2018). Network analyses in systems biology: new strategies for dealing with biological complexity. *Synthese*, 195 (4):1751–1777.

Guay, A. and Sartenaer, O. (2016). A new look at emergence. Or when after is different. *European Journal for Philosophy of Science*, 6(2):297–322.

Gunton, C. (1993). *The One, the Three and the Many: God, Creation and the Culture of Modernity, The 1992 Bampton Lectures*. Cambridge University Press, Cambridge.

Ha, M. J. et al. (2020). Compositional zero-inflated network estimation for microbiome data. *BMC Bioinformatics*, 21:581.

Haag, R., Hugenholtz, N. M., and Winnink, M. (1967). On the equilibrium states in quantum statistical mechanics. *Commun. Math. Phys.*, 5:215–236.

Haag, R., Kastler, D., and Trych-Pohlmeyer. E. B. (1974). Stability and equilibrium states. *Commun. Math. Phys.*, 38:173–193.

Hacking, I. (1983). *Representing and Intervening: Introductory Topics in the Philosophy of Natural Science*. Cambridge University Press, Cambridge.

Hansson, T. H., Oganesyan, V., and Sondhi. S. L. (2004). Superconductors are topologically ordered. *Annals of Physics*, 313(3):497–538.

Harbecke, J. and Atmanspacher, H. (2012). Horizontal and vertical determination of mental and neural states. *Journal of Theoretical and Philosophical Psychology*, 32: 161–179.

Hardcastle, V. G. and Stewart, M. C. (2002). What do brain data really show? *Philosophy of Science*, 69(S3):341769

Harrington, A. (ed.). (1997). *The placebo effect: An interdisciplinary exploration*. Harvard University Press, Cambridge, MA.

Harrist, R. S. and Richardson, F. C. (2014). Pathologizing the normal, individualism, and virtue ethics. *Journal of Contemporary Psychotherapy*, 44:201–211.

Hart, C. L. (2021). *Drug Use for Grown Ups: Chasing Liberty in the Land of Fear*. Penguin Press, New York.

Hartmann, S. (2001). Effective field theories, reductionism and scientific explanation. *Studies in History and Philosophy of Modern Physics Part B*, 32(2), 267–304.

Hawthorne, J. and Silberstein, M. (1995). For whom the bell arguments toll. *Synthese*, 102:99–138.

Hayashi, Y. and Liu, C. (2017). The evolution and function of sleep. In Shigeno, S., Murakami, Y., and Nomura, T. (eds.), *In Brain Evolution by Design: From Neural Origin to Cognitive Architecture*. Springer, pp. 243–366.

Healey, R. (1994). Nonseparable processes and causal explanation. *Studies in History and Philosophy of Science Part A*, 25(3), 337–374.

Heitman, N. Sennett, R., Mok, K.-W., Saxena, N., Srivastava, D., Martino, P., Grisanti, L. G., Wang, Z., Ma'ayan, A., Rompolas, P., and Rendl, M. (2020). Dermal sheath contraction powers stem cell niche relocation during hair cycle regression. *Science*, 367:161–166.

Held, C. (2013). The Kochen–Specker theorem. In Zalta, E.N. (ed.), *The Stanford Encyclopedia of Philosophy* (Winter 2014 Edition), http://plato.stanford.edu/archives/win2014/entries/kochen-specker/.

Hellman, G. (1989). *Mathematics without Numbers: Towards a Modal-Structural Interpretation*. Clarendon Press, Oxford.

Hendry, R. F. (2019). Emergence in Chemistry: Substance and Structure. In Gibb, S., Hendry, R. F., and Lancaster, T. (Eds.), *Emergence and Complexity*. Routledge Handbook of Emergence, pages 339–351. Routledge, Abingdon, UK.

Hett, B. C. (2018). *The Death of Democracy: Hitler's Rise to Power and the Downfall of the Weimar Republic*. Henry Holt and Co, New York.

Hettema, H. (2017). *The Union of Chemistry and Physics: Linkages, Reduction, Theory Nets and Ontology*. Springer, Cham.

Hilgetag, C. C. and Goulas, A. (2020). Hierarchy in the organization of brain networks. *Philosophical Transactions of the Royal Society B*, 375(1796):20190319.

Hilgetag, C. C. and Goulas, A. (2015). Is the brain really a small-world network? *Brain Structure and Function*, 221:2361–2366.

Hill, R. (1967). Instantaneous action-at-a-distance in classical relativistic mechanics. *Journal of Mathematical Physics*, 8:201–220.

Hoekstra, A., Chopard, B., and Coveney, P. (2014). Multiscale modelling and simulation: a position paper. *Philosophical Transactions of the Royal Society A*, 372:20130377.

Hoffmann-Kolss, V. (2014). Interventionism and higher-level causation. *International Studies in the Philosophy of Science*, 28(1): 49–64.

Holt, J. (2012). Physicists, stop the churlishness. *New York Times Online* (June 8, 2012).

Hooker, C. (2011). Conceptualising reduction, emergence and self-organization in complex dynamical systems. In Hooker, C. (ed.), *Philosophy of Complex Systems*, Vol. 10. Elsevier, Amsterdam, pp. 195–222.

Horst, S. (2007). *Beyond Reduction: Philosophy of Mind and Post-Reductionist Philosophy of Science*. Oxford University Press, New York.

Hossenfelder, S. (2018). *Lost in Math: How Beauty Leads Physics Astray*. Basic Books, New York.

Howard, D. (2007). Reduction and emergence in the physical sciences: some lessons from the particle physics and condensed matter debate. In Murphy, N., and Stoeger W. R. (eds.), *In Evolution and Emergence: Systems, Organisms, Persons*, Oxford University Press, Oxford, pp. 141–157.

Huang, Z., Zhang, J., Wu, J., Mashour, G. A., and Hudetz, A. G. (2020). Temporal circuit of macroscale dynamic brain activity supports human consciousness. *Cognitive Neuroscience*, 6(11):eaaz0087.

Huemer, M. and Kovitz, B. (2003). Causation as simultaneous and continuous. *The Philosophical Quarterly*, 53:556–565.

Humphreys, P. (1997). How properties emerge. *Philosophy of Science*, 64:1–17.

Humphreys, P. (2016). *Emergence. A Philosophical Account*. Oxford University Press, Oxford.

Huneman, P. (2008). Emergence made ontological? Computational versus combinatorial approaches. *Philosophy of Science, Proceedings of the 2006 Biennial Meeting of the Philosophy of Science Association* Part II: Symposia Papers Edited by Cristina Bicchieri and Jason Alexander (December 2008), 75(5):595–607.

Huneman, P. (2010). Topological explanations and robustness in biological sciences. *Synthese*, 177:213–245.

Huneman, P. (2018a). Diversifying the picture of explanations in biological sciences: ways of combining topology with mechanisms, *Synthese*, 195:115–146.

Huneman, P. (2018b). Outlines of a theory of structural explanations. *Philosophical Studies*, 175:665–702.

Hüttemann, A. (2005). Explanation, emergence, and quantum entanglement. *Philosophy of Science*, 72:114–127.

Hüttemann, A. and Love, A. C. (2011). Aspects of reductive explanation in biological science: intrinsicality, fundamentality, and temporarily. *British Journal for the Philosophy of Sciences*. 62(3):519–549.

Hüttemann, A. and Love, A. C. (2016). Reduction. In Humphreys, P. (ed.), *The Oxford Handbook of Philosophical Science*. Oxford University Press, New York, pp. 460–484.

Ingram, J., Hand, C. J., and Maciejewiski, G. (2021). Social isolation during Covid-19 lockdown impairs cognitive function. *Applied Cognitive Psychology*, 35(4)935–947.

Jablonka, E. and Lamb, M. (2005). *Evolution in Four Dimensions: Genetic, Epigenetic, Behavioral, and Symbolic Variation in the History of Life*. MIT Press, Cambridge, MA.

Jaeger, L. and Calkins, E. R. (2012). Downward causation by information control in micro-organisms. *Interface Focus*, 2:26–41.

Jain, J. K. (1989). Composite-fermion approach for the fractional quantum hall effect. *Physical Review Letters*, 63(2):199.

James, W. (1898). *Human Immortality*. Houghton Mifflin and Company, Cambridge, MA.

James, W. (1904). *Does consciousness exist? Journal of Philosophy, Psychology and Scientific*. Methods, 1(18):477–491.

James, W. (1905a). The notion of consciousness: Communication made at the 5th international congress of psychology, Rome, 30 April 1905. *Journal of Consciousness Studies*, 12(7):55–64 (2005).

James, W. (1905b). The place of affectional facts in a world of pure experience. *The Journal of Philosophy, Psychology and Scientific Methods*, 1(18):477–491.

James, W. (1912). *Essays in Radical Empiricism*. Longmans, Green and Co, New York.

James, W. (1988). *Manuscript Lectures*. Harvard University Press, Cambridge, MA.

Jammer, Max (1961). *Concepts of Mass in Classical and Modern Physics*. Harvard University Press, Cambridge, MA.

Jansson, L. (2020). Network explanations and explanatory directionality. *Philosophical Transactions of the Royal Society B*, 375:20190318.

Joos, E. and Zeh, H. D. (1985). The emergence of classical properties through interaction with the environment. *Z. Phys.*, B59:223–243.

Kahn, J. (2017). *Race on the Brain: What Implicit Bias Gets Wrong About the Struggle for Racial Justice*. Columbia University Press, New York.

Kaiser, M. (2020). *Changing Connectomes: Evolution, Development, and Dynamics in Network Neuroscience*. MIT Press, Cambridge, MA.

Kaiser, M. I. and Krickel, B. (2017). The metaphysics of constitutive mechanistic phenomena. *The British Journal for the Philosophy of Science*, 68(3):745–779

Kanaya, T., Williams, I. R., and Ohno, H. (2020). Intestinal M cells: tireless samplers of enteric microbiota. *Traffic*, 21(1):34–33.

Kant, I. (1998). *The Critique of Pure Reason*. Cambridge University Press, Cambridge.

Kaplan, D. M. (2005). How to reconcile physicalism and antireductionism about biology. *Philosophy of Science*. 72(1):43–68.

Kaplan, D. M. (2018). Mechanics and dynamical explanation. In Glennan, S. (ed.), *The Routledge Handbook of Mechanisms and Mechanical Philosophy*. Oxford University Press, Routledge, pp. 267–280.

Kaplan, J. (2008). Review of genes in development: rereading the molecular paradigm. In Neumann-Held, E. M., and Rehmann-Stutter, C. (eds.), *Biology and Philosophy*, 2:427–429.

Kellert, S. H., Longino, H. E., and Waters, C. K. (2006). *Scientific Pluralism*. University of Minnesota Press, Minneapolis.

Kendi, I. X. (2016). *Stamped from the Beginning: The Definitive History of Racist Ideas in America*. Bold Type Books, New York.

Kershenbaum, A. (2020). *The Zoologist's Guide to the Galaxy: What Animals on Earth Reveal About Aliens—and Ourselves*. Penguin Press.

Keshavarzi, A., Nomura, D., and Teubner, T. (2018). The muon g-2 and α(M2Z): a new data-based analysis. *Physical Review D*, 97:114025.

Kim, D. J., Davis, E. P., Sandman, C. A., Glynn, L., Sporns, O., O'Donnell, B. F., and Hetrick, W. P. (2019). Childhood poverty and the organization of structural brain connectome. *Neuroimage*,1:84:409–416.

Kim, J. (1998). *Mind in a Physical World*. MIT Press, Cambridge, MA.

Kim, J. (1999). Making sense of emergence. *Philosophical Studies*, 95:3–36.

Kim, J. (2007). *Physicalism, or Something Near Enough*. Princeton University Press.

Kim. S., Thapa, I., Zhang, L., and Ali, H. (2019). A novel graph theoretical approach for modeling microbiomes and inferring microbial ecological relationships. *BMC Genomics*, 20:945.

Kleckner, D. and Irvine, W. T. M. (2013). Creation and dynamics of knotted vortices. *Nature Physics*, 9:53–258.

Koonin, E. V. (2011). *The Logic of Chance: the Nature and Origin of Biological Evolution. Financial Times*, Prentice Hall.

Koperski, J. (2001). Has chaos been explained? *British Journal for the Philosophy of Science*, 52:683–700.

Kostic, D. (2018). The topological realization. *Synthese*, 195(1):79–98.

Kostic, D. (2020). General theory of topological explanations and explanatory asymmetry. *Philosophical Transactions of the Royal Society B*, 375:20190321.

Kotler, S. et al. (2021). Direct observation of deterministic macroscopic entanglement. *Science*, 372(6542):622–625.

Kraemer, M. U. G. et al. (2021). Spatiotemporal invasion dynamics of SARS-CoV-2 lineage B.1.1.7 emergence. *Science*, 373:889–895.

Krall, N. A. and Trivelpiece, A. W. (1986). *Principles of Plasma Physics*. San Francisco Press, San Francisco.

Kraman, M., Bambrough, P. J., Arnold, J. N., Roberts, E. W., Magiera, L., Jones, J. O., Gopinathan, A., Tuveson, D. A., and Fearon, D. T. (2010). Suppression of antitumor immunity by stromal cells expressing fibroblast activation protein-alpha. *Science*, 330(6005):827–830.

Kronz, F. and Tiehen, J. (2002). Emergence and quantum mechanics, *Philosophy of Science*, 69(2):324–347.

Kryachko, E. S. and Ludña, E. V. (1990). *Energy Density Functional Theory of Many-Electron Systems*. Kluwer Academic, Dordrecht.

Kubo, R. (1957) Statistical-mechanical theory of irreversible processes. I. General theory and simple applications to magnetic and conduction problems *J. Phys. Soc. Jpn.*, 12:570–586.

Law, Y.-H. (2021). Long live the queen. *Science* 371:1302–1305.

Ladyman, J. (2017). An apology for naturalized metaphysics. In Slater, M., and Yudell, Z. (eds.), *Metaphysics and the Philosophy of Science: New Essays*, Oxford University Press, Oxford, Chapter 7.

Ladyman, J. and Ross, D. (2007). *Everything Must Go: Metaphysics Naturalized*. Oxford University Press, Oxford.

Lancaster T. and Blundell S. J. (2012). *Quantum Field Theory for the Gifted Amateur*. Oxford University Press, Oxford.

Lancaster, T. and Pexton, M. (2015). Reduction and emergence in the fractional quantum Hall state. *Studies in History and Philosophy of Science Part B: Studies in History and Philosophy of Modern Physics*, 52:343–357

Landau, L.D. and Lifshitz, E.M. (1977). *Quantum Mechanics (Non-Relativistic Theory)*, 3rd Edition. Pergamon Press, Oxford.

Lanse, N. (2021). Gut bacteria may 'talk' to the brain, mouse study suggests. *Live Science* (June 30, 2021). https://www.livescience.com/microbiota-shape-social-behavior-mice.html

Laughlin, R. B. (1983). Anomalous quantum Hall effect: an incompressible quantum fluid with fractionally charged excitations. *Physical Review Letters*, 50(18): 1395–1398.

Laughlin, R. B. (2005). *A Different Universe: Reinventing Physics from the Bottom Down*. Basic Books, New York.

Laughlin, R. B. and Pines, D. (2000). The theory of everything. *Proceedings of the National Academy of Sciences*, 97(1):28–31.

Levin, M. and Wen, X. G. (2005). Colloquium: photons and electrons as emergent phenomena. *Reviews of Modern Physics*, 77(3):871.

Levy, A. and Bechtel, W. (2016). Towards mechanism 2.0: expanding the scope of mechanistic explanation. *In the 25th Biennial Meeting of the Philosophy of Science Association in Atlanta, GA on November 3–5, 2016*.

Levy, N. (2013). Addiction is not a brain disease. *Front Psychiatry*, 4:24.

Lewis. D. 1986. *On the Plurality of Worlds*. Blackwell Publishers, Oxford.

Lewis, D. (1999). *Papers in Metaphysics and Epistemology*. Cambridge University Press, Cambridge.

Lewis, G. N. (1923). *Valence and the Structure of Atoms and Molecules*. The Chemical Catalog Company, New York.

Lewis, P. (2016). *Quantum Ontology: A Guide to the Metaphysics of Quantum Mechanics*. Oxford University Press, Oxford.

Lewton, T. (2021). Does life use quantum effects? *New Scientist*, 251:40–41.

Liu, C. (2019). Infinite idealization and contextual realism. *Synthese*, 196(5): 1885–1918.

Loewer, B. (2001). The rise of physicalism. In Gillett, C. and Loewer, B. (eds.), *Physicalism and Its Discontents*. Cambridge University Press, Cambridge, pp. 3–36.

Love, A. C. (2012). Hierarchy, causation and explanation: ubiquity, locality, and pluralism. *Interface Focus*, 2:115–125.

Love, A. C. (2018). Developmental mechanisms. In Glennan, S., and Illari, P. (eds.), *The Routledge Handbook of the Philosophy of Mechanisms and Mechanical Philosophy*. Routledge, New York, pp. 332–347.

Love, A. C. and Hüttemann, A. (2011). Comparing part-whole explanations in biology and physics. In Dieks, D., Gonzalex, W. J., Hartmann, S., Uebel, T., and Weber, M. (eds.), *Explanation, Prediction, and Confirmation*, Springer, Berlin, pp. 183–202.

Lower, B. (2008). Why there is anything except physics. In Jakob Hohwy and Jesper Kallestrup (eds.), *Being Reduced: New Essays on Reduction, Explanation, and Causation*. Oxford University Press, New York, pp. 149–163.

Lundqvist, S. (1992). *Nobel Lectures, Physics 1971–1980*. World Scientific, Singapore.

Lyre, H. (2008). Does the Higgs mechanism exist? *International Studies in the Philosophy of Science*, 22(2):119–133.

Machamer, P., Darden, D., and Craver, C. F. (2000). Thinking About Mechanisms. *Philosophy of Science* 67:1–25.

Machon, T. (2020). The Godbillon–Vey invariant as a restricted Casimir of three-dimensional ideal fluids. *J. of Phys. A*, 53:235701.

Mackie, J. L. (1965). Causes and conditions. *American Philosophical Quarterly*, 2: 245–264.

Mackie, J. L. (1974). *The Cement of the Universe*. Clarendon Press, Oxford.

Maillé, S. and Lynn, M. (2020). Reconciling current theories of consciousness. *Journal of Neuroscience*, 40(10):1994–1996.

Mainwood, P. (2006). *Is More Different? Emergent Properties in Physics*. Oxford University Press, Oxford.

Markov, N. T., Ercsey-Ravasz, M., Van Essen, D. C., and Knoblauch, K. (2013). Cortical high-density counterstream architectures. *Science*, 342:1238406.

Marletto, C. (2021). *The Science of Can and Can't: A Physicist's Journey through the Land of Counterfactuals*. Viking Publishing, New York.

Martin, P. C. and Schwinger J. (1959) Theory of many-particle systems. I. *Phys. Rev.*, 115:1342–1373.

Massimini, M. and Tononi, G. (2018). *Sizing Up Consciousness*. Oxford University Press, Oxford.

Massimo, P. (2014). Between holism and reductionism: a philosophical primer on emergence. *Biological Journal of the Linnean Society*, 112(2):261–267.

Matthiessen, D. (2017). Mechanistic explanation in systems biology: cellular networks. *The British Journal for the Philosophy of Science*, 68:1–25.

Maudlin, T. (2007). *A Modest Proposal Concerning Laws, Counterfactuals, and Explanations: The Metaphysics Within Physics*, Oxford University Press, Oxford.

McComb, W. D. (2004). *Renormalization Methods: A Guide for Beginners*. Clarendon Press, Oxford.

McLaughlin, B. (1982). British emergentism. In Emergence or reduction? In Beckermann A., Flohr, H, and Kim, J. (eds). *Emergence or Reduction? Essays on the Prospects of Nonreductive Physicalism*. (De Gruyter, Berlin, pp. 49–93.

McLeish, T., Pexton, M., and Lancaster, T. (2019). Emergence and topological order in classical and quantum systems. *Studies in History and Philosophy of Science Part B: Studies in History and Philosophy of Modern Physics*, 66:155–169.

McMullin, E. (1985). Galilean Idealization. *Stud. Hist. Phil. Sci.*, 16(3):247–273.

McNally, R. J. (2016). Can network analysis transform psychopathology? *Behaviour Research and Therapy*, 86:95–104.

Melnyk, A. (2003). *A Physicalist Manifesto: Thoroughly Modern Materialism*. Cambridge University Press, Cambridge.

Menon, T and Callender, C. (2013). Turn and face the strange... Ch-ch-changes: Philosophical questions raised by phase transitions. In Batterman, R. W. (ed.), *The Oxford Handbook of Philosophy of Physics*. Oxford University Press, New York, chapter 5.

Mercier de Lépinay, L., Ockeloen-Korppil, C. F., Woolley, M. J., and Sillanpää, M. A. (2021). Quantum mechanics–free subsystem with mechanical oscillators. *Science* 372: 625–629.

Mermin, N. (1981). Bringing home the atomic world: Quantum mysteries for anybody. *Am. J. Phys.*, 49:940–943.

Mitchell, S. D. (2002). Integrative pluralism. *Biology and Philosophy*, 17:55–70.

Montessori, A., Tiribocchi, A., Bonaccorso, F., Lauricella, M., and Succi, S. (2020). Lattice Boltzmann simulations capture the multiscale physics of soft flowing crystals. *Phil. Trans. R. Soc.* A, 378: 20190406.

Monthoux, P., Balatsky, A. V., and Pines, D. (1991). Toward a theory of high-temperature superconductivity in the antiferromagnetically correlated cuprate oxides. *Physical Review Letters*, 67(24):3448.

Moreno, A., Ruiz-Mirazo, K., and Barandiaran, X. (2011). The impact of the paradigm of complexity on the foundational frameworks of biology and cognitive science. In Hooker, C. (ed.), *Philosophy of Complex Systems*. Elsevier, pp. 311–333.

Morgan, L. (1923). *Emergent Evolution*. Williams and Norgate, London.

Moroz, L. L., Kocot, K. M., Citarella, M. R., Dosung, S., et al. (2014). The ctenophore genome and the evolutionary origins of neural systems. *Nature*, 510:109–114.

Morrison, M. (2003). Spontaneous symmetry breaking: theoretical arguments and philosophical problems. *Symmetries in Physics: Philosophical Reflections*, 347–363.

Morrison, M. (2012). Emergent physics and micro-ontology. *Philosophy of Science*, 79(1):141–166.

Morrison, M. (2015). *Reconstructing Reality: Models, Mathematics, and Simulations*. Oxford University Press, New York.

Müller-Herold, U. (1980). Disjointness of β-KMS states with different chemical potential. *Lett. Math. Phys.*, 4:45–48.

Murthy, G. and Shankar, R. (2003). Hamiltonian theories of the fractional quantum Hall effect. *Reviews of Modern Physics*, 75:1101.

Nagel, E. (1961). *The Structure of Science: Problems in the Logic of Scientific Explanation*. Hackett, New York.

Najafi, M., McMenamin, B., Simon, J., and Pessoa, L. (2016). Overlapping communities reveal rich structure in large-scale brain networks during rest and task conditions. *NeuroImage*, 135:92–106.

Nand, A., Zhan, Ye., Salazar, O., Aranda, M., Voolstra, C. R., and Dekker, J. (2021). Genetic and spatial organization of the unusual chromosomes of the dinoflagellate Symbiodinium microadriaticum, *Nature Genetics*, 53(5):618–629.

Nayak, C., Simon, S. H., Stern, A., Freedman, M., and Sarma, S. D. (2008). Non-Abelian anyons and topological quantum computation. *Reviews of Modern Physics*, 80(3):1083.

von Neumann, J. (1955/1932). *Mathematical Foundations of Quantum Mechanics*. Princeton University Press, Princeton.

Newen, A., de Bruin, A., and Gallagher, S. (2018). *The Oxford Handbook of 4E Cognition*. Oxford University Press, Oxford.

Niquil, N., Haraldsson, M., Sime-Ngando, T., Huneman, P., and Borrett, S. R. (2020). Shifting levels of ecological network's analysis reveals different system properties. *Philosophical Transactions of the Royal Society B*, 375:20190326.

Noble, D. (2008). *The Music of Life: Biology Beyond Genes*. Oxford University Press, Oxford.

O'Connor, T. (1994). Emergent properties. *American Philosophical Quarterly*. 31(2): 91–104.

O'Conner, T. and Wong, H. Y. (2006). The metaphysics of emergence. *Noûs*, 39(4): 658–678.

O'Malley, M. A., Brigandt, I., Love, A. C., Crawford, J. W., Gilbert, J. A., Knight, R., Mitchell, S. D., and Rohwer, F. (2014). Multilevel research strategies and biological systems. *Philosophy of Science*, 81:811–828.

Omnès, R. (1999). *Understanding Quantum Mechanics*. Princeton University Press, Princeton, NJ.

Paolini, M. P. and Orilia, F. (2017). *Philosophical and Scientific Perspectives on Downward Causation*. Routledge.

Papineau, D. (2001). The rise of physicalism. In Gillett, C., and Loewer, B. (eds.), *Physicalism and Its Discontents*. Cambridge University Press, Cambridge, pp. 3–36.

Papineau, D. (2002). *Thinking About Consciousness*. Clarendon Press, Oxford.

Parrington, J. (2021). *Mind Shift: How Culture Transformed the Human Brain*. Oxford University Press, Oxford.

Pathria, R. (1972). *Statistical mechanics*. Oxford: Pergamon Press.

Paul, M., Chiam, K.-H., Cross, M., and Greenside, H. (2003). Pattern formation and dynamics in Rayleigh–Bénard convection: numerical simulations of experimentally realistic geometries. *Physica D: Nonlinear Phenomena*, 184:114–126.

Pauli, W. (1933). Allgemeinen prinzipien der wellenmechanik. In Geiger, H. and Scheel, K. (eds.), *Handbuch der Physik*, vol. 24. Springer, Berlin, pp. 83–272.

Pearl, J. (2009). *Causality: Models, Reasoning, and Inference*, 2nd edn. Cambridge University Press, Cambridge.

Pedersen, M. and Omidvarnia, A. (2016). Further insight into the brain's rich-club architecture. *Journal of Neuroscience*, 36(21):5675–5676.

Peleg, O. (2021). Living Orbs of Light: Solving the Mystery of How and why Fireflies Flash in Time Can Illuminate the Physics of Complex Systems. *Aeon*. (21 September 2021). https://aeon.co/essays/how-firefly-flashes-illuminate-the-physics-of-complex-systems

Pessoa, L. (2014). Understanding brain networks and brain organization. *Physics of Life Reviews*, 11:400–435.

Pexton. M. (2014). Can asymptotic models be explanatory? *European Journal for Philosophy of Science*, 4(2):233–252.

Pexton, M. (2015). Emergence and fundamentality in a pancomputationalist universe. *Minds and Machines*, 25(4):301–320.

Pigliucci, M. (2014). Between holism and reductionism. *A Philosophical Primer on Emergence. Biological Journal of the Linnean Society*. 112:261–267.

Piomelli, U. (1999). Large-eddy simulation: achievements and challenges. *Prog. in Aero. Sci.* 35 (1999) 335–362.

Polger, T. W. and Shapiro, L. A. (2016). *The Multiple Realization Book*. Oxford University Press, Oxford.

Povich, M. and Craver, C. F. (2018). Mechanistic levels, reduction, and emergence. In Glennan, S., McKay. P., and Illari, P, *The Routledge Handbook of Mechanisms and Mechanical Philosophy*. Routledge, pp. 185–197.

Price, H. (1996). *Time's Arrow and Archimedes Point: New Directions for the Physics of Time*. Oxford University Press, Oxford.

Primas, H. (1977). Theory reduction and non-Boolean theories. *Journal of Mathematical Biology*, 4:281–301.

Primas, H. (1983). *Chemistry, Quantum Mechanics and Reductionism*. Springer, Berlin.

Primas, H. (1988). Emergence in exact natural sciences. *Acta Polytech. Scand.*, 91: 83–98.

Pustilnik, A. (2012). *Racey, Racey Neuro-hype! Can a Pill Make You Less Racist?* https://neuroethics16.rssing.com/chan-16594202/article6.html.

Putnam, H. (1975). Philosophy and our mental life. In Mind, *Language and Reality: Philosophical Papers*, Vol. 2. Cambridge University Press, Cambridge, pp. 291–303.

Qi, X.-L. and Zhang, S.-C. (2011). Topological insulators and superconductors. *Reviews of Modern Physics*, 83:1057–1110.

Raja, V. and Anderson. M. L. (2021). Behavior considered as an enabling constraint. In Calzavarini, F., and Viola, M. (eds.), *Neural Mechanisms: New Challenges in the Philosophy of Neuroscience*. Springer Neuroscience, pp. 209–232.

Ranade, R., Hill, C., and Pathak, J. (2021). DiscretizationNet: A machine-learning based solver for Navier–Stokes equations using finite volume discretization. *Computer Methods in Applied Mechanics and Engineering*, 378:113722.

Rathkopf, C. (2018). Network representation and complex systems. *Synthese*, 195: 55–78.

Redhead, M. (1999). Quantum field theory and the philosopher. In Cao, T. Y. (ed.), *Conceptual Foundations of Quantum Field Theory*. Cambridge University Press, Cambridge, pp. 34–40.

Reed, M. and Simon B. (1972). *Methods of Modern Mathematical Physics I: Functional Analysis*, Revised and Enlarged Edition. Academic Press, San Diego, CA.

Rescher, N. (1993). *Pluralism*. Clarendon Press, Oxford.

Richardson, F. C., Fowers, B., and Guignon, C. (1999). *Re-envisioning Psychology: Moral Dimensions of Theory and Practice*. Jossey-Bass Publishers, San Francisco.

Richerson, P. J., Boyd, R., and Henrich, J. (2010). Gene-culture coevolution in the age of genomics. *Proceedings of the National Academy of Sciences*, 107(Supplement 2):8985–8992.

Rohrlich, F. (2001). Cognitive scientific realism. *Philosophy of Science*, 68(2):185–202.

Rommelfanger, K. S. (2012). Placebo for psychogenic illnesses: Why "It's all in my head" does and doesn't matter. *Soapbox Science: A community guest blog from Nature.com*, http://blogs.nature.com/soapboxscience/2012/07/18/placebo-for-psychogenic-illnesses-why-its-all-in-my-head-does-and-doesnt-matter.

Ross, L. N. (2021). Distinguishing topological and causal explanation. *Synthese*, 198:9803–9820.

Rosaler, J. and Harlander, R. (2019). Naturalness, Wilsonian renormalization, and "fundamental parameters" in quantum field theory. *Studies in History and Philosophy of Science Part B: Studies in History and Philosophy of Modern Physics*, 66: 118–134.

Ross, L. (2015). Dynamical models and explanation in neuroscience. *Philosophy of Science*, 82:32–54.

Rovelli, C. (2021). *Helgoland: Making Sense of the Quantum Revolution*. Penguin Books, Riverhead.

Rueger, A. (2000a). Robust supervenience and emergence. *Philosophy of Science*, 67(3):466–489.

Rueger, A. (2000b). Physical emergence, diachronic and synchronic. *Synthese*, 124(3):297–322.

Russell, B. (1927). *An Outline of Philosophy*. George Allen and Unwin, London.

Russell, B. (1927). *The Analysis of Matter*. Kegan Paul, London.

Ruiz-Mirazo, K. and Moreno, A. (2012). Autonomy in evolution: from minimal to complex life. *Synthese*, 185(1):21–52.

Sachdev, S. (2011). *Quantum Phase Transitions*, 2nd edn. Cambridge University Press, Cambridge.

Safron, A. (2020). An integrated world modeling theory (iwmt) of consciousness: combining integrated information and global neuronal workspace theories with the free energy principle and active inference framework; toward solving the hard problem and characterizing agentic causation. *Frontiers in Artificial Intelligence*, 3(30):3389.

Salazar, V. W., Tscheoke, D. A., Swings, J., Consenze, C. A., Mattoso, M., Thompson, C. C., and Thompson, F. L. (2020). A new genomic taxonomy system for the Synechococcus collective. *Environmental Microbiology*, 22(11):4557–4570.

Santos, G. C. (2015). Ontological emergence: how is that possible? Towards a new relational ontology. *Foundations of Science*, 20:429–446.

Santos, G. C. (2020). Integrated-structure emergence and its mechanistic explanation. *Synthese*, 198(9):8687–8711.

Sarfati, R., Hayes, J. C., and Peleg, O. (2021). Self-organization in natural swarms of photinus carolinus synchronous fireflies. *Science Advances*, 7(28):1126.

Scheinin, A. (2020). Foundations of human consciousness: imaging the twilight zone. *Journal of Neuroscience*, 41(8):1769–1778.

Schirner, M., McIntosh, A. R., Jirsa, V., Deco, G., and Ritter, P. (2018). Inferring multi-scale neural mechanisms with brain network modelling. *Elife*, 8(7):28927.

Seki, R., Li, C., Hayashi, S., Fang, Q., Egawa, S., et al. (2017). Functional roles of aves class-specific cis-regulatory elements on macroevolution of bird-specific features. *Nature Communications*, 6(8):14229.

Serban, M. (2020). Exploring modularity in biological networks. *Philosophical Transactions of the Royal Society B*, 375:20190316.

Sewell, G. (2002) *Quantum Mechanics and Its Emergent Macrophysics*. Princeton University Press, Princeton, NJ.

Shaffer, J. (2010). Monism: The priority of the whole. *Philosophical Review*, 119(1):31–76.

Shang, J. J. S. (2019). Landmarks and new frontiers of computational fluid dynamics. *Advances in Aero*. 1:5.

Shine, J. M., Breakspear, M., Bell, P. T., Martens, K. A. E., Shine, R., Koyejo, O., Sporns, O., and Poldrack, R. A. (2019). Human cognition involves the dynamic integration of neural activity and neuromodulatory systems. *Nature Neuroscience*, 22: 289–296.

Shoemaker, S. (1980). *Causality And Properties*. Springer, Netherlands.

Shrapnel, S. and Costa, F. (2018). Causation does not explain contextuality. *Quantum*, 2:63.

Shyer, A. E., Rodrigues, A. R., Schroeder, G. G., Kassianidou, E., Kumar, S., and Harland, R. M. (2017). Emergent cellular self-organization and mechanosensation initiate follicle pattern in the avian skin. *Science*, 357:811–815.

Silberstein, M. (1998). Emergence and the mind–body problem. *Journal of Consciousness Studies*, 5(4):464–482.

Silberstein, M. (2001). Converging on emergence: consciousness, causation and explanation. *Journal of Consciousness Studies*, 8(9–10):61–98.

Silberstein, M. (2006). In defense of ontological emergence and mental causation. In Davies, P. (ed.), *The Re-emergence of Emergence*. Oxford University Press, Oxford, pp. 203–226.

Silberstein, M. (2009a). Quantum nonseparability and mereology. In Gerogiorgakis, S. Seibt, S., and Imaguire, G. (eds.), *Handbook of Mereology*. Philosophia, Munich, pp. 300–320.

Silberstein, M. (2009b). Emergence and consciousness. In T. Bayne, T., Cleeremans, A., and Wilken, P. (eds.), *Oxford Companion to Consciousness*. Oxford University Press, Oxford, pp. 254–57.

Silberstein, M. (2011). Metaphysics or science: the battle for the soul of philosophy of mind. *Philosophical Psychology*, 24(4):561–573.

Silberstein, M. (2012). Emergence and reduction in context: philosophy of science and/or analytic metaphysics. *Metascience*, 21:627–642.

Silberstein, M. (2015). Extending neural monism to the hard problem. *Journal of Consciousness Studies*, 22(3–4):181–194.

Silberstein, M. (2016). The implications of neural reuse for the future of cognitive neuroscience and the future of folk psychology. *Brain and Behavioral Sciences*, 120: 1–45.

Silberstein, M. (2017). Strong emergence no, contextual emergence yes. *Special Issue of Philosophica on Emergence*, 91:145–192.

Silberstein, M. (2019). Neural monism reborn: breaking the gridlock between emergence and inherence. In Seager, W. (ed.), *The Routledge Handbook of Panpsychism*, Routledge, Chapter 18.

Silberstein, M. (2020). Constraints on localization and decomposition as explanatory strategies in the biological sciences 2.0. In Calzavarini, F., and Viola, M. (eds.), *Neural Mechanisms: New Challenges in the Philosophy of Neuroscience*. Springer Neuroscience, Chapter 15.

Silberstein, M. and Chemero, A. (2012). Complexity and extended phenomenological-cognitive systems. *Topics in Cognitive Science*, 4(1):35–50.

Silberstein, M. and Chemero, A. (2013). Constraints on localization and decomposition as explanatory strategies in the biological sciences. *Philosophy of Science*, 80:958–970.

Silberstein, M. and Machamer, P. (2002). Reduction, emergence, and explanation. In *The Blackwell Guide to the Philosophy of Science*. Blackwell Publishers, pp. 80–107.

Silberstein, M. and McGeever, J. (1999). The search for ontological emergence. *Philosophical Quarterly*, 49(195):201–214.

Silberstein, M. and Stuckey, W. M., and McDevitt, T. (2020). Re-thinking the world with neutral monism: removing the boundaries between mind, matter, and spacetime. *Entropy*, 22(5):551.

Silberstein, M. (ed.) (2008), *The Blackwell Guide to the Philosophy of Science*. Blackwell, Oxford.

Silberstein, M., Stuckey, W. M., and McDevitt, T. (2021). Beyond causal explanation: Einstein's principle not Reichenbach's. *Entropy*, 23(1):114.

Silberstein, M., and Stuckey, W. (2022). The completeness of quantum mechanics and the determinateness and consistency of intersubjective experience: Wigner's friend and delayed choice. In *Quantum Mechanics and Consciousness*, Gao, S. (ed.), Oxford University Press: Oxford, forthcoming 2022.

Silberstein. M. (2014). *Dynamics, Systematicity and Extended Cognition in Systematicity and the Post-Connectionist Era*. MIT Press, Cambridge, MA.

Silberstein, M., Stuckey, W., and McDevitt, T. (2018). *Beyond the Dynamical Universe: Unifying Block Universe Physics and Time as Experienced*. Oxford University Press, Oxford.

Singleton J. (2001). Band theory and electronic properties of solids. *Physics Today*, 55(12):61.

Sklar, L. (1967) Types of inter-theoretic reduction. *British Journal for the Philosophy of Science*, 18:109–124.

Skow, B. (2014). Are there non-causal explanations (of particular events)? *British Journal for the Philosophy of Science*, 65(3):445–467.

Slater, M. H. and Yudell, Z. (2017). *Metaphysics and the Philosophy of Science: New Essays*. Oxford University Press, Oxford.

Slezov, V. V. (1978). Formation of the universal distribution function in the dimension space for new-phase particles in the diffusive decomposition of the supersaturated solid solution. *J. Phys. Chem. Solids*, 39:367–374.

Smeenk, C. (2006). The elusive Higgs mechanism. *Philosophy of Science*, 73(5): 487–499.

Smith, P. (1998). *Explaining Chaos*. Cambridge University Press, Cambridge.

Smolin, L. (2013). *Time Reborn: From the Crisis in Physics to the Future of the Universe*. Houghton Mifflin Harcourt.

Solé, R. and Valverde, S. (2020). Evolving complexity: how tinkering shapes cells, software and ecological networks. *Philosophical Transactions of the Royal Society B*, 375:20190325.

Soto, A. M. and Sonnenschein, C. (2012). Is systems biology a promising approach to resolve controversies in cancer research? *Cancer Cell International*, 12(1):1–5.

Spekkens, R. (2005). Contextuality for preparations, transformations, and unsharp measurements. *Phys. Rev. A*, 71:052108.

Spence, S. (2009). *The Actor's Brain: Exploring the Cognitive Neuroscience of Free Will*. Oxford University Press, Oxford.

Spivey, M. J. (2020). *Who You Are: The Science of Connectedness*. MIT Press, Cambridge, MA.

Sprons, O. (2011). The human connectome: a complex network. *Annals of the New York Academy of Sciences*, 1224(1):109–125.

Stern, A. (2008). Anyons and the quantum Hall effect a pedagogical review. *Annals of Physics*, 323(1):204–249.

Stern, F. (1974). *The Politics of Cultural Despair: A Study in the Rise of the Germanic Ideology*. University of California Press, Berkeley.

Stinson, C. (2016). Mechanisms in psychology: ripping nature at its seams. *Synthese*, 193(5):1585–1614.

Strogatz, S. H. (2001). Exploring complex networks. *Nature*, 410:268–276.

Strogatz, S. H. (2004). *Sync: The Emerging Science of Spontaneous Order*. Hachette Books, New York.

Stephan, A. (1992). Emergence – a systematic view on its historical facets. In Beckermann, A., Flohr, H., and Kim, J. (eds.), *Emergence or Reduction? Essays on the Prospects of Nonreductive Physicalism*. De Gruyter, Berlin.

Strogatz, S. H. and Stewart, I. (1993). Coupled oscillators and biological synchronization. *Scientific American*, 269(6):102–109.

Stuckey, W. M., Silberstein, M., McDevitt, T., and Le, T. D. (2020). Answering Mermin's challenge with conservation per no preferred reference frame. *Nature Scientific Reports*, 10:15771.

Sukoriansky, S, Chekhlov, A., Orszag, S. A., Galperin, B., and Staroselsky (1996). Large eddy simulation of two-dimensional isotropic turbulence. *J. of Sci. Comp.*, 11:13–45.

Sutcliffe, B. T. and Woolley, G. R. (2012). Atoms and molecules in classical chemistry and quantum mechanics. In Hendry, FR. F., Needham, P., and Woody, A. I., *Handbook of the Philosophy of Science*, vol 6. Elsevier, Amsterdam, pp. 388–426.

Symons, J. (2018). Brute facts about emergence. In Elly Vintiadis and Constantinos Mekios (eds.), *Brute Facts*. Oxford University Press. Chapter 10.

Takesaki, M. (1970) Disjointness of the KMS-states of different temperatures. *Commun. Math. Phys.*, 17:33–41.

Takesaki, M. (2002). *Theory of Operator Algebras I*. Springer: Berlin.

Taylor, C. (1985a). *Philosophical Papers Vol. 1: Human Agency and Language*. Cambridge University Press, Cambridge.

Taylor, C. (1985b). *Philosophical Papers Vol. 2: Philosophy and the Human Sciences*. Cambridge University Press, Cambridge.

Taylor, C. (1995). *Philosophical Arguments*. Harvard University Press.

Taylor, C. (2002). Foundationalism and the inner-outer distinction. In Smith, N. H. (ed.), *Reading McDowell: On Mind and World*. Routledge, London and New York, pp. 106–119.

Taylor, E. (1996). *William James on Consciousness Beyond the Margin*. Princeton University Press.

Taylor, R. (1963). Causation. *The Monist*, 47(2):287–313.

Taylor, R. (1966). *Action and Purpose*. Prentice-Hall, Englewood Cliffs, NJ.

Teller, P. (1986). Relational holism and quantum mechanics. *British Journal for the Philosophy of Science*, 37(1):71–81.

Thalos, M. (2013). *Without Hierarchy: The Scale Freedom of the Universe*. Oxford University Press. Oxford.

Tsui, D. C., Stormer, H. L., and Gossard, A. C. (1982). Two-dimensional magnetotransport in the extreme quantum limit. *Physical Review Letters*, 48:1559–1562.

Uscinski, J. E., Enders, A. M., Seelig, M. I., Klofstad, C. A., Funchion, J. R., Everett, C., Wuchty, S., Premaratne, K., and Murthi, M. N. (2021). American politics in two dimensions: partisan and ideological identities versus anti-establishment orientations. *American Journal of Political Science*, 65(4):877–895.

Valencia, A. L. And Froese, T. (2020). What binds us? Inter-brain neural synchronization and its implications for theories of human consciousness. *Neuroscience of Consciousness*, 2020(1):1093.

Van den Heuvel, M. P, and Sporns, O. (2011). Rich-club organization of the human connectome. *Journal of Neuroscience*, 31(44):15775–15786.

Van Gulick, R. (2001). Reduction, emergence and other recent options on the mind/body problem: a philosophic overview. *Journal of Consciousness Studies*, 8: 1–34.

Van Vechten, C. (1926). *Nigger Heaven and the Harlem Renaissance*. Alfred A. Knopf, New York.

Varela, F. J. and Thompson, E. T. (2001). *The Embodied Mind: Cognitive Science and Human Experience*. MIT Press, Cambridge, MA.

Venturelli, N. A. (2016). A cautionary contribution to the philosophy of explanation in the cognitive neurosciences. *Minds and Machines*, 26 (3):259–285.

Vernier, C. L., Krupp, J. J., Marcus, K., Hefetz, A., Levine, J. D., and Ben-Sharar, Y. (2019). The cuticular hydrocarbon profiles of honey bee workers develop via a socially-modulated innate process. *eLife*, 8:e41855.

Vintiadis, E. (2018). There is nothing (really) wrong with emergent brute facts. In Elly Vintiadis and Constantinos Mekios (eds.), *Brute Facts*. Oxford University Press. Chapter 11.

Vuilleumier, P., Chicherio, C., Assal, F., Schwartz, S., Slosman, D., and Landis, T. (2001). Functional neuroanatomical correlates of hysterical sensorimotor loss. *Brain*, 124(6):1077–1090.

Wada, K., Chen, C., and Jarvis, E. D. (2017). Molecular profiling reveals insight into avian brain organization and functional columnar commonalities with mammals. In *Brain Evolution by Design*. Springer, Japan, p. 285.

Wallace, D. (2012) *The Emergent Multiverse: Quantum Theory According to the Everett Interpretation*. Oxford University Press, Oxford.

Weinan, E. (2011). *Principles of Multiscale Modeling*. Cambridge University Press, Cambridge.

Weinan, E. and Jianfeng, L. (2011). Multiscale modeling. *Scholarpedia*, 6(10):11527.

Weinberg, S. (1974). Gauge and global symmetries at high temperature. *Physical Review D*, 9(12): 3357.

Weinberg, S. (1995). Reductionism redux. The New York Review, October 5, 1995.

Weinberg, S. (2014). *Conceptual Foundations of Quantum Theory*. Cambridge University Press.

Weinberg, S. (2017). The trouble with quantum mechanics. http://quantum.phys.unm.edu/466-17/QuantumMechanicsWeinberg.pdf.

Weiskopf, D. A. (2016). Integrative modeling and the role of neural constraints. *Philosophy of Science*, 83:674–685.

Weissman, D. (1978). Dispositions as geometrical–structural properties. *The Review of Metaphysics*, 32(2):280.

Wen, X.-G. (2004). *Quantum Feld Theory of Many-Body Systems*. Oxford University Press, Oxford.

Wen, X.-G. (2013). ISRN condensed matter physics. *Physical Review B*, 100(24):198710.

Williams, R. (2001). The biologization of psychotherapy: understanding the nature of influences. In Slife, B., Williams, R., and Barlow, S. (eds.), *Critical Issues in Psychotherapy: Translating New Ideas into Practice*. Sage Publications, Thousand Oaks, CA, pp. 51–73.

Wilson, J. (2010). Non-reductive physicalism and degrees of freedom. *British Journal for the Philosophy of Science*, 61(2):279–311.

Wilson, J. (2013). A determinable-based account of metaphysical indeterminacy. *Inquiry*, 56:359–385.

Wilson, J. (2015). Metaphysical emergence: weak and strong. In Bigaj, T., and Wüthrich, C (eds.), *Metaphysics in Contemporary Physics*. Brill, Leiden, pp. 345–402.

Wilson, M. (1993). Honorable intensions. In Wagner, S. J., and Warner, R. (eds.), *Naturalism: A Critical Appraisal*. Notre Dame University Press, pp. 53–94.

Wilson, M. (2017). *Physics Avoidance: Essays in Conceptual Strategy*. Oxford University Press, Oxford.

Wimsatt, W. C. (2007). *Re-Engineering Philosophy for Limited Beings: Piecewise Approximations to Reality*. Harvard University Press, Cambridge, MA.

Windt, J. (2015). *Dreaming: A Conceptual Framework for Philosophy of Mind and Empirical*. MIT Press, Cambridge, MA.

Windt, J. M. (2016). Just in time—dreamless sleep experience as pure subjective temporality: a commentary on Evan Thompson. *Open MIND*, 2(1): 1571–1604.

Winning, J. (2020). Mechanistic causation and constraints: perspectival parts and powers, non-perspectival modal patterns. *British Journal for the Philosophy of Science*, 71(4):1385–1409.

Winning, J. and Bechtel, W. (2018). Rethinking causality in biological and neural mechanisms: constraints and control. *Minds and Machines*, 28(2): 287–310.

Wittmann, M. (2016). *Felt Time: The Psychology of How We Perceive Time*. MIT Press, Cambridge, MA.

Wolf, Y. I., Karev, G., and Koonin, E. V. (2002). Scale-free networks in biology: new insights into the fundamentals of evolution? *Bioessays*, 24(2):105–109.

Wolley, R. G. (1976). Quantum theory and molecular structure. *Adv. in Phys.*, 25(1):27–52.

Wong, H. Y. (2006). *Emergents from Fusion*. Philosophy of Science Association, University of Chicago Press.

Woodson, C. B., Schramski, J. R., and Joye, S. B. (2020). Food web complexity weakens size-based constraints on the pyramids of life. *Proceedings of the Royal Society B*, 287:20201500.

Woodward, J. (2003). *Making Things Happen*. Oxford University Press, Oxford.

Wu, W.-L. et al. (2021). Microbiota regulate social behaviour via stress response neurons in the brain. *Nature*, 595:409–414.

Wyss, P. (2018). Emergence: Inexplicable but explanatory. In Elly Vintiadis and Constantinos Mekios (eds.), *Brute Facts*. Oxford University Press. Chapter 12

Xia, Z. (1992). The existence of noncollision singularities in Newtonian systems. *Annals of Mathematics, Second Series*, 135:411–468.

Yamashita, W. and Nomura, T. (2017). The neocortex and dorsoventricular ridge: functional convergence and underlying developmental mechanisms. In Shigeno S., Murakami Y., and Nomura T. (eds.), *Brain Evolution by Design*. Springer, Tokyo, pp. 146–165.

Zbili, M., Rama, S., and Debanne, D. (2016). Dynamic control of neurotransmitter release by presynaptic potential. *Front Cell Neurosci*, 10:278.

Zednik, C. (2011). The nature of dynamical explanation. *Philosophy of Science*, 78 (2):238–263.

Zednik, C. (2014). Are systems neuroscience explanations mechanistic? Presented at Philosophy of Science Association 24th Biennial Meeting (Chicago, IL), November 2014, http://philsci-archive.pitt.edu/10859/.

Zednik, C. (2015). Heuristics, descriptions, and the scope of mechanistic explanation. *In Explanation in Biology*. Springer, pp. 295–318.

Zednik, C. (2018). Models and mechanisms in network neuroscience. *Philosophical Psychology*, 32 (1):23–51.

Zeh, H. D. (2000). The meaning of decoherence. In Blanchard, P. et al. (eds.), *Decoherence: Theoretical, Experimental, and Conceptual Problems*. Springer, Berlin, pp. 19–42.

Zerilli, J. (2021). *The Adaptable Mind: What Neuroplasticity and Neural Reuse Tell us About Language and Cognition*. Oxford University Press, Oxford.

Zhang, L., Wang, H., & Weinan, E. (2018). Reinforced dynamics for enhanced sampling in large atomic and molecular systems. *J. Chem. Phys.* 148(12):124113.

Zhou, Y., Xu, X., Wei, Y., et al (2021). A widespread pathway for substitution of adenine by diaminopurine in phage genomes. *Science*, 372(6541):512–516.

Zimmer, C. (2018). *She Has Her Mother's Laugh: The Powers, Perversions and Potential of Heredity*. Dutton Press, New York.

Zurek, W. H. (1991). Decoherence and the transition from quantum to classical. *Physics Today*, 44(10):36–44.

Zurn, P. and Bassett, D. S. (2020). Network architectures supporting learnability. *Philosophical Transactions of the Royal Society B*, 375:20190323.

Index

A

abstraction 55, 56, 59–64
  and algebra of observables 63–64
  and contextual topology 63–64
  dimensional analysis 61–63
  distinct from idealization or approximation 63
  Ising model 60–61
  and modality 60, 63
addiction 241–242
algebras of observables
  quantum-classical 103, 136–137
  See also observables
Anderson, Michael 227
Anderson, P. W. 3–4, 10, 23, 24, 150, 159, 162, 165, 174, 189, 209, 261
anyons 180, 186, 188
approximation 50, 52, 58, 60, 63, 74, 75, 82, 89, 120, 198, 201, 204–208
artificial intelligence xvi–xvii. *See also* machine learning
asymptotic expansions 57. *See also* singular limits
Atmanspacher, Harald 57, 200
attractors xiv, 92, 243, 245
  functional convergence 245
  strange 177
autonomy 218

B

Bain, Jonathan 205–207
Ball, Philip 136
Barnes, Elizabeth 208–209
Batterman, Robert 8, 33, 52, 67, 95–97, 102, 114, 144–146, 147
Bechtel, William 236–237
Becker, Adam 9
Bedau, Mark 5, 7
Bell inequalities 130
Bell's theorem. *See* no-go theorems
Berry, Michael 67, 114
biology. *See* autonomy; plasticity; robustness; systems biology
biopsychosocial xxi, 225, 274, 300
Bohm, David 133
Boi, Luciano 215, 216
Boids xvi
Bokulich, Alisa 114, 177
Boltzmann, Ludwig 168
Born, Max 120
Born–Oppenheimer procedure. *See* clamped nucleus assumption
bosons 109, 173, 180, 186
boundary conditions 21
bridge laws or principles xvii, 33, 42, 66, 67, 71, 72, 73, 77, 113, 285
British emergentists 65, 129–130. *See also* Broad, C. D.
Broad, C. D. 16, 36, 261, 277, 285, 319
Butterfield, Jeremy 13, 58, 68, 71–76, 97
  meshing dynamics 74–75

C

Cacioppo, Stephanie and John 224, 235
Callender, Craig 33, 146
Cao, Tian Yu 200
Cartwright, Nancy 57, 282–283, 284, 285, 319
Castellani, Elena 201, 205
Caudy, William 176
causal closure 35, 290, 294–296, 301–302
  of physics 115, 292–294, 312
causal cooperation 291
causal exclusion 291, 301. *See also* causal over-determination
causal inheritance 290
causal over-determination 290
causal powers 35
  causal-power actuality principle 41–42
causation 25–26, 266, 283–286, 302
  as contextual 283–284, 302
  difference making 111, 238–239, 257
  downward 12–13, 237
  fundamentalism 283
  intervention/manipulability 25–26
  multilevel 78
  property compatibility relations 284
  retrocausal 133
  synchronous 42

See also INUS condition; mental causation; possibility space
Chalmers, David 12, 13
Charcot, Jean-Martin 296
Chern–Simons field 185, 186, 189, 190, 191, 192, 194, 204
clamped nucleus assumption 120–121. *See also* stability conditions
classical mechanics 71, 286–287, 293
relationship to quantum mechanics 113–115
See also Newtonian mechanics
coarse graining 28, 74, 75, 88–89, 105
cognitive science 226, 231, 232, 233, 239, 310
4E 239–242, 300, 303
internalist 239, 241, 242, 305
See also neuroscience
collective effects xix, xx, 38, 39, 41, 92, 106, 141, 147–148, 149, 151, 171, 178, 179, 189, 191, 192, 195, 210, 244–245. *See also* quasiparticles
compatibility relations 32, 160–161, 162, 167, 169, 179, 195, 210, 262–263, 264, 265–267, 279, 284–285, 288
complexity 2–3
compositionality principle 120, 130, 135. *See also* quantum entanglement
condensed-matter physics 159, 162–170, 202, 209–210
and contexts 159–162
See also fractional quantum Hall state; superconductivity
constraints xii, 26, 81, 106
as downward 39, 40, 96–97, 147, 283–284
See also stability conditions
contexts xx, 59, 63, 65, 69–70, 74, 76, 82–83, 101, 132–134, 135–136, 226, 243–245, 247–248, 249, 254–255, 256, 260, 278, 288
and condensed-matter physics 159–162, 169, 171
cultural 219–220
defined by stability conditions xvii, 27, 31, 33, 41, 43, 52, 58–59, 72, 77, 93–94, 97, 105, 106, 112, 138, 164, 169, 172, 197–198, 205, 208, 227–228, 237, 238, 241–242, 254, 255, 259, 261, 263, 266, 271, 275–276, 277, 278, 282
endogenous 164, 165, 170, 171–172, 194–195, 197
exogenous 164, 165, 170, 172, 193–194, 195–196, 197
and networks 197
and semantic content 70
and theoretical vocabulary 21–22, 67–68
See also multiscale contexts; stability conditions
contextual emergence 8, 14, 24, 27, 28, 31, 33, 35, 42, 57–58, 69, 97, 101, 138, 240, 251, 257, 259, 277
and appendage regeneration 248
and bee colonies 249
and biology 78, 79, 81, 82–83
and causation 283
and chemical potential 106–107
and cognitive science 235–236, 241
compared with approximations 57–58
compared with Barnes, Elizabeth 209
compared with Batterman, Robert 52, 95–97, 102
compared with Butterfield, Jeremy 71–76
compared with Cartwright, Nancy 283
compared with deductive nomological explanations 34
compared with emergence intuitions 34–35
compared with French, Steven 324–326
compared with functionalism 274
compared with Humphreys, Paul 196, 263, 316, 317–320
compared with Ladyman, James and Ross, Don 169–170, 196, 322–324
compared with localization and decomposition 213, 220, 236–237, 238–239, 244, 246, 276
compared with mereological supervenience 40, 276
compared with Morrison, Margaret 179, 195
compared with Nagelian reduction 67–70
compared with ontological structural realism 321–326
compared with priority monism 102
compared with realization 112, 268, 275–276
compared with Santos, Gil 320–321
compared with Schaffner, Kenneth 76–78
compared with transformational emergence 316–321
compared with Wilson, Jessica 33, 263–264
and condensed-matter physics 158, 160, 162, 170, 197–198
and convection 28–29, 37–42, 84, 101, 128, 197, 256
and determinism 286–288
and dispositions 259, 320, 323, 324, 325–326
and ecology 249–251
and effective field theories 200, 204–205, 207–208, 209–210
and emergence of life 310
as eschewing metaphysical grounding 32
and evolution 220, 271–272, 273, 278, 310
as explanatory 14, 24, 98, 284, 285
and fireflies xxi, 254

contextual emergence (*Continued*)
and fluid dynamics 64
and follicles 15, 211
and fractional quantum Hall states 184, 192–198
and fundamentality 16–17, 121, 138–139, 209–210, 313–315, 318, 319–320, 321–322, 325
and genetics 69–70
and human fireflies 91–93
and ideal fluids 153–154
and intertheoretic relations 64–83
and lasers 31, 152–153
and laws 36, 40, 152–153, 279–281, 282, 283, 293, 319
and metastable liquids 154
and microbial community assembly 245–246
and molecular shape 115–124
as multiscale 17, 24, 36, 64, 137–138
and multiscale models 88, 94–97
and neuroscience 222–223, 225–226, 227–228, 235–236, 241
and neutral monism 308–309
as ontological 14, 24, 134
and ontology 288–290
and physicalism 303, 310
practical consequences of 17–19
and pressure 107–108
and quantum electrodynamics 122
and quantum mechanics 58, 113–115, 124–125, 133–134, 135–140
as relational 16, 23, 32, 132, 257, 288
and renormalization 142, 149–152
and scanning tunneling microscopy 153
and superconductivity 170–179
and systems biology 218–219, 220
and temperature 27, 102–106
and topological defects 168
and whisky xxii–xxiii, 253–254
and white dwarf stars 108–112
See also possibility space
contextual topology 46, 50–55, 196, 201, 206, 262
and abstractions and idealizations 59–64
and intertheoretic relations 55–59, 74, 75, 206–207
control hierarchy 39
conversion disorders 295–301, 302–303
Craver, Carl 216, 243
criticality 102, 141, 143, 146
critical exponent 143
critical phenomena 141
critical point 38, 43, 73, 92, 99, 143, 144–145, 150–151, 187–188

D

Davidson, Eric 215
Dawkins, Richard 212
Debye shielding 141, 170
decoherence. *See* quantum mechanics
degeneracy pressure 108, 109–111, 156–157
as acausal stability condition 111
degrees of freedom 25, 33, 39, 57, 110, 134–135, 141, 147–148, 190, 198, 200, 203, 204, 205–207, 209, 262–264
descriptions. *See* scientific descriptions; vocabulary
determination 288
as bottom-up 10, 171, 212, 242–243
failure of strictly bottom-up determination 15, 33, 40, 97, 114, 149, 151, 173, 215–216, 227–228, 230–231, 250, 276, 283–284, 288
larger scale to smaller scale 15, 49–50, 93–97, 133, 137, 171, 218, 236–237, 238
See also mereological supervenience
determinism 74, 77, 286–288
genetic 214–215, 240
unique evolution 286–287
diachronic emergence. *See* emergence
diffeomorphism 181, 184
Dirac, Paul 115–116
disjoint states 105, 107, 114–115
dispositions 255, 257–261, 262–263, 324, 325–326
as contextual 172
dispositionalism and contexts 260
as extrinsic 35, 36, 260
as intrinsic 258–259
and laws 265, 319
and quantum mechanics 258
relationality 32, 172, 191–192, 258, 259–260, 263, 288
and scientific investigation 259
Dorato, Mauro 257–258
dressed states 116, 141, 187–188
Duan, Feng 159
dynamical systems xiv, 92, 148

E

Earman, John 175
effective field theories 142, 198–201
as approximations 204–208
decoupling theorem 203
fundamentality 208–210
underdetermination 199, 203, 204
Einstein, Albert 128–129, 131
Elitzur's theorem 176
Ellis, George 32
Embryogenesis 216
emergence 6, 65, 72, 159, 213
as diachronic 7, 13, 316

and fractional quantum Hall states 187–193
and metaphysics 8, 251, 254
as synchronic 7, 41
See also contextual emergence; epistemic emergence; ontological emergence; strong emergence; transformational emergence; weak emergence
entropy 105
environment xix, xx, xxii, 16, 21, 24, 27, 59
microenvironment 17
See also stability conditions; system/environment distinction
epigenetics 215, 217, 225, 249
epistemic emergence 3, 5, 7, 11, 33, 79, 159, 213
equivalence classes 28, 41, 45–46, 57, 102, 106, 121, 122–123, 140, 144, 218, 244. *See also* universality classes
ergodicity 104, 168
Euler, Leonard 22
evolution 69, 72, 219, 220, 224, 247–248, 273, 277–278, 310
convergent 270–271, 272, 273
natural selection 8, 10, 251, 271–272
sexual selection 81
explanatory indispensability 177–178, 296–297, 315

F

false forced choice 14, 17, 212, 213, 279, 285, 296
Feldman Barrett, Lisa 223–224, 227, 234–235, 270
fermions 109, 111, 155, 173, 180, 181, 186
"composite" 185–186, 188–189
wavefunction 112, 155–156
filling factor 179, 180
fireflies xii–xiii, xiv, xvii–xxii, 254
fixed points 144, 147–148, 151
fluid dynamics 38, 42, 53, 60, 68–69, 71, 150, 153–154
bulk flow 39, 41, 44, 46–47, 51, 58, 59, 64, 84, 85, 86–88, 93, 94
eddies 85–88, 93, 94
See also contextual emergence; multiscale models; Rayleigh-Bénard convection; turbulence
Foster, Brett 253, 254
foundationalism 10, 12, 35, 200, 260, 279, 288
fractional quantum Hall state 167, 168, 170, 179–180, 181–182, 185, 186–187, 191–194, 197–198, 204, 210, 266
multiple realizability 41, 106, 144–145, 183–186, 218, 246, 255, 268–269, 273–275
not derivable from microstates and observables 183
and quasiparticle realism 187–189
Franklin, Alexander 149, 159
French, Stephen 324–326
Freud, Sigmund 296
Friederich, Simon 176
functionalism 274
fundamentality xvi, xxi, xxii, xxxiii, 1, 14, 29, 31–33, 36, 59, 74, 115, 121, 139, 142, 151, 159, 162, 198, 199–200, 313, 314–315
contexts as fundamental xxii, 16–17, 32, 138, 158, 161, 164, 210, 257, 314, 322, 323
fundamental ontology 288–303
fundamental theory 9, 33, 49, 55–56, 58, 64, 125, 204, 209
stability conditions as fundamental 33, 74, 138
See also effective field theories; foundationalism

G

Game of Life xiv, xvi, xxii, 11, 15, 32–33, 312–314
gauge symmetry 172–174
physical significance 174–179
gene regulatory networks 77, 215–216, 248
general relativity 65, 71, 143, 207, 281
generalized rigidity 38, 167–170, 171–172, 180
and symmetry breaking 167, 178
generative atomism 316–317. *See also* reductive atomism
genetics 80–81
classical genetics relationship to molecular genetics 69–70
See also gene regulatory networks
geophysical modeling 88–90, 93, 94–95. *See also* multiscale models
Georgi, Howard 199
Gibbs state 105, 107–108
Gillett, Carl 5
glia 219, 222
Godfrey-Smith, Peter 223, 232, 272
good coordinates 166, 189–191, 192, 196, 208, 264
and stability conditions 190, 197–198
Goujin, Jin 159
Gould, Stephen J. 215
graph theory 22, 228
Greensite, Jeff 176
Gunton, Colin 285

H

Haag, Rudolf 104
Hacking, Ian 70

Hardcastle, Valerie Gray 294–295
Hartmann, Stefan 202, 203
Heisenberg uncertainty principle 71, 122
as structural stability condition 112
Hempel, Carl 35
Hettema, Hinne 119
hierarchy. *See* control hierarchy
Higgs, Peter 174
Higgs mechanism 174–175, 176
Hiley, Basil 133
Hoff, Jacobus Henricus van't 117
Howard, Don 178
Human Genome Project 214–215
Hume, David 278
Humean supervenience 127–128
Humphreys, Paul 193, 196, 263, 290, 316–320
fusion of properties 160, 193, 196, 263, 320

I

idealization 52, 59–64, 201
integer quantum Hall effect 179, 180, 188
intertheoretic relations
and contexts 68, 72–74, 77
contrasted with multiscale modeling 88, 89
as logical relations 65, 66, 68–69, 70–71, 72–74, 76
See also contextual emergence; contextual topology; scientific descriptions
INUS condition 40, 195, 284
Ising, Ernst 60–61
Ising model 60–61

J

James, William 285, 303, 304–308, 309

K

Kadanoff, Leo 33
Kant, Immanuel 306
stability conditions for subjectivity 304–305
Kim, Jaegwon 35, 41–42, 128
Kochen-Specker theorem. *See* no-go theorems
Kronz, Fred 129–130
Kubo-Martin-Schwinger states 103, 122
conditions for 28, 103–104
as stability condition 103–104, 105, 106, 107–108, 110

L

Ladyman, James 169, 269, 322–324
Landau, Lev 165
Landsman, N. P. 33
language 220, 221, 224
Laughlin, Robert 4–5, 10–11, 23, 159, 180
laws xv, xvii, 4, 10, 11, 15, 29, 36, 57, 106, 123, 206, 278–283, 286, 293
adynamical 280–281
and compatibility relations 265, 279
conservation 26, 28, 37, 39, 293
as constraints 151, 280–281, 282, 319
as contextual 265
dynamical 26, 279, 287, 316, 318
emergent 282, 319
false forced choice 279
fundamental 208
fundamentalism 278
Humean 278
as nomic relations 66
nomological realism 278–279, 280
and possibility space 31, 43, 138, 279–280, 282
power or scaling laws 148, 151, 231
regularities 278–279, 280, 282
universal 283
See also bridge laws or principles; determinism
Leggett, Anthony 13
levels 23–24, 35, 196, 205
of description 46, 48, 52, 77, 84, 88
macroscopic not governed by microscopic 113
of organization 13, 39, 69, 78–80, 82–83, 225, 257, 324
of reality 15
Levin, Michael 187–188
Lewis, David 110–111, 127–128, 189
Lewis, Peter 131–132
locality principle 127, 128 *See also* quantum nonlocality
localization and decomposition 35, 212–213, 215, 233, 235–239, 241, 243, 276, 277, 315
contextual decomposition 238
failure of 213–214, 216, 220, 231, 244–245
Love, Alan 236, 276–277, 285
Lower, Barry 292
Lyre, Holger 175

M

machine learning 47, 113, 244, 245
macroscale independence 95, 113, 133, 143, 149, 151
macrostates 74, 144, 146, 149, 151, 171
Mainwood, Paul 145–146, 261
Maxwell, James Clerk 61
McMullin, Ernan 59–60
mechanisms 212–218, 224–228, 231–248, 270–278
mechanistic explanation 80, 82
Menon, Tarun 33, 146
mental causation 13, 292–295, 300–302
mereological supervenience 40, 128
failure of 128, 136, 185, 191–193, 276
metaphysics
non-Humean 266

and science 8–9, 251
See also determinism; dispositions; laws; neutral monism; ontology; possibility space; properties
microstates 74, 146, 149, 183
modal constraints. *See* possibility space; stability conditions
modal facts. *See* possibility space
modality. *See* possibility space
Morrison, Margaret 60, 147–148, 158, 171, 172–175, 176–177, 178–179
multilevel explanation 78, 80, 82–83, 284
multiple realizability 41, 46, 74, 106, 144, 173–174, 179, 183–186, 228, 246, 268–278. *See also* fractional quantum Hall state
multiscale contexts xxii, 14, 15, 16–17, 32, 33, 35, 41, 47, 64, 69, 82–83, 85, 93–94, 124–125, 136, 138, 159, 169–170, 196, 215–216, 225, 230, 248, 277, 282, 284, 309, 314
multiscale interactions 23, 41, 43, 102, 233
multiscale models 64, 76, 83–85, 163, 221, 223
and contexts 85–90, 95–96, 112
data analysis 90–93
subgrid models 86–87, 89, 93, 94–95, 96–97
multiscale properties 25, 82, 196
multiscale relations 20, 64, 70, 90, 101, 209, 233
muon anomalous magnetic moment 259–260

N

Nāgārjuna 20
Nagel, Ernst 58, 65–66, 67, 97
natural philosophy 9, 10
natural selection 8, 10, 36, 273, 277, 291. *See also* evolution
Navier–Stokes equations 46, 47, 52, 58, 64
network neuroscience 228–235, 240, 278
network topology 214
networks xvii, xix, 15, 16, 22–23, 70, 118, 216, 218–220, 245–246, 251, 274, 275–276, 287, 321
and compatibility relations 32
as constraints 28
and dispositions 32, 259
and laws 280
multiscale 16, 217, 223, 225, 228, 232, 242, 275
"Rich-Club" 229–230, 231
signal centers 247, 248
small world 25, 222, 228–230, 231
social 234–235
topological 26, 214, 228
See also contexts; gene regulatory networks
neural correlates 295, 297, 298–299, 308–309
neural degeneracy 227, 270
neural redundancy 227
neural reuse 227
neuroscience 221–223, 297–299, 310
multiscale 222–223
RNA regulatory networks 15, 215, 222, 232
See also cognitive science; network neuroscience; social neuroscience; systems neuroscience
neutral monism 303–310
contrast with panpsychism 307
contrast with primary/secondary distinction 308
Newton, Isaac 61, 287
Newtonian mechanics 129, 280, 293, 302
relationship to general relativity 65, 281
relationship to quantum mechanics 56, 65
relationship to special relativity 65, 67
Nickels, Thomas 66–67
no-go theorems 131–135, 139–140
independence 131–132
locality 132, 133
realism 131
nonseparability 41, 135–136
novelty 34, 35–36, 42, 57, 58, 165, 261–262

O

observables 45, 48–49, 63–64, 65, 69, 106, 127, 144, 149, 183
algebras of 49–50, 51–52, 67, 103, 105, 113, 121, 136, 155
classical 105, 106, 107, 114–115, 119, 122, 123, 137
contextual 105, 106, 256, 264
and different scales 113, 137, 150, 200
quantum 119–120, 122, 130, 137
and singular limits 50–51, 121, 147
and stability conditions 51, 68, 121–122, 195, 197, 200, 208
See also good coordinates
O'Connor, Tim 13, 319
ontological emergence xvii, xx, 7, 13–14, 16, 33, 42, 120–122, 158–159, 251–252, 303, 319
ontological structural realism 257, 259, 321–325
ontology 196, 203, 206
and algebras of observables 121
of molecules 136
as multiscale 24
ontological dependence 209
ontological independence 209
and quantum entanglement 135, 136–137
as relational 192, 257, 288
See also relative onticity

Oppenheimer, Robert 120
order 9, 15, 24, 32, 44, 61–62, 164–165
  partial order 58, 78
order parameter 38, 61, 143, 165, 173, 175, 176, 190, 275

P

Pauli, Wolfgang 155
Pauli exclusion principle 109, 110, 112, 155–156
persistent homology 90–91, 92, 99
  as stability condition 90–91, 93
phase transition xiv, 143, 144, 146, 154, 168, 170–171, 173, 174, 176, 178, 180, 195
  contrast with fractional quantum Hall effect 186–187
  quantum phase transition 180
  second-order 165
Pigliucci, Massimo 212
Pines, David 4–5
plasticity 217–218, 226–227, 245, 273–274, 275, 277
possibility space 30–33, 35, 42, 43, 49, 55, 94, 111, 112, 123, 138, 147, 151, 152–153, 160–162, 165, 166–167, 168, 169–170, 196–197, 208, 209–210, 220, 237, 250–251, 256–257, 262, 263, 264, 265, 282, 287, 288, 317–318, 323, 324
  and compartmentalization 163–164, 168, 197
  and degrees of freedom 263
  as dispositions 258
  modal facts 255, 256, 262, 264
  nested 267–268
  See also compatibility relations; stability conditions
Povich, Mark 216
pressure 107–108, 110, 157. *See also* degeneracy pressure
Primas, Hans 55, 56, 67, 74, 76, 116, 148
priority monism 102
properties 261–265, 317
  causes 284
  fusion of 160, 193, 196, 263
  gatekeeper 268
  incompatibility relations 265–267
  intensional 189–190, 192
  network 217, 275
  property exclusion argument 262
  relational 192–193, 195, 262, 289–290
  See also multiscale properties
property dispositionalism 32
protein folding 2, 216
Proteome 15, 215, 235, 240
Putnam, Hilary 145–146

Q

quantum contextuality 135
quantum electrodynamics 122, 180, 199, 281
quantum entanglement 115, 117, 126–131, 135, 137–138, 185–186, 251
  and emergence 7, 320
  entangled states 126, 130
  long-range entanglement 185–186, 191–192, 194, 210
quantum mechanics 71, 138, 258, 289
  basis problem 114
  decoherence 115, 122–123, 125–126, 134, 136
  Ghirardi–Rimini–Weber theory 293
  Hamiltonian 116, 118–119, 181, 183
  Many-Worlds 133
  mixed states 117
  and molecular shape 115–124
  pure states 114, 117
  relationship to macroscopic physics 56, 58, 113–115, 125–126
  stability conditions for allowable states 112
quantum nonlocality 126, 130. *See also* quantum entanglement
quantum protectorates 5
quantum superposition 124–125, 126, 135
quasiparticles 166, 185, 187–189, 191

R

radical ontological emergence 11, 12–13, 14. *See also* strong emergence
rainbow realism 169, 196
Rayleigh–Bénard convection 27–28, 37–42, 43, 44, 84, 101–102, 127, 128, 143, 146–147, 164, 194, 197, 256
realization 35, 112, 135, 146, 268–278, 290
Redhead, Michael 201–202, 203
reductionism 3, 5, 58, 72, 100, 116, 122, 125, 145, 201, 203, 211–212, 233, 239, 241, 264–265, 269
  as contextual 203–204
  failure of 67–68, 79, 81–82, 83, 105–106, 113, 154, 198
  horizontal 312
  intertheoretic 58, 65, 66–67
  as ontological 4, 9, 11, 14, 27, 33, 79, 116, 208–209, 213, 252, 290, 310, 311, 312, 314, 315
  vertical 311–312
reductive atomism 40, 127, 135
relative onticity 200, 315
renormalization 95, 140–151, 199, 203
  and equivalence classes 141–142, 150–151
  renormalization group explanation 144–147, 149–151
  See also fixed points
robustness 217–218, 245, 273–274, 275

Ross, Don 169, 322–324
Rovelli, Carlo 260–261, 289–290, 323
Russell, Bertrand 303, 304, 307–308

S

Santos, Gil 316, 320–321
Schrödinger, Erwin 116
Schweber, Sam 200
scientific descriptions 44–48, 55, 105, 166, 200, 285, 287
  and abstractions and idealizations 59–64
  context-free descriptions not possible 58–59
  contrast with intertheoretic reductionism 47–48, 57–58, 67–68, 71
  higher-level not implied by lower-level 54, 56–58, 205–206
  as multiscale 44–45
self-organization xiii, xvii, xix, 152, 216, 283
separability 127
  separability principle 127, 128
  See also nonseparability
sets 49–50, 207. *See also* topology
Shakespeare 100
singular limits 50–51, 59, 66–67, 75, 76, 113–114, 120
  and changes of ontology 121, 147
  and contextual topologies 51
  See also asymptotic expansions
Skow, Bradford 110
Smeenk, Christopher 175
Smolin, Lee 36, 282
social neuroscience 223–226, 234, 240, 273
spatial dimensionality 28–30, 289
  as global stability condition 30, 110, 181–182, 193–194, 314
  as modal gatekeeper 31, 193–194
special relativity 65, 67, 126, 129, 135, 280–281
Spence, Sean 296, 297
Spivey, Michael 240–241
spontaneous symmetry breaking. *See* symmetry breaking
Sporns, Olaf 231, 232, 234–236
stability conditions 14, 15, 17, 18, 27–34, 35, 38, 43, 46, 51, 58–59, 63, 69–70, 74, 77, 85, 103, 107, 112, 119, 123, 161, 187, 190, 200, 208, 235, 257, 265, 284, 291
  as acausal 25, 26, 27, 28, 29, 106, 110, 122, 181–182, 281
  as adynamical 25, 31, 42, 135, 138, 280
  as affordances 240, 245
  as causal 25
  and collective effects 41
  as compatibility relations 162, 266
  as constraints 16, 17, 28, 36, 37–38, 39, 40, 81–82, 88, 96–97, 101, 123, 147, 149, 151, 152, 153–154, 161, 163–164, 172, 186, 194, 196, 205, 217, 231, 236–237, 245, 250–251, 255, 263, 272, 276, 281, 282, 305, 310, 313
  contrasted with bridge laws 71
  as cultural 220, 225, 230, 241–242
  and data analysis 90–91
  and determinism 287
  as difference makers 28, 238–239, 261–262, 268
  as dynamical 25, 26, 28, 31, 39, 40–41, 42, 161, 194, 230
  as endogenous 194–195
  as environment 24, 27, 271
  as exogenous 179, 193–194, 195, 196
  and gene regulatory networks 216
  as global 16, 25, 26, 28, 29, 106, 110, 122, 178, 181–182, 217–218, 237, 239, 276, 277, 283
  and good coordinates 192
  as larger scale 15, 111–112, 124–125, 132–133, 136–138, 146, 147, 149, 151, 152, 192–193, 195, 211, 220, 225, 226, 227–228, 232, 236–237, 238, 245–246, 248, 249–251, 255, 263, 267, 289
  and lasers 152–153
  and metastable liquids 154
  as modal gatekeepers 14, 30–31, 34, 40–42, 112, 123, 138, 151, 172, 193–194, 250–251, 255, 256, 267, 280, 284
  and molecular shape 121
  as multiscale 16, 81–83, 93–94, 123–124, 140, 151, 193–194, 217, 220, 237, 238, 254, 255, 284, 310, 314, 320
  and multiscale models 88, 89, 93–94, 97
  as networks 22, 32, 34, 118, 197, 214, 216, 219, 223, 225, 232, 235, 237, 246, 250, 275
  as nomological 25
  as organizational principles 217, 242, 276, 277
  and phase space 168–169
  and scanning tunneling microscopy 153
  as social 18, 19, 33, 219, 223–224, 225, 235, 240, 241, 249, 254, 295, 296, 301
  as statistical 111
  as structural 25, 112, 122, 149, 167, 179, 283
  and superconductivity 174–179
  as top-down information control 218
  as topological 25, 26, 153–154, 184, 192–193, 196, 197, 231
  and wave-particle duality 124–125, 156
  See also abstraction; spatial dimensionality
state space 45, 63, 110, 142, 144, 146, 147, 148, 190, 256, 280
  of Hamiltonians 144, 148–149, 150, 151

state space (*Continued*)
microstate space 74, 75, 151
statistical mechanics 28, 47, 53, 57, 102–103, 105, 148, 168. *See also* ergodicity; temperature; thermodynamics
Stewart, Matthew 294–295
Stoppard, Tom 44
Strogatz, Steven xiii–xv, xviii
strong emergence 7, 12, 13, 16, 212. *See also* radical ontological emergence
superconductivity 170–179
superfluidity 178
superselection sectors 105, 107, 122
supervenience 151, 184–185. *See also* Humean supervenience; mereological supervenience
symmetry 11, 38, 112, 118, 150, 165–167, 172, 263
and possibility space 165
pressure 110
symmetry breaking 4, 38, 165, 166, 168, 173, 186
spontaneous 105, 172, 174, 176, 177–178, 190, 195, 268
See also generalized rigidity
sync xii–xiv, xviii, xix, 251. *See also* synchronization
synchronization xii–xiv, xix
in fireflies xiii–xiv, xviii–xix
system/environment distinction 20–21, 59, 64, 69, 70, 161, 169, 171, 195, 196, 197, 198, 200–201, 240
and contextual topology 201, 262
as epistemic 198
and modality 264
systems biology 214–220, 231, 240
systems neuroscience 223–226, 228, 231

T

Takesaki, Masamichi 105
temperature 102–106, 110, 150, 268
theoretical vocabulary 21, 65–66, 67–69, 70, 77, 84, 255
conceptual embedding 21–23, 70
and context 68, 84, 85
and system/environment distinctions 22, 69
thermodynamics
relation to statistical mechanics 51–52, 74, 79, 103, 105–106, 107
second law 105
thermodynamic equilibrium 103
zeroth law 103, 104
See also temperature
topological defects 167
and context 167–168
topological equivalence 75
topological order 180, 184, 186, 210
topological phase 184
topological states 167
topology 49–50, 55, 63
coarse 51, 53, 57
fine 51, 53, 57
and fractional quantum Hall states 181–182, 193–194, 197
strong 51
weak 51
See also contextual topology; network topology
transformational emergence 315–321
turbulence 85–87, 93
twin-slit experiment 124–125, 136, 138, 139

U

universality xiv, 11, 52, 60, 87, 95, 142, 285
in biology 143, 214, 218, 231, 245–246
and renormalization 143–149
in superconductivity 172
universality classes 44–45, 57, 151
physical connections 150

V

VanderMeer, Jeff 211
Vishik, Inna 158
vocabulary 21–23, 29, 48, 255. *See also* levels, of description; scientific descriptions; theoretical vocabulary
von Goethe, Johann Wolfgang 1
Vuilleumier, Patrik 298

W

Wallace, David 114–115, 187
weak emergence 5, 12, 14. *See also* epistemic emergence
Weinberg, Stephen 9–10, 11, 125, 174, 176, 209
Weiskopf, Daniel 231
Weissman, David 266
Wen, Xiao-Gang 175, 187–188
whisky xxii–xxiii, 253–254
white dwarf stars 108–112, 156–157
Wilson, Jessica 7, 33, 58, 205, 263–264
Wilson, Kenneth 142
Wilson, Mark 84, 88, 97, 189, 196

Wimsatt, William 78–83
winding number 181–182
  braiding 182, 184
Windt, Jennifer 309–310
Winning, Jason 237
Wittman, Marc 309
Wong, Hong Yu 13
Woodward, James 25, 34, 265
Woolley, Guy 116

Z

Zerilli, John 227, 269–270
Zimmer, Carl 215
Zurek, Wojciech 114